Linnaeus, natural history and the circulation of knowledge

Oxford University Studies in the Enlightenment
– formerly *Studies on Voltaire and the Eighteenth Century* (*SVEC*),
is dedicated to eighteenth-century research.

http://www.voltaire.ox.ac.uk/our-books/for-authors-studies-enlightenment

Senior publishing manager
Lyn Roberts

The Voltaire Foundation is a department of the University of Oxford.
It furthers the University's objective of excellence in research,
scholarship and education by publishing worldwide.

Linnaeus, natural history and the circulation of knowledge

Edited by

HANNA HODACS,
KENNETH NYBERG
and
STÉPHANE VAN DAMME

VOLTAIRE FOUNDATION
OXFORD

www.voltaire.ox.ac.uk

ISBN 978 0 7294 1205 6
Oxford University Studies in the Enlightenment 2018:01
ISSN 0435-2866

Voltaire Foundation
99 Banbury Road
Oxford OX2 6JX, UK
www.voltaire.ox.ac.uk

A catalogue record for this book is available from the British Library

The correct style for citing this book is
H. Hodacs, K. Nyberg and S. Van Damme (ed.), *Linnaeus, natural history and the circulation of knowledge*, Oxford University Studies in the Enlightenment (Oxford, Voltaire Foundation, 2018)

Cover illustration: frontispiece from Hermann Boerhaave, *Index plantarum, quæ in Horto Academico Lugduno Batavo reperiuntur* (Leiden, Cornelis Boutestein, 1710).

MIX
Paper from responsible sources
FSC® C013056
www.fsc.org

FSC® (the Forest Stewardship Council) is an independent organization established to promote responsible management of the world's forests.

This book is printed on acid-free paper

Printed in the UK by TJ International Ltd, Padstow, Cornwall

OXFORD UNIVERSITY STUDIES IN THE ENLIGHTENMENT

LINNAEUS, NATURAL HISTORY AND THE CIRCULATION OF KNOWLEDGE

The name of Carl Linnaeus (1707-1778) is inscribed in almost every flora and fauna published from the mid-eighteenth century onwards. However, his rich legacy has been largely overlooked in recent scholarship. In this book a group of specialists argue for the need to re-centre Linnaean science and de-centre Linnaeus the man by exploring the ideas, practices and people connected to his taxonomic innovations.

Contributors examine the various techniques, materials and methods that originated within the 'Linnaean workshop': paper technologies, publication strategies, and markets for specimens. Fresh analyses of the reception of Linnaeus's work in Paris, Königsberg, Edinburgh and beyond offer a window on the local contexts of knowledge transfer, including new perspectives on the history of anthropology and stadial theory. The global implications and negotiated nature of these intellectual, social and material developments are further investigated in chapters tracing the experiences and encounters of Linnaean travellers in Africa, Latin America and South Asia.

Through focusing on the circulation of Linnaean knowledge and placing it within the context of eighteenth-century globalisation, authors provide innovative and important contributions to our understanding of the early modern history of science.

History of science / history of ideas / cultural history

Histoire de la science / histoire des idées / histoire culturelle

Contents

List of illustrations and tables

Illustrations

Table

Preface

This edited volume began its life in Stockholm, at the Centre for History of Science at the Royal Swedish Academy of Sciences in October 2012, during the workshop 'A global history of Linnaean science, 1750-1820'. A generous grant from Åke Wibergs stiftelse and Stiftelsen Lars Hiertas minne enabled the Director of the Centre, Karl Grandin, Hanna Hodacs and Kenneth Nyberg to invite a number of junior and senior scholars currently working on different aspects of natural history and Carl Linnaeus to attend. The Stockholm workshop gave rise to plans to put together an edited volume representing new research on Linnaeus and natural history. A second workshop, 'Empire's nature: a global history of Linnaean sciences in the long eighteenth century', hosted by Stéphane Van Damme at the European University Institute, took place in Florence in November 2014. The chapters that make up this book represent the main bulk of the papers presented at the Florence workshop.

We would like to thank Karl Grandin and the Centre for History of Science for their continuous support in the work with this volume. We would also like to thank the Chairs at the workshop in Stockholm, Marie-Christine Skuncke and Otto Sibum, both at Uppsala University, as well as Sverker Sörlin (Royal Institute of Technology, Stockholm), who provided conclusions and comments both in Stockholm and in Florence.

We would also like to express our gratitude towards Kungliga patriotiska sällskapet for supporting the project and enabling us to employ Dr Alexia Grosjean as a language editor. Finally, to Lyn Roberts and the Voltaire Foundation for the invaluable help we have received in turning our manuscript into a book, thank you.

Hanna Hodacs, Kenneth Nyberg and Stéphane Van Damme

Notes on naming conventions

Carl Linnaeus (1707-1778) is a frequently occurring name in this book. Carolus Linnaeus and Carl von Linné are variant names, referring to the same person, that occur in some quotations from contemporary sources and in bibliographical references. Linnaeus's son, also named Carl (1741-1783), is usually referred to as Carl Linnaeus the Younger in the text, but in quotations and references the form Linnaeus fil. (Latin for 'Linnaeus the son') may also occur. 'Linnaeus' on its own always refers to the father, not the son.

List of abbreviations

AFSt/M	Online catalogue of the archive of Die Franckeschen Stiftungen zu Halle
AGS	Archivo General de Simancas
ALMW/DHM	Online catalogue of the archive of Die Franckeschen Stiftungen zu Halle
AN	Archives nationales, Paris
AS	Asiatic Society in Calcutta
BCMNHN	Bibliothèque centrale du Muséum national d'histoire naturelle, Paris
CRC	Centre for Research Collections, Edinburgh
DAC	Danish Asiatic Company
DNA	Danish National Archives
EIC	English East India Company
EUL	University of Edinburgh, Library
GNFB	Gesellschaft naturforschender Freunde zu Berlin
KDVS	Det kongelige danske videnskabernes selskab
KNVS	Det kongelige norske videnskabernes selskab
LSL	Linnean Society of London
MNCN	Museo Nacional de Ciencias Naturales, Madrid
NHM	Natural History Museum, London
RJB	Archivo del Real Jardín Botánico, Madrid
RSAS	Royal Swedish Academy of Sciences, Stockholm
SEIC	Swedish East India Company
VOC	Vereenigde Oost-Indische Compagnie

Introduction: de-centring and re-centring Linnaeus

HANNA HODACS, KENNETH NYBERG
and STÉPHANE VAN DAMME

Carl Linnaeus (1707-1778) represents a challenging departure point for a book which aims to explore eighteenth-century scholarship concerned with nature. On the one hand, his reform of scientific nomenclature for species of plants and animals helped to ensure that Linnaeus's name was inscribed on almost every general flora and fauna published after the mid-eighteenth century. In this respect the *princeps botanicorum*, or the Prince of Botany, almost became immortalised in the process. On the other hand, few traces of Linnaeus are to be found in many recent explorations of natural history in early modern science histories which focus on the making of knowledge and include considerations of social, spatial and global processes.[1]

Consequently this book has two somewhat contradictory ambitions: it wants to both *de-centre* and *re-centre* Linnaeus. One step towards de-centring Linnaeus is by broadening the focus; in the following, the term 'Linnaean science' is used with reference to scholarship and practices which are connected to Linnaeus's taxonomic innovations in one way or another. The starting point is that Linnaean natural history was a collective and collaborative enterprise and not the work of one man. Aside from ideas of how to order, describe and name species, it was also a set of practices embedded in a materiality, and in social contexts. Applied, acted out and materialised globally, practices and materialities were constantly renegotiated. The term 'Linnaean science' is deliberately vague; in fact, one objective of this book is to explore what it can refer to, and the possibility of using it to outline the contours of a field of studies we tentatively call 'Linnaean studies'.[2] If the term 'Linnaean science' is

1. In the pioneering work *Cultures of natural history*, ed. Nicholas Jardine, James A. Secord and Emma C. Spary (Cambridge, 1996), for instance, most of the references to Linnaeus are casual or marginal. Two out of twenty-six chapters (Koerner's and Schiebinger's) discuss Linnaean botany more closely, but considering that it is a book for specialists it is interesting to note how both present relatively basic facts about Linnaeus as if they are assumed to be unknown to the readers. Most notable of all, perhaps, is how much of this influential volume on natural history that does *not* deal with Linnaean science.
2. This notion was suggested by Sverker Sörlin at the workshop 'Nature's empire' at the European University Institute in Florence, Italy, on 14 November 2014.

about de-centring Linnaeus, the aim of the term 'Linnaean studies' is to re-centre Linnaeus, by mapping out the impact of Linnaeus's work on a wide range of fields of studies, inspired by recent explorations of early modern scholarly techniques for organising information, consumption, cosmopolitanism, political economy, enlightenment ideas, public spheres and individual career building.

The recent upsurge in the history of early modern natural history, the tradition to which this book seeks to connect Linnaean science, is surveyed in the first part of this introduction. Linnaeus's position in different historiographical discussions provides the theme for the second part, while the third introduces the different contributions that together provide a mosaic exemplifying the many ways in which we can trace the application and development of Linnaean science, and trace the contours of the field of Linnaean studies. The historical circumstances that can help us frame and explain Linnaean studies in a broader chronology in the form of global history are discussed in the final section of the introduction.

Natural history in the history of sciences

Natural history, encompassing the study of plants, animals and minerals, the three kingdoms of nature, has recently regained its lustre for historians of science.[3] In the period leading up to this comeback, from the 'triumphal' days of the nineteenth century, natural history was seen as somewhat disconnected from contemporary scientific disciplines such as mathematics, physics and chemistry. In recent scholarship, which largely focused on the early modern period, the paradigm of natural history – whether in the form of the *historia* of the Renaissance, or the field-science of the eighteenth century – has gradually eclipsed other early modern sciences, supplanting astronomy and mathematics in historical research, and questioning the centrality of experimental culture.[4] We cannot understand one without the other; natural history in the form of gemstone collections was, for example, central to physical experiments investigating light.[5]

Natural history opens up avenues by the help of which we can study change on its own terms, predating nineteenth-century modernity. The

3. Katharine Park and Lorraine Daston (ed.), *Early modern science* in *The Cambridge history of science*, ed. David C. Lindberg and Ronald L. Numbers, 8 vols (Cambridge, 2003-2013), vol.3 (Cambridge, 2006).
4. Gianna Pomata and Nancy G. Siraisi (ed.), *Historia: empiricism and erudition in early modern Europe* (Cambridge, MA, 2005); Silvia Sebastiani, *The Scottish Enlightenment: race, gender and the limits of progress* (New York, 2013).
5. Michael Trevor Bycroft, 'Physics and natural history in the eighteenth century: the case of Charles Dufay', unpublished doctoral dissertation, University of Cambridge, 2013.

ancient heritage of Pliny and other classical writers of natural history formed a central reference point in early modern natural history, even in the wake of the early 'discoveries' imposed on Europe by the New Worlds. Throughout the sixteenth century, a chain of discussions enabled scholars to distance themselves from the legacy of antiquity. However, even Linnaeus, whose aim was to systematically accommodate species from across the globe, made use of numerous references to ancient authors, thereby underlining the continued importance of the scholarly past. In fact, antiquarians and naturalists worked collaboratively to interpret traditions of investigating the natural world.[6] By holding on to the classical heritage while also aiming at understanding nature from across the globe, early modern natural history diverged from the trajectory typically associated with modernity. Progress entailed encompassing the world and scholarship into a single whole rather than just moving on from what had gone before by fundamentally shifting the parameters or methods of scholarship.

Beyond the complex linkage to traditions, early modern natural history also took the form of a composite social space where many practices and groups coexisted. The world of plants and botanists was also inhabited by gardeners, horticulturists and flower lovers who used different and not always compatible systems and terms of definition for species and their varieties, for example.[7] The complexity of this world is captured to some degree in Linnaeus's own attempt to classify those contributing to the production of botanical knowledge. In his *Philosophia botanica* Linnaeus mapped out the epistemic community by opposing botanists (professional scholars) to 'botanophiles' (amateurs).[8] The first group contained 'methodisers' like Linnaeus and 'collectors', including those 'concerned primarily with the number of species of vegetables'; Linnaeus listed them as 'fathers, commentators, illustrators, describers, monographers, the meticulous, Adonises, compilers of Floras, and travellers'.[9] The 'amateurs of Botany', the other group, Linnaeus described as having 'produced various [works] of botany which do not properly pertain to botanical science, including anatomists, gardeners, physicians, and miscellaneous persons'.[10]

6. Marie-Noëlle Bourguet, Christian Licoppe and Heinz Otto Sibum (ed.), *Instruments, travel and science: itineraries of precision from the seventeenth to the twentieth century* (London, 2002).
7. Jean-Marc Drouin, *L'Herbier des philosophes* (Paris, 2008). Paula Findlen, 'Natural history', in *The Cambridge history of science*, ed. David C. Lindberg and Ronald L. *Numbers*, 8 vols (Cambridge, 2003-2013), vol.3, *Early modern science*, ed. Katharine Park and Lorraine Daston (Cambridge, 2006), p.435-68.
8. Carl Linnaeus, *Linnaeus's 'Philosophia botanica'*, translated by Stephen Freer (Oxford, 2003).
9. Linnaeus, *Linnaeus's 'Philosophia'*, p.18.
10. Linnaeus, *Linnaeus's 'Philosophia'*, p.26.

Of course, Linnaeus was not the only one attempting to order the world of naturalists, and neither were his demarcations accepted by everyone. The author of the article on 'Natural history' in the *Encyclopédie* asserted, for example, that 'the culture of flowers and of trees belongs to Botany' – botanical connoisseurship was something that flower merchants and botanists shared.[11] As numerous studies have shown, the social world of European natural history was also highly diverse, involving many types of actors. With regard to developments outside Europe, Creole and indigenous people also played a crucial role in the production of natural knowledge.[12] While the diversity of natural history challenged Linnaeus, to historians it has offered an opportunity to write a more inclusive history of science.

While maintaining proximity with other knowledge traditions, such as classical culture and law, the observation of nature did more than generate textual comments that could be restricted to libraries and books. Natural history was foremost an empirical science; collections in cabinets of curiosity and public museums helped to make natural history and the systems it generated in order to classify nature tangible to the wider public. In 1780, Dezallier d'Argenville listed 695 cabinets of curiosity throughout Europe, of which 450 were in Paris.[13] In this respect the history of natural history is to no small extent embedded in the histories of various institutions, museums and collections.

The paradigm of collecting and accumulating things largely took place en route, while travelling, walking and botanising. This 'mobilisation' of natural history, which, according to Brian Ogilvie, dates back to at least the sixteenth century, meant that next to observation the senses of smell and taste became core to the work of the naturalist.[14] To Linnaeus travelling was central to the discipline of natural history; he argued this in his inaugural professorial speech at Uppsala University in 1741, *Oratio*

11. Quote from Drouin, *L'Herbier des philosophes*, p.38.
12. Kapil Raj, 'Surgeons, fakirs, merchants and craftsmen: making L'Empereur's *Jardin* in early modern South Asia', in *Colonial botany: science, commerce and politics in the early modern world*, ed. Londa Schiebinger and Claudia Swan (Philadelphia, PA, 2005), p.252-69; Miles Ogborn, 'Discriminating evidence: closeness and distance in natural and civil histories of the Caribbean', *Modern intellectual history* 11:3 (2014), p.631-53.
13. Aline Pelletier, 'Cabinet', in *1740, un abrégé du monde: savoirs et collections autour de Dezallier d'Argenville*, ed. Anne Lafont (Lyon, 2012), p.60-74. More broadly, Barbara M. Benedict, *Curiosity: a cultural history of early modern inquiry* (Chicago, IL, 2002); Lorraine Daston and Katharine Park, *Wonders and the order of nature, 1150-1750* (New York, 1998); Robert John Weston Evans and Alexander Marr (ed.), *Curiosity and wonder from the Renaissance to the Enlightenment* (Aldershot, 2006); Neil Kenny, *The Uses of curiosity in early modern France and Germany* (Oxford, 2004).
14. Brian W. Ogilvie, *The Science of describing: natural history in Renaissance Europe* (Chicago, IL, 2006).

qua peregrinationum intra patriam asseritur necessita, and in later writings such as *Instructio peregrinatoris* (1759). Autopsy, seeing for yourself, was key to Linnaeus and one reason why he favoured travelling, while observation, description, nomenclature, classification and pictorial representation constituted a continuum of activities.[15] Mobility was essential in this economy of naturalistic enterprise, and allowed the inventorying of ever more species. During the Enlightenment, the notion of 'the field' also became linked to the idea of measurement and quantification; fields in the abstract sense became areas of investigation, in the exploration of the systematic relationship between the life and geography of plants.[16] By the end of the eighteenth century, the physical environment – locality, distribution, temperature, altitude, climate – generated a bio-geographical approach to plants most famously connected to the work of Alexander von Humboldt.

Much of the work on collections, mobility and knowledge circulation owes its current popularity to the spatial turn, paired with an interest in European trade and colonisation. A culture had existed in Europe, since the Renaissance, of accommodating plant transfer and acclimatising exotic products, such as tomatoes, potatoes and strawberries. Whether in England, Spain, France, or the Dutch Republic, early modern interest in colonies was largely motivated by opportunities for long-distance trade.[17] As mobility became an element of the naturalist's basic practice in the modern era, botanical gardens and academies' collections expanded; the new imperative was collections representing nature globally. In the eighteenth century, the description of nature also became linked to colonial and commercial impulses as a result of the great European monarchies' searches for exotic products and natural resources.[18]

Linnaeus, too, used the backdrop of global trade to highlight the potential role of natural history. In his treatise on the tea plant, Linnaeus wrote:

15. Sachiko Kusukawa, *Picturing the book of nature: image, text and argument in sixteenth-century human anatomy and medical botany* (Chicago, IL, 2011); Sue Ann Prince (ed.), *Of elephants and roses: French natural history, 1790-1830* (Philadelphia, PA, 2013).
16. Marie-Noëlle Bourguet, 'Measurable difference: botany, climate and the gardener's thermometer in eighteenth-century France', in *Colonial botany: science, commerce and politics in the early modern world*, ed. Londa Schiebinger and Claudia Swan (Philadelphia, PA, 2005), p.270-86.
17. Sarah Easterby-Smith, *Cultivating commerce: cultures of botany in Britain and in France, 1760-1815* (Cambridge, 2017).
18. Richard H. Grove, *Green imperialism: colonial expansion, tropical island Edens and the origins of environmentalism, 1600-1860* (New York, 1995); Jordan Kellman, 'Nature, networks and expert testimony in the colonial Atlantic: the case of cochineal', in *Itineraries of Atlantic science: new questions, new approaches, new directions*, ed. Neil Safier, *Atlantic studies* 7:4 (2010), p.373-95.

> It is amusing to note that the Europeans, the wisest of all, conquer the most distant regions of America, and there with the greatest difficulty excavate precious s i l v e r, and with greatest danger repatriate it to Europe, in order to, with no less danger, carry it to another part of the world, to the East Indies, only to exchange it for the leaves of a certain shrub.[19]

Linnaeus had two solutions for this specific problem: one was trying to move tea production from China to Sweden; the second was to find Swedish leaves to infuse in hot water. Needless to say, neither his tea transplant project nor his import substitution plan were ever realised. Other contemporary and later European projects, to move and substitute, and shift productions across the globe, from Asia to the Atlantic world, were, however, successful. The organisation of Kew Gardens is emblematic of this imperial projection of applied botany.[20] In the wake of environmental history, historians have also begun to pay attention to colonial bioprospecting and natural resources.[21] This approach gave weight to a new avenue of research based on practices of mobility and artefacts, one that considers scientific work as a result of different tasks, and of different types of actions.

The connection between natural history and European expansion is also something that has become a central focal point over the last fifteen years, in the much revitalised study of 'science and empire'.[22] These new research perspectives are the result of a radical questioning of the prevailing diffusionist representation, and the desire to subject this field to more complex forms of analysis. These are to a large extent a legacy of anthropological inquiry. By criticising the earlier research agenda that focused on the 'centre/periphery' relationship, historians have sought to

19. 'Det är roande att tänka att europeerna, de klokaste bland alla, erövra de avlägsnaste trakter i Amerika, därstädes med största möda utgräva det kostbara s i l v r e t, under största fara hemforsla det till Europa, och slutligen med icke mindre fara föra det till en annan del av jorden, till Ostindien, endast för att hämta bladen av en viss buske'; Carl von Linné, *Linnés avhandling Potus theae 1765* (Uppsala, 2002), p.14.
20. Richard Drayton, *Nature's government: science, imperial Britain and the 'improvement' of the world* (New Delhi, 2005). Recently Skuncke has pointed out that Swedish plant transfer failed due to being of insufficient scale, while plant transfers within the British Empire could operate on a global scale with similar climate in several very different locations. See Marie-Christine Skuncke, *Carl Peter Thunberg, botanist and physician: career-building across the oceans in the eighteenth century* (Uppsala, 2014).
21. See Grove, *Green imperialism*; Fredrik Albritton Jonsson, *Enlightenment's frontier: the Scottish Highlands and the origins of environmentalism* (New Haven, CT, 2013). See also Londa Schiebinger, *Plants and empire: colonial bioprospecting in the Atlantic world* (Cambridge, MA, 2004).
22. For example, Patrick Petitjean, Catherine Jami and Anne Marie Moulin (ed.), *Science and empires: historical studies about scientific development and European expansion* (Dordrecht, 1992); for the collection of articles edited by Londa Schiebinger in the special forum section, see 'Focus: colonial science' in *Isis* 96:1 (2005), p.52-87; Benedikt Stuchtey (ed.), *Science across the European empires 1800-1950* (Oxford, 2005).

place the construction of knowledge within an entangled, albeit asymmetric, historical narrative. At the same time, they have tried to demonstrate that colonial processes are marked by interaction and exchange according to the dictates of networks and circuits, at the centre of which exchange is produced.

This new area of research has, from the beginning, emphasised the importance of its learned and predominantly scientific elements as privileged agents of imperial domination. As a result, it has promoted the analysis of places and spaces beyond Europe, and has integrated them into the general inquiries of the history of science. Furthermore, it has demonstrated the limits of the Eurocentric definition of science and knowledge in the existing studies of the field. Initially focusing on nineteenth- and twentieth-century studies, research has extended back in time to earlier European empires and urban areas, which are generally missing from the broad maps of scientific knowledge as inherited from the classic historiography of the Scientific Revolution. Studies devoted to the history of science in the Spanish or Portuguese empires have in recent years emphasised the part played by the extensive networks of information that structured the worldly exchange of knowledge from the sixteenth through the eighteenth centuries.[23] At the same time, they have underlined the need for a better analysis of 'areas of contact', their specificity, and their localisation.[24]

In the last decade several books on 'nature and empire' have reshaped the earlier debate on the role of science in the formation of colonial phenomena. New approaches, with a periodisation more familiar to historians of the modern period, have rejuvenated the approach to the study of cities and capitals – especially in the colonies. The model of the 'colonial machine' represents one way to explain the integration of faraway spaces via scientific activity.[25] By focusing on each sequence of the naturalist's work, historians have underlined the multi-faceted operations utilised by naturalists to transform travelling practices into

23. Kellman, 'Nature, networks', p.373-95.

24. On Portugal, see the crucial contributions by Luís Saraiva (ed.), *History of mathematical sciences, Portugal and East Asia II: scientific practices and the Portuguese expansion in Asia (1498-1759)* (Hackensack, NJ, 2004); Luís Saraiva and Catherine Jami (ed.), *History of mathematical sciences, Portugal and East Asia III: the Jesuits, the Padroado and East Asian science (1552-1773)* (Hackensack, NJ, 2008). On Spain, beyond the handbook by Jorge Cañizares-Esguerra, *Nature, empire and nation: explorations of the history of science in the Iberian world* (Stanford, CA, 2006), see Jesús María Carrillo Castillo, *Naturaleza e imperio: la representación del mundo natural en la Historia general y natural de las Indias de Gonzalo Fernández de Oviedo* (Madrid, 2004), and Víctor Navarro Brotóns and William Eamon (ed.), *Más allá de la Leyenda Negra: España y la revolución científica* (Valencia, 2007).

25. James E. McClellan III and François Regourd, *The Colonial machine: French science and overseas expansion in the old regime* (Turnhout, 2011); Schiebinger and Swan (ed.), *Colonial botany.*

scientific publications and reputations.[26] The scientific book is therefore now considered to be a final link in a chain of operations and activities including travel, collecting, corresponding, seeing, classifying, settling and publishing.[27]

The historiography of Linnaean natural history

Taken as a whole, what is conveyed by the historiographical renewal we have just described is an increasingly clear understanding of the history of natural history in the long eighteenth century as a globalisation of knowledge about nature. There were competing strategies – French, Dutch and English – for this process of globalisation from the 1660s to the 1730s, and eventually it was the attempt based on Linnaean systematics that proved most effective and influential. For much of the latter half of the eighteenth century it achieved a dominant position in botany and zoology, and although it was, to a large degree, supplanted by other taxonomic traditions around the turn of the nineteenth century it has had a lasting impact on both science and society up to the present day.

This significance is frequently acknowledged, yet the work of Linnaeus and his students has played a relatively minor role in the lively scholarly discussion on the history of science in recent decades. The literature contains a paradoxical 'metropolitan' focus, as it were, on England, France and the Netherlands, not fully taking into account the important role that Linnaean natural history, broadly conceived, has played.[28] However, while Linnaean natural history may not always have been 'central' in recent international scholarship, when it has been discussed it has often been viewed from within the diffusionist model of the history of science that has been challenged and rejected elsewhere. There has been a strong focus on Linnaeus himself as a person and the centralised, coherent and monolithic nature of the whole Linnaean enterprise has been emphasised in a way strongly reminiscent of Bruno Latour's concepts of 'centres of calculation' and 'cycles of accumulation'. In fact, these concepts align very closely with Linnaeus's own view of himself

26. Neil Safier, *Measuring the new world: Enlightenment science and South America* (Chicago, IL, 2008); Jim Endersby, *Imperial nature: Joseph Hooker and the practices of Victorian science* (Chicago, IL, 2008).
27. Endersby, *Imperial nature.*
28. See Sverker Sörlin's comment: 'Centrality and peripherality are, in many ways, later constructs that have followed from the patterns of recent scholarly priorities and patterns of intellectual, if not linguistic, hegemony'; Sverker Sörlin, 'Science, empire and enlightenment: geographies of northern field science', *European review of history/Revue européenne d'histoire* 13 (2006), p.455-72 (466). This can be seen as an example of how smaller states such as Sweden are especially interesting when analysing the connections between politics, economics and science.

as the centre and the master of a vast machinery of taxonomic activity associated with his name. One might say that Linnaeus, especially as he presented himself, is the very archetype of a 'centre of calculation' in Latour's sense and that the international literature on Linnaean natural history – not least due to the influence that Linnaeus himself has had on the creation of his own legacy – with few exceptions has tended to reflect this kind of perspective.

We will return to those exceptions, but until very recently, then, Linnaean natural history has not extensively been situated within or inscribed into the kind of history of science that has come to the fore in recent years, and that is what we hope to do with this book. We believe that Linnaean science, in a broad sense of the term, represents an important set of empirical data for understanding how knowledge moves and is negotiated in different contexts; how it consists of practices as well as ideas; how it evolves not through isolated individuals but through collaborative and competitive networks; and how a tradition such as Linnaeanism does not suddenly appear and then go away but has threads reaching backwards and forwards through time. Knowledge does not move in a simple, one-way pattern of dissemination, but rather through complex and ever-evolving processes of circulation and (re)negotiation.

The 'Linnaean tradition' and its critics[29]

This view of history of science as a circulation of knowledge shaped by collaboration, negotiation and materiality is in some respects quite different from what one might call 'the Linnaean tradition', which is deeply embedded within a Swedish historical context.[30] There, Linnaeus was long made to represent a specific Swedish approach to the world, an approach that although far reaching was free of any colonial legacy. The Linnaean model of scientific travel, epitomised by his 'apostles' – young students who travelled the world to collect and describe plants and animals according to Linnaean principles – has likewise played an important role in defining a particular Swedish type of global engagement and scientific development.[31] More broadly, Linnaeus himself was

29. It should be stressed that this is in no way a comprehensive overview of the existing literature, which is vast; rather, what we offer here are some very broad outlines, exemplified by a few more influential works. For a somewhat more detailed treatment in Swedish, see Hanna Hodacs and Kenneth Nyberg, *Naturalhistoria på resande fot: om att forska, undervisa och göra karriär i 1700-talets Sverige* (Lund, 2007), p.17-35.
30. See Sverker Sörlin, 'Scientific travel: the Linnean tradition', in *Science in Sweden: the Royal Swedish Academy of Sciences 1739-1989*, ed. Tore Frängsmyr (Canton, MA, 1989), p.96-123.
31. Critically discussed in Sverker Sörlin, 'Apostlarnas gärning: vetenskap och offervilja i Linné-tidevarvet', *Svenska Linnésällskapets årsbok* (1990-1991), p.75-89 (77-84). See the argument about Linnaeus's apostles in Sverker Sörlin and Otto Fagerstedt, *Linné och hans apostlar*

seen and used as a symbol of Swedish identity, for example in America where Swedish Americans erected statues of him as a reminder of 'the old country'.[32]

As a field of scholarly literature, the 'Linnaean tradition' resides in the intersection of botany, history of ideas and biographical writing. Its focus has long been Linnaeus as a person and a scientist, which has often resulted in detailed studies of his life or of people and phenomena associated with him. On the one hand this has meant rather atheoretical work, but on the other it also represents a substantial body of empirical work that in some respects points forward to the notion of a field of 'Linnaean studies' referred to in the beginning of this chapter. The perspectives which characterise this tradition go back to the time of Linnaeus himself – the large amount of material he left behind, including several autobiographies and lists of his accomplishments, came to serve as foundations of the tradition – but it really emerged around the time of the bicentenary of Linnaeus's birth in 1907.[33] A few years earlier, Thore Magnus Fries had published what is still, rather surprisingly, the only major biography in Swedish of Linnaeus. While Fries voiced scepticism about some of the Linnaean legends circulating at the time, as a whole his book was instrumental in cementing a heroic, idealistic and idolising image of Linnaeus as the 'Prince of Botanists' and, not least, beloved teacher of a large number of devoted students. Among them, the travelling 'apostles' took pride of place in creating a powerful image of Linnaeus as a man larger than life.

While there were dissident voices, it was really only after World War Two that the legend of Linnaeus was seriously challenged by historians of ideas such as Sten Lindroth and Olle Franzén.[34] Both Lindroth and Franzén critically examined the history of this legend and how it had been reproduced over the years, but in some respects scholarship still remained heavily centred on Linnaeus as a person – only as a more complex or even unsympathetic individual rather than the selfless scientific hero that had dominated previously. Around the same time,

(Stockholm, 2004), and Hodacs and Nyberg, *Naturalhistoria på resande fot*, p.17-35 (20-27).

32. Dag Blanck, 'Att bli svensk-amerikan. Framväxten av en svensk-amerikansk identitet i USA', in *Främlingar: ett historiskt perspektiv*, ed. Anders Florén and Åsa Karlsson (Uppsala, 1998), p.59-69 (62-64).

33. Linnaeus was also a prolific correspondent who left behind a collection of several thousand letters; for these, and many other writings directly connected to him, see *The Linnaean correspondence*, http://linnaeus.c18.net/ (last accessed 3 March 2015).

34. See Sten Lindroth, 'Linné: legend och verklighet', *Lychnos* 1965-1966, p.56-122; Olle Franzén, 'Hur Linnébilden formades 1778-1850', *Svenska Linnésällskapets årsskrift* (1963), p.5-41 and Gunnar Eriksson, *Botanikens historia i Sverige intill år 1800* (Uppsala, 1969), to just give a few examples.

however, a more theoretically grounded history of ideas and science emerged, investigating the ideas, methods, results and institutions of Linnaean natural history and early modern history of science more broadly. These scholars included Lindroth, again, in his accounts of the history of the Royal Swedish Academy of Sciences, but also Gunnar Eriksson's work on the history of botany, and Gunnar Broberg's studies of Linnaeus's anthropology and view of nature.[35] Tore Frängsmyr should also be mentioned in this context, as well as the botanist Bengt Jonsell who focused on the scientific contributions of the travelling students, and, towards the end of the 1980s, Sverker Sörlin's work on scientific travel as a Swedish speciality.[36] The Linnaean tradition was placed under greater critical analysis and contextualisation in these works than it had been before, resulting in a more nuanced understanding of the development of Linnaean natural history.

Eriksson, for instance, was among the first to emphasise that there were precursors to some of Linnaeus's ideas – they did not arise out of nothing – and that the dominant position of Linnaean systematics in the latter half of the eighteenth century also had some negative effects by inhibiting or delaying the general development of botany and biology after Linnaeus's time.[37] In the same vein, Jonsell has increased our understanding of the role Linnaeus's travelling students played in his work and to what extent they in turn were dependent on the infrastructure of East India companies and colonial empires. As we will see below, Sörlin, meanwhile, has been instrumental in analysing the claims that Linnaeus and others made of Linnaean travel as a Swedish speciality, and in situating the travelling students in a global context where their work became entangled with processes of globalisation and colonialism.[38]

35. Sten Lindroth, *Kungl. svenska vetenskapsakademiens historia 1739-1818* 1:1, *Tiden intill Wargentins död (1783)* (Stockholm, 1967); Sten Lindroth, *Svensk lärdomshistoria* 3, *Frihetstiden* (Stockholm, 1978); and Eriksson, *Botanikens historia.*

36. Tore Frängsmyr (ed.), *Linnaeus: the man and his work*, new edn (Canton, MA, 1994); Bengt Jonsell, 'Thunberg, Linné och det linneanska', in *Carl Peter Thunberg: linnean, resenär, naturforskare 1743-1828*, ed. Bertil Nordenstam (Stockholm, 1993), p.41-55, and Bengt Jonsell, 'Apostlarnas resor och gärningar: Linnélärjungarnas roll i upptäckten av världen', in *Ljus över landet? Upplysningen som drivkraft i 1700-talets svenska vetenskap och vitterhet*, ed. Paul Hallberg (Gothenburg, 2005), p.79-98; Sörlin, 'Scientific travel', p.96-123; Sörlin, 'Apostlarnas gärning', p.75-89; Sörlin and Fagerstedt, *Linné och hans apostlar*; Sörlin, 'Science, empire', p.455-72; and Sverker Sörlin, 'Globalising Linnaeus: economic botany and travelling disciples', *TijdSchrift voor Skandinavistiek*, vol.29:1-2 (2008), p.117-43.

37. Eriksson, *Botanikens historia*, p.176-78.

38. Jonsell, 'Apostlarnas resor och gärningar', p.79-86. See also Kenneth Nyberg, 'Linnaeus's apostles, scientific travel and the East India trade', *Zoologica scripta* 38, Supplement 1 (2009), p.7-16; and Hanna Hodacs, 'Linnaeans outdoors: the transformative role of studying nature "on the move" and outside', *The British journal for the history of science* 44:2 (2011), p.183-209; Sörlin, 'Apostlarnas gärning', p.75-89; and Sörlin and Fagerstedt, *Linné och hans apostlar.*

More recently, these lines of inquiry have been continued and extended through more detailed studies of travel as a scientific and educational practice and the contributions of Linnaeus's 'apostles' to some of his most influential works. Several of these studies focus on both a quite specific, 'internal' history of science issues and a broader type of contextualisation of that history within larger historical processes that foreshadows the current discussion on the global circulation of knowledge.[39]

To sum up, Swedish Linnaean literature has gone from person-centred scholarship to one focused on ideas, theories, methods, and institutions of history of science, and more recently broader historical issues, of which we will see more below. It is in many ways a fundamental change of perspective, corresponding to the shift in recent international scholarship from diffusionist models to more complex questions about the circulation and materiality of knowledge.

The general perspective of the Swedish Linnaean tradition, apart from, perhaps, some specifically national aspects, for a long time also influenced English-language literature on Linnaeus. If nothing else, the notion of the ideas of a lone genius spreading through the world has proved pervasive; it is clearly visible, for instance, in the major English-language biography of Linnaeus by Wilfrid Blunt from 1971 where the outlook and interpretation of events closely follow that in Linnaeus's own autobiographical writings.[40] And while the author of another, more scholarly, work argues that it is important 'to relate the events in the then so small world of science to those in human society and culture in general', the account in Frans Stafleu's *Linnaeus and the Linnaeans: the spreading of their ideas in systematic botany, 1735-1789* mostly focuses on internal scientific developments and a similar interpretation of how Linnaeus's ideas conquered the European Republic of Letters.[41]

To some extent, then, Lisbet Koerner's widely cited book *Linnaeus: nature and nation* was something of a Linnaean comeback in the international scholarship on history of science. Her study focuses on the political economy of Linnaean botany, and places it in the context of growing global trade where cameralist thinkers of states without colonies

39. Pär Eliasson, *Platsens blick: vetenskapsakademien och den naturalhistoriska resan 1790-1840* (Umeå, 1999); Mariette Manktelow and Kenneth Nyberg, 'Linnaeus's apostles and the development of the *Species plantarum*', in *'Species plantarum' 250 years: proceedings of the 'Species plantarum' symposium*, held in Uppsala 22-24 August 2003, ed. Inga Hedberg (Uppsala, 2005), p.73-80 (76-80); Hodacs and Nyberg, *Naturalhistoria på resande fot*; Hodacs, 'Linnaeans outdoors', p.183-209.

40. Wilfrid Blunt (with William T. Stearn), *The Compleat naturalist: a life of Linnaeus* (London, 1971).

41. Frans A. Stafleu, *Linnaeus and the Linnaeans: the spreading of their ideas in systematic botany, 1735-1789* (Utrecht, 1971), quote from the preface (p.v).

emphasised the role of import substitution for strengthening national economies. One way of achieving this was plant transfer, another to find domestic substitutes for items such as tea that could be cultivated more extensively to reduce dependence on foreign trade. In this context natural history, especially botany, became very important and that is how Linnaeus could attract much of the necessary public support for his and his students' work.[42] Recently the intersection of political economy and natural history has also been the object of great scholarly interest with contributions from Staffan Müller-Wille, Sverker Sörlin and Fredrik Albritton Jonsson.[43]

An issue related to political economy and utility is that of local vs universal knowledge, which in turn is connected with the question of how Linnaean natural history was able to travel from the Swedish and European context as analysed by Koerner, to the global scale where it became an essential tool for European expansion and colonial enterprises. This discussion is complex and contains rather divergent strands, where some have emphasised the universalising ambitions of Linnaeanism and its lack of interest in local plant knowledge, while others have argued almost the opposite.[44] Koerner, for instance, has

42. Within the Swedish literature the utilistic dimension of Linnaean botany more generally has long been an established theme; see, for example, Eli F. Heckscher, 'Linnés resor: den ekonomiska bakgrunden', *Svenska Linnésällskapets årsskrift* 25 (1942), p.1-11; and Lindroth, *Kungl. svenska vetenskapsakademiens* 1:1, p.30-35, and 1:2, p.557. However, Koerner has driven the argument further and done so in a major English-language study which has become very influential since its publication. It should also be noted that some Swedish Linnaean scholars such as Jonsell have criticised Koerner for exaggerating the role of such motives for Linnaeus's work, arguing that while these were important, curiosity and basic research were also integral to the whole Linnaean enterprise; see Jonsell, 'Apostlarnas resor och gärningar', p.79-86 (84-85).

43. Staffan Müller-Wille, 'Nature as a marketplace: the political economy of Linnaean botany', in *Oeconomies in the age of Newton*, ed. Margaret Schabas and Neil De Marchi (Durham, NC, 2003), p.154-72, and Staffan Müller-Wille, 'Walnuts at Hudson Bay, coral reefs in Gotland: the colonialism of Linnaean botany', in *Colonial botany: science, commerce and politics in the early modern world*, ed. Londa Schiebinger and Claudia Swan (Philadelphia, PA, 2007), p.34-48; Sörlin, 'Globalising Linnaeus', p.117-43; Fredrik Albritton Jonsson, 'Rival ecologies of global commerce: Adam Smith and the natural historians', *The American historical review* 115:5 (2010), p.1342-63.

44. See Daniela Bleichmar, 'Training the naturalist's eye in the eighteenth century: perfect global visions and local blind spots', in *Skilled visions: between apprenticeship and standards*, ed. Cristina Grasseni (New York, 2007), p.166-190 (186); Daniela Bleichmar, 'The geography of observation: distance and visibility in eighteenth-century botanical travel', in *Histories of scientific observation*, ed. Lorraine Daston and Elizabeth Lunbeck (Chicago, IL, 2011), p.373-95 (392); and Hanna Hodacs, 'Circulating knowledge on nature: travellers and informants, and the changing geography of Linnaean natural history', in *Travel, agency and the circulation of knowledge*, ed. Gesa Mackenthun, Andrea Nicholas and Stephanie Wodianka (Münster, 2017), p.75-97.

stressed the importance to Linnaeus of local knowledge, such as vernacular names.[45] She also points out that:

> For Linnaeus's circles believed that new natural knowledge could form by way of a cross-cultural mediation between high and folk knowledge, and they believed that this syncretic new science was both an epistemology and a technology, that is, both a way to know and a material tool. Hence the Linnaeans' emphasis on vernacular names for flora and fauna, which they argued indicated salient characteristics and local uses.[46]

This view of Linnaean natural history as a form of mediation has later been explored by Müller-Wille, in his attempts to answer precisely the question of how Linnaeanism was able to move from the local and national context in which it was formulated to the truly global context where it later would have such an impact. Focusing in particular on binomial nomenclature, Müller-Wille's conclusion is that it was 'designed to operate in or, rather, to mediate between different cultures rather than to serve the interests of a particular one. [This] made it possible for binomial nomenclature to come into existence in advance of the colonialist conquest for which it later proved to be instrumental.'[47]

Another scholar who has recently engaged the question of Linnaean natural history in a global perspective is Sverker Sörlin. He has mainly worked on Linnaean travel and the ideology that underpinned it, not least the function it filled in Linnaeus's own rhetoric about the value of his work in both a national and global context. Of particular interest here are, perhaps, two articles from 2000 and 2006 respectively, where Sörlin discusses the entanglement of natural history and Linnaean travel with colonialism and global trade. He was among the first to also discuss the commodification of scientific objects, the value, not least commercial, of collections, and problematised our notions of centres and peripheries in the history of science and elsewhere.[48] When situating Linnaeanism in a global context, Sörlin points out that Sweden and its neighbouring countries were both colonial states and colonised, through the indigenous Sami population in the far north. More generally, he has

45. Lisbet Koerner, *Linnaeus: nature and nation* (Cambridge, MA, 1999), p.46; and Lisbet [Koerner] Rausing, 'Underwriting the oeconomy: Linnaeus on nature and mind', in *Oeconomies in the age of Newton*, ed. Margaret Schabas and Neil De Marchi (Durham, NC, 2003), p.173-203.

46. Rausing, 'Underwriting the oeconomy', p.197; see Alix Cooper, *Inventing the indigenous: local knowledge and natural history in early modern Europe* (Cambridge, 2007).

47. Müller-Wille, 'Walnuts at Hudson Bay', p.48; see also Müller-Wille, 'Nature as a marketplace', p.154-72 (168-69).

48. Sverker Sörlin, 'Ordering the world for Europe: science as intelligence and information as seen from the northern periphery', in *Nature and empire: science and the colonial enterprise*, *Osiris* 15 (2000), p.51-69; and Sörlin, 'Science, empire', p.455-72. See also Sörlin and Fagerstedt, *Linné och hans apostlar*; Sörlin, 'Globalising Linnaeus', p.117-43.

emphasised the situatedness of scientists both nationally and internationally, and that they are driven by a range of motivations. Even so, like Koerner he has argued that 'it was economic botany that made the project possible'.[49]

Towards a connected history of Linnaean sciences

Over the last fifteen years the history of the science of the modern era – and particularly of the eighteenth century – has taken a global turn, expanding the discipline's horizon of interests. While this turn was initially manifested in a 'science and empires' framework and the history of the globalisation of the Scientific Revolution, an increasing degree of sophistication led to a greater focus on local contexts, trans-imperial trajectories, connections and the local and transnational intermediaries involved in the mobility of Linnaean sciences. The histories of Linnaean science outlined above are embedded in a global history in multiple ways. Linnaeus was clearly a central figure in what is usually labelled Linnaean natural history or Linnaean science. But the latter was also a system, an infrastructure, or a movement that soon outgrew him and became something much, much larger than a single individual. And perhaps that was also the point. By establishing a set of principles that in themselves constituted an institutionalisation that went beyond specific empirical facts or conclusions, Linnaeus ensured that his name lived on. But what that name, used as a shorthand or label, meant or what it stood for is still open to debate today, 250 years later. The chapters comprising this book suggest a number of methods, processes, ideas and approaches that can help outline what constitutes Linnaean science beyond the sexual system and the binominal names.

In the Linnaean workshop

The first two chapters in this volume, written by Isabelle Charmantier and Bettina Dietz, are concerned with techniques, materials and methods that originated within what we might want to call the 'Linnaean workshop'. The condition that dictated how work was organised in the workshop was the accumulation of information about the natural world, generating what has been called the 'first bio-information crisis'; a development essential for an understanding of Linnaeus's rise to prominence.[50]

49. Sörlin, 'Science, empire', p.455-56; and Sörlin, 'Globalising Linnaeus', p.125 (quote).

50. Staffan Müller-Wille and Sara T. Scharf, *Indexing nature: Carl Linnaeus (1707-1778) and his fact-gathering strategies, Working papers on the nature of evidence: how well do 'facts' travel?* 36:08, http://www.lse.ac.uk/economichistory/pdf/factspdf/3909muellerwillescharf.pdf (last accessed 10 October 2015).

Charmantier's contribution constitutes an excellent first chapter by helpfully providing an overview of Linnaeus's publications. Her work belongs to an emerging research field concerned with early modern practices of recording, storing, organising, retrieving and disseminating information.[51] Charmantier's contribution synthesises research she conducted together with Müller-Wille.[52] Her chapter also provides new and exciting perspectives on developments after Linnaeus's death (1778) and the application of new paper technologies in the emerging centre of Linnaean natural history, London. The chapter draws on the surviving paperwork which reflects the work methods of Linnaeus's son (Carl Linnaeus the Younger, 1741-1783) and some of Linnaeus's most prominent students, such as the London-based Daniel Solander (1733-1782) and Jonas Carlsson Dryander (1748-1810), and Carl Peter Thunberg (1743-1828) who succeeded Linnaeus in Uppsala.

The collaborative aspect of natural history is a theme that has been developed in multiple publications by the author of the second chapter, Bettina Dietz. Previously, Dietz has looked at the role of Linnaeus's correspondents in sending Linnaeus 'error reports' which the latter used when preparing forthcoming new editions of his main publications.[53] In the light of this Dietz has helped underline the 'contribution-based epistemology' (see p.63) and provisional nature of many of Linnaeus's publications, which accords with the conclusion drawn by Charmantier and Müller-Wille. This chapter features Pehr Osbeck (1723-1805), one of Linnaeus's earliest 'apostles', and his collaboration with Linnaeus, but also the afterlife of Osbeck's published travel account in the form of new German and English editions. When comparing the editions, Dietz focuses on additions which take the form of supplementary editorial notes as well as new and revised information, and outlines how this continuous work helped to redefine the boundaries of the species

51. See for example Sara T. Scharf, 'Identification keys, the "natural method" and the development of plant identification manuals', *Journal of the history of biology* 42:1 (2009), p.73-117.
52. Staffan Müller-Wille and Isabelle Charmantier, 'Natural history and information overload: the case of Linnaeus', *Studies in history and philosophy of science*, part C, *Studies in history and philosophy of biological and biomedical sciences* 43:1 (2012), p.4-15; Isabelle Charmantier and Staffan Müller-Wille, 'Lists as research technologies', *Isis* 103:4 (2012), p.743-52; Isabelle Charmantier and Staffan Müller-Wille, 'Carl Linnaeus's botanical paper slips (1767-1773)', *Intellectual history review* 24:2 (2014), p.215-38. See also Isabelle Charmantier, 'Carl Linnaeus and the visual representation of nature', *Historical studies in the natural sciences* 41:4 (2011), p.365-404.
53. Bettina Dietz, 'Linnaeus's restless system: translation as textual engineering in eighteenth-century botany', *Annals of science* 73:2 (2016), p.143-56; Bettina Dietz, 'Contribution and co-production: the collaborative culture of Linnaean botany', *Annals of science* 69:4 (2012), p.551-69. See also Bettina Dietz, *Das System der Natur: die kollaborative Wissenskultur der Botanik im 18. Jahrhundert* (Cologne and Vienna, 2017).

mapping project – from global to regional. Dietz's research also has implications for our understanding of scientific authorship and identity in the shift from early modern to modern history.

Global, local and urban reconfigurations

The shift away from a scholarly system in which one man, situated in the small university town of Uppsala, in the northern part of the Baltic world, could function as a clearing house for a natural history of global parameters forms an important backdrop in the next two contributions. Hanna Hodacs and Stéphane Van Damme's chapters situate Linnaeus's work within the context of the two emergent mega-centres of European and global natural history in the second half of the eighteenth century, London and Paris. However, the chapters in which these two cities take central stage tell quite different stories about how natural history shaped the metropole.

Hodacs follows the money as she traces the shifting valuations and movements of natural history collections with a Linnaean provenance in the north-western part of Europe in the second half of the eighteenth century. Price tags, attached to Linnaeus's own collections before and after his own and his son's death in 1778 and 1783 respectively, are the starting point for the first section of Hodacs's contribution. In the second and third part of the paper she discusses the relocation of the Linnaean collections and its implication by focusing on the material history of natural history collections – their more or less perishable nature – but also the changing styles and fashions associated with organising and displaying specimens in Europe. In the final part of Hodacs's chapter the perspective broadens, offering an outline of a North European market for natural history specimens more generally, as well as the rise of London as a staple market in the specimen trade in the final decades of the eighteenth century more specifically. The chapter ends with a discussion of the emerging use of binomial names in the trade of natural history specimens.

The traces of Linnaean natural history in Paris form the focal point of Stéphane Van Damme's contribution, as he compares overlapping local histories of natural history with those of a global extent. Paris as the bastion of resistance to Linnaean taxonomy is not the main theme of the history outlined here. Instead, chapter 4 illuminates the many ways in which natural history shaped the notion of Paris, its geography and inhabitants in a broader sense. Van Damme starts with an overview of research into the history of natural history in Paris. He draws attention to both its role as a colonial power centre, and to the city's relations with other centres of learning in France, several of which had strong Linnaean connections, and networks of well-known *and* forgotten scholars. Different groups and individuals associated natural history with different ambitions – Van Damme uses the notion of the 'invisible empire' to identify naturalists operating outside the realm of the state and institutions in the capital, independently of the universalising ambitions of the latter sphere of power. Zooming in on Paris as a locality, Van Damme also investigates different communities and areas in which natural history became an important reference point. These include a community of Parisian chemists and apothecaries as well as naturalists who were busy making inventories of the local Parisian landscape.

History of mankind: the Linnaean lineage

Chapters 5 and 6 relate to Linnaeus's influence on intellectual history and philosophy. Jonas Gerlings's, and Linda Andersson Burnett and Bruce Buchan's contributions intersect in that they both outline the effects of Linnaeus's system in allocating a place for mankind in natural history. Together the two papers serve to highlight the many dialogues in which Linnaeus's taxonomy was invoked, and the implications shaping a distinct European understanding of progress, power and the explanatory role given to race in the forthcoming century.

Gerlings's main focal point is how Linnaeus's taxonomy, as it was represented in the publications of several generations of German naturalists, influenced Immanuel Kant – foremost his *Critique of the power of judgment* – and particularly in relationship to the science of man. Gerlings approaches the topic by mapping the status of natural history and the growing influence of Linnaean taxonomy in the southern and eastern parts of the Baltic area, Prussia and particularly Königsberg, from the 1720s to the 1790s. This scope allows Gerlings to bridge as well as problematise two divides in intellectual history and post-colonial studies: firstly, one traditionally employed to chronologically separate Linnaeus from Kant; and secondly, one which typically presupposes Europe as one unproblematic entity engaged in defining and exploiting the rest of the world. The case of Kant and Königsberg helps to illuminate the importance of peripheries in providing alternative perspectives to central strands in European intellectual developments.

Andersson Burnett and Buchan's paper also studies a process of intellectual interleafing, examining the interplay between Linnaeus's work and ideas central to the Scottish Enlightenment, foremost stadial history. However, the authors also shift the focus outwards, zooming in on Edinburgh-trained scholars travelling in western and southern Africa, studying the formation of a discourse in which the notion of human races became increasingly significant, creating divides between and legitimising exploitation of different groups. Starting in Britain, chapter 6 provides a backdrop to the introduction of Linnaeus to the English-speaking world, complementing the histories told by Charmantier and Hodacs discussed previously. The first part of the paper does not, however, focus on London but Edinburgh and academics engaged in natural as well as stadial and conjectural history.[54] This overlapping of different intellectual traditions forms the background to the second part of the paper which focuses on the African journeys of Mungo Park (1771-1806) and William Somerville (1771-1860), both students in Edinburgh in the 1790s. Taken together Somerville's and Park's accounts provide a bridge, highlighting not only how Linnaean taxonomy and the tradition of ethnographic writing became entangled in ideas about stadial theory central to the Scottish Enlightenment, but also, moving forwards in history, to the inscription of race in the development of different social organisations.

Tracing knowledge: writing lives and empires

Applying one's knowledge of Linnaean taxonomy to the natural world was not exclusively a collective information-gathering project or part of a broader intellectual movement; it also reflects how individuals in the past used their knowledge searching for opportunities and means to further their positions in a world of changing political and social circumstances. The last chapters in this book follow the lives of two students of Linnaeus and how they negotiated nature as well as states, empires, trading companies, institutions and networks which provided them with opportunities and access to areas across South Asia and Spanish Latin America.

The history of Johann Gerhard König (1728-1785) told by Niklas Thode Jensen captures how an individual life could be shaped by engaging in natural history, taking into account conditions both local to the Baltic area and South and South East Asia.[55] König was of German descent, born in what was then Polish Latvia. His working life was, however, fundamentally shaped by the changing fortunes of the Danish and the British empires in Asia. König arrived in South Asia in 1768, when political powers were fluid and unstable: the Dutch, French, Danish and Asian rulers, such as the Nawab of Bengal, were losing out to the

54. See also Linda Andersson Burnett, 'Northern noble savages? Edward Daniel Clarke and British primitivist narratives on Scotland and Scandinavia *c.*1760-1822', unpublished doctoral dissertation, University of Edinburgh, 2012.

55. For more work on Danish-Indian scholarship see also Niklas Thode Jensen, 'Making it in Tranquebar: science, medicine and the circulation of knowledge in the Danish-Halle Mission, *c.*1732-1744', in *Beyond Tranquebar: grappling across cultural borders in South India*, ed. Esther Fihl and A. R. Venkatachalapathy (Delhi, 2014), p.325-51.

increasingly powerful English East India Company (EIC) which, in 1778, employed König as its first stationary naturalist in South Asia. Until his death in 1785, König travelled extensively, reporting on natural history and potentially profitable ways to exploit nature in different parts of present-day India, Sri Lanka, Indonesia and Thailand. Jensen captures how König negotiated the changing political circumstances, drawing on his local and regional networks, but also those centred in north west Europe, in London and Copenhagen. Zooming in on König's field work Jensen highlights how communication problems, social and work hierarchies, and the ambulatory dimension erased the explicit acknowledgement of local people who provided König with crucial information.

Studying the life and legacy of Linnaeus's student Pehr Löfling (1729-1756), Kenneth Nyberg and Manuel Lucena Giraldo explore how Linnaean natural history travelled and how it became an integrated part of different contexts in northern and southern Europe, and in Spanish America.[56] Löfling became a student of Linnaeus in Uppsala in the 1740s; in 1751 he left Sweden for Spain and Madrid, where he was to remain for two years. These first two stages of Löfling's life provide a basis for investigating both agency and dependency as the relationship between Löfling and Linnaeus changed and as the former developed his own networks of scholarly contacts. In the last two parts of the chapter the focus shifts, taking in Löfling's participation in a Spanish expedition to the Orinoco basin of present-day Venezuela, the aim of which was to settle border disputes between European powers and to do an inventory of natural resources. Fragmented sources provide only a few glimpses of Löfling in this context, how he operated and the local knowledge he drew on. The general history of the expedition and its legacy do, however, provide a telling example of how Linnaean natural history became incorporated into a colonial project.

Linnaean studies as part of a global history

The histories of Linnaean science outlined above are embedded in a global history, and by considering the geography and chronology of this history more carefully it is possible to outline the potential scope and relevance of Linnaean studies. A series of global historical connections and conjunctures needs to be taken into account; the Iberian expansion across the Atlantic basin at the end of the fifteenth century provides an early departure point. Northern Europeans responded to the vast Latin American silver deposits, looted by the Spanish, by envisioning similar

56. See also Kenneth Nyberg, 'Från Tolvfors till Orinoco: Linnélärjungen Pehr Löflings resa och den globala historien', *Historielärarnas förenings årsskrift* (2015), p.23-39.

amounts of wealth hidden underground. The physic-theological idea of a divine and symmetrical distribution of natural wealth across the globe came to elevate natural history and particularly mineralogy in a context of mining, as the search began for riches undergrown at home. The roots of the political economy underpinning Linnaeus's work can be found here.[57] While Linnaeus followed the tradition of land-locked continental naturalists seeking natural riches at home rather than abroad, his followers did not remain local. As suggested by Löfling's employment as an imperial naturalist in Spanish service, the Spanish Empire was still powerful in the eighteenth century. By this point, the centre of the European expansion had however been gradually shifting away from Madrid, to north west Europe. Not only did the rising Northern European maritime powers, foremost the French, the British and the Dutch Republic, start to compete with the Portuguese and the Spanish in the Atlantic basin, they also moved eastwards into the Indian Ocean. The Dutch took the early lead in the latter development, as the Vereenigde Oost-Indische Compagnie (VOC) established itself in East Asia.

While the latter part of the eighteenth century saw the rise of the second British Empire with its focus on Asia and subsequently Africa, the legacy of Dutch dominance in the East forms an important strand in Linnaeus's own life and that of his students. Take, for example, Linnaeus's Dutch employer in the 1730s, George Clifford III (1685-1760). Linnaeus worked for three formative years as Clifford's in-house naturalist, writing and studying exotic plants that arrived from across the world at the large botanical garden and natural history collections in Hartekamp, in the Dutch Republic. Clifford himself sat on the board of the VOC. Although the VOC's influence was in decline, and in 1799 the Company went bankrupt, it still provided Linnaeus's student Thunberg with an infrastructure which helped him to explore Asian and African flora and fauna. Thunberg travelled on Dutch ships for nine years, most famously to Japan, before he returned to take over Linnaeus's chair in Uppsala.[58]

Several other East India companies provided vessels and opportunities for Linnaean travellers. Osbeck was just one of several students of Linnaeus who worked for the Swedish East India Company (SEIC), which almost exclusively traded with the still-powerful Chinese Empire. A late player in the East India trade, the SEIC operated under different premises to the older companies which combined trade with colonialisation. Nonetheless, it was the budding European taste for Asian goods, mag-

57. Cooper, *Inventing the indigenous*; Alix Cooper, '"The possibilities of the land": the inventory of "natural riches" in the early modern German territories', in *Oeconomies in the age of Newton*, ed. Margaret Schabas and Neil De Marchi (Durham, NC, 2003), p.129-53.

58. Skuncke, *Carl Peter Thunberg*.

nified by the trade of the seventeenth-century Dutch and English East India companies, that stimulated the establishment of the Swedish Company in 1731. By the early 1750s, the Company had provided transport for at least four of Linnaeus's students eastwards.[59]

Operating out of Tranquebar, the Danish Asiatic Company's Indian centre, König's early South Asian career was closely connected to Danish East India trade. The Danish colonial role in South Asia was relatively minor in scale and, more importantly, was soon overshadowed by British expansion as it harnessed an ever-growing share of the vast resources of India. König's change of employers reflects this shift in European power and influence over South Asia. Although Russia, too, employed Swedish naturalists to explore and expand its eastern borders, Britain did provide the main source of employment for Linnaeus's students. While König was the first of them to explore South and South East Asia by land on behalf of the EIC, Solander and Dryander were busy organising libraries, archives and collections across London according to Linnaean taxonomy. Solander also travelled with Joseph Banks and James Cook on the latter's first circumnavigation of the world in 1768, marking a shift in the British pursuit of natural history and colonial expansion in the Pacific. Other students of Linnaeus followed suit, joining Cook on his second expedition, or taking part in West Africa projects under the British flag. By extending the scope to include students who, like Park and Somerville, were educated by keen Linnaean scholars in Edinburgh, the number of naturalists with Linnaean training who travelled the realm of the expanding British Empire multiplied.

France was the only European superpower in the eighteenth century in whose service none of Linnaeus's 'apostles' found employment, although of course natural history was a core interest to the expanding French Empire, and a central factor in turning Paris into a tribunal of science in the eighteenth century. In a parallel movement London became another centre, a process which also reflected the shifting economics and fashions in north west Europe, including the downturn Amsterdam suffered as a staple market for natural history specimens. More importantly, embedded here is the shift from a poly-centred world of European scholarship in the form of the Republic of Letters to one in which national universities, the nation state, and centres of empires became the dominant formats in which science was undertaken. Ironically, perhaps, Linnaeus, or at least his legacy in Sweden, suffered in this process, as Linnaeus's own collections were moved from Uppsala to London, becoming part of an ever-growing assembly of collections from across Europe and the world that were housed in the British capital.

59. Nyberg, 'Linnaeus's apostles, scientific travel', p.12.

This re-location mirrors a growing asymmetry of power within Europe, which also explains the momentum Linnaean natural history lent to the expansion of European and particularly, perhaps, British influence over the rest of the world. Linnaean science was an important part of the curriculum in universities in the Baltic region, and in European states such as Prussia, which remained somewhat peripheral to the maritime and colonial expansion. In this Baltic context Linnaean natural history largely served to educate future servants of the State and the Church, provincial doctors, clergymen, 'projectors', preparing them for exploring local resources and the economic interests of their states, alongside an economic programme which reflected an ambition to locate the source of wealth at home rather than in the long-distance trade with exotic goods or colonising ventures.

Linnaeus's universal scope, aiming at a global systematisation of nature, was integral to a broad-spectrum education which also opened up opportunities for students to travel outside the Baltic region, to areas targeted in long-distance trade and by European colonial powers. Löfling and König are only two among many naturalist students from the northern shores of the European continent who undertook long and complex journeys, socially, professionally as well as geographically. In this respect we can compare Linnaean science to a pair of lenses which sharpened the vision of both the eye that was turned inwards, scanning domestic landscapes for substitutes and richness, *and* the eye directed outwards, scrutinising nature outside Europe. Linnaean natural history created a bridge between an educational system largely situated in northern Europe, and long-distance trade and European colonialism. Over time and embedded in academic environments like Edinburgh and Königsberg, Linnaean sciences also came to fertilise theories which hoped to explain and legitimise European expansion, by producing theories about human races and developments.

The publication strategies and paper technologies discussed above are suggestive in that they indicate one precondition for the success of Linnaeus's sciences worldwide, namely their capacity to process and communicate vast sets of information. Remains in the form of correspondence and publications between and by naturalists, as well as paper trails in the form of notes and index cards, pay witness to this evolving capacity of Linnaeus's system. Even more daunting, albeit largely hidden, is the idea of the amount of communication that *must* have taken place in the field between naturalists trained in Linnaean sciences and those providing information on indigenous names and uses of species of animals and plants, information that subsequently became re-labelled and reconfigured to fit with Linnaeus's universal system. As the cases of König and Löfling illustrate there are only a few tangible records that can help us reconstruct the exchange between Linnaean naturalists and those carrying indigenous knowledge. More significantly these exchanges leave relatively little to work with in order to reassemble local knowledge systems existing prior to or parallel with the one Linnaeus developed. One great problem for historians to overcome is of course that local experts on nature outside Europe left few paper trails which we can use mapping knowledge outside the realm of Linnaeus. The largely 'invisible' knowledge about nature outside Europe, essential for the success of Linnaeus's system, is something our book at best only can indicate the scope of – as a known unknown.

Today's historiographical landscape of Linnaean sciences in a global context is far more nuanced and complex than that of twenty years ago, when research was still obsessed with the model of centre and periphery and the Needham paradigm of the delayed development of science outside Europe. Today globalisation is viewed as discontinuous, with strong and weak phases in a world formed of mosaics and archipelagos, local encounters and go-betweens, initiating a new dialogue with specialists in different cultural regions. For a long time the grand narrative of Linnaean science was Eurocentric and heroic. The global history of Linnaean sciences has become a critical field abounding in debates that do not shy away from controversy and reflexivity. Its perspectives have become more solid and credible, even though they are regularly called into question in approaches ranging from the post-colonial to those of environmental studies, which regard the expansion of the world as the starting point of the anthropocene period. But while the analyses of historians of science avoid any whiff of teleology, globalisation tends to be described as a monolithic, irresistible and globally positive phenomenon linked to the growth of a scientific knowledge seen as universal.

To sum up, the development of Linnaean science can be explained through a history of global connections and conjunctures. It is with the help of this global history we can both de- and re-centre Linnaeus, hopefully evoking continued interest so as to instil further meaning and content into the field of Linnaean studies.

1. Notebooks, files and slips: Carl Linnaeus and his disciples at work

ISABELLE CHARMANTIER

In the past few years, research has shed light on the Swedish naturalist Carl Linnaeus's (1707-1778) idiosyncratic ways of working. Using well-known, contemporary paper technologies such as commonplace books and interleaved books, Linnaeus successfully adapted them to his own work, and developed a publication strategy which contributed to his immense success as a botanist and a naturalist. For most of his life, Linnaeus's paper technologies suited his mode of working, allowing him to publish successive editions of his main works quickly to get the information out, and trigger feedback from his numerous correspondents. The system of nature with Linnaeus's main classificatory systems for the three kingdoms of nature, entitled *Systema naturae* and first published in 1735, went through twelve editions in Linnaeus's lifetime; his catalogue of plant genera, *Genera plantarum*, first published in 1737, was edited six times; the universal catalogue of plant species, *Species plantarum* was first published in 1753 and was re-edited in 1762-1763. Further editions of *Systema naturae* and *Species plantarum* were published after Linnaeus's death. Each edition was bigger than the last one, including more species and more genera. Towards the end of his career, faced with ever-increasing information on new plants, Linnaeus resorted to publishing supplements of his major works, rather than full new editions. At the same time he experimented with a new methodology (for him): that of paper slips. This new methodology proved so successful that it became a filing system, rather than simply another preparatory strategy for a specific publication.

Linnaeus had a reputation for being a charismatic teacher. His lectures were well attended, and he attracted many students who roamed the globe and collected specimens, not only for themselves, but also for

Material for the first part of this article has been compiled from previously published articles: Staffan Müller-Wille and Isabelle Charmantier, 'Natural history and information overload: the case of Linnaeus', *Studies in history and philosophy of science*, part C, *Studies in history and philosophy of biological and biomedical sciences* 43:1 (2012), p.4-15; and Isabelle Charmantier and Staffan Müller-Wille, 'Carl Linnaeus's botanical paper slips (1767-1773)', *Intellectual history review* 24:2 (2014), p.215-38. The author thanks the editors of this volume for organising two successful workshops in Stockholm (2012) and in Florence (2014).

their patrons and correspondents, amongst whom Linnaeus figured prominently. Linnaean botany was characterised by its simple sexual system of identifying and classifying plants, and by its clear binomial nomenclature. Linnaeus's main works had established a precise way of naming and describing a plant genus or species. The layout in *Genera plantarum* and in *Species plantarum* followed a strict order of description, while the terms botanists should use to describe a plant were meticulously determined by Linnaeus in his 1762 *Termini botanici*.[1] The nomenclature and the order in which a plant genus or species should be described became Linnaeus's established norm for describing plants, a convention followed by his students. They learnt his descriptive methods not only from Linnaeus's published works, but also from the lectures and private tutorials Linnaeus gave, and from the field expeditions undertaken around Uppsala.[2] Information concerning new plants that Linnaeus's so-called 'apostles' encountered in the far-flung corners of the Earth was therefore more easily communicated by the simplicity and the rigidity of Linnaean botany. The increased mobility of Linnaeus's disciples and the profusion of newly discovered exotic plants were counterbalanced by an immutable way of recording information about plants. Whilst the exchange of plants and letters between Linnaeus and his 'apostles' has been extensively studied, and Linnaeus's paper and writing technologies are now relatively well known, the potential legacy of these paper technologies in Linnaeus's disciples' work, and the extent to which these technologies were part and parcel of the application of Linnaean science, has not been studied at all.

In a first instance, this chapter will provide a very broad overview of Linnaeus's working technologies, from his student days to the end of his working life. Secondly, the chapter will bring to light the extent to which Linnaeus's writing practices were emulated by his students, including not only the way a plant was formally described, but also the paper technologies which were used to convey that description. It will focus on Linnaeus's son Carl Linnaeus the Younger (1741-1783), and on his 'apostles' Daniel Solander (1733-1782) and Carl Peter Thunberg (1743-1828).

Linnaeus's paper-based information technologies

The search for a system

Between 1727, when he enrolled at the University of Lund, and spring 1735, when he left Sweden for continental Europe, Carl Linnaeus's

1. Carl Linnaeus, *Termini botanici* (Uppsala [no name], 1762).
2. See Carl Linnaeus, *Herbationes Upsalienses* (Uppsala, Laur. M. Hojer, 1753); Hanna Hodacs, 'In the field: exploring nature with Carolus Linnaeus', *Endeavour* 34:2 (2010), p.45-49.

university years were varied and eventful. They comprised studying medicine and natural history at both Lund and Uppsala universities; demonstrating in Uppsala's botanical garden; attracting the patronage of eminent professors Olof Celsius (1670-1756) and Olof Rudbeck the Younger (1660-1740), acquiring the privilege to use their libraries; travelling to Lapland (1732) and to the mining province of Dalecarlia (1733/1734); and last but not least, elaborating his own sexual system of classifying plants. More reliable than Linnaeus's own, often self-glorifying *Vitae*, the early manuscripts provide an alternative window into Linnaeus's university years. They contain a wealth of information regarding his learning, memorising and practising in a wide variety of subjects which included anatomy, medicine, botany, zoology, mineralogy and assaying, which he studied extensively while in Falun. They are an invaluable source through which to understand Linnaeus's studying and learning, which he did through note-taking but also composing plant catalogues, and teaching botany to fellow students. They indicate which books he read, which naturalists and physicians he considered important enough to take notes on, and how he went about taking these notes. Lastly, they show how his own sexual system of classifying plants, and his ways of conveying it to the students he was teaching as demonstrator, were forged during these university years.

An important set of student notebooks were entitled 'Manuscripta medica' by Linnaeus. The Linnean Society holds volumes I and III, volume II having disappeared. Volume I is of special interest and was filled between 1727 and 1730.[3] Compiled on separate bundles of pages, or fascicles, before being sewn together in order according to subject areas, it contains a wealth of excerpts (250 folios) from various different authors, mostly copied from books read in his mentors' libraries. This early notebook is a jumble of notes on various subjects, although botany already takes pride of place. There, Linnaeus copied and also tinkered with the classification systems of botanists such as Joseph Pitton de Tournefort (1656-1708), Augustus Quirinus Rivinus (1652-1723), Johann Jacob Dillenius (1684-1747), and John Ray (1627-1705), to name but a few. The early manuscripts, and 'Manuscripta medica' especially, indicate that Linnaeus was accumulating and processing information by turning it into two-dimensional arrangements of words, using three principle techniques: lists, dichotomous charts, and tables. These techniques used paper space exhaustively and expediently, and conveyed a wealth of information diagrammatically, as in a map. But most importantly, they show that Linnaeus, early in his student years, had begun

3. Carl Linnaeus, 'Manuscripta medica I', Library of the Linnean Society of London, LM/LP/BIO/3/2. Volume III contains many excerpts not in Linnaeus's hand.

experimenting with his data, by playing around with various methods of classification and their visual representations. Linnaeus did not simply copy the information found in previous naturalists' works, but sought to arrive at a clearer understanding of it by means of writing it out, and rearranging it. The clearest example of his tentative application of various systems is found in a manuscript entitled 'Spolia botanica' (1729), in which he classified the flora of three different Swedish regions according to three different systems: that of Tournefort, Ray and Rivinus.[4]

The outcome of playing around with paper technologies and botanical classifications was the publication of his own sexual system, initially written out in manuscript form for the students he was teaching in the botanical garden. After writing out two catalogues of plants growing in gardens around Uppsala and especially those growing in the local botanical garden according to Tournefort's system, which he entitled 'Hortus uplandicus', Linnaeus struck out on his own and produced a further three catalogues, arranged according to his own system of classification.[5] In the fifth catalogue, entitled 'Adonis uplandicus' (1731), Linnaeus combined different representational strategies to illustrate his sexual system: a dichotomous diagram lays out the (then twenty-three) classes, facing a table listing plant genera under the respective classes of his system. This table is in fact a schematic map of flower beds that were planted according to the sexual system, and it fits neatly into the bigger plan of the botanical garden that served as a frontispiece to the manuscript. If Linnaeus's project of digesting botanical information can be compared to a kind of map-making – as he himself repeatedly did – here was an instance of him modelling an actual 'territory' according to a preconceived map.[6] 'Adonis uplandicus' was in some ways the pinnacle of Linnaeus's early years' experimentation with various modes of representing the natural world, and it was written out specifically for teaching purposes.

4. Carl Linnaeus, 'Spolia botanica', Library of the Linnean Society of London, LM/LP/BOT/1/2.
5. The various manuscript versions of Linnaeus's 'Hortus Uplandicus' can be found in the following libraries: 'Hortus Uplandicus I' (1730), Library of the Linnean Society of London, LM/LP/BOT/2/1; 'Hortus Uplandicus II' (1730), Uppsala University Library, Leufsta MS; 'Hortus Uplandicus III' (1730), Staatsbibliotek, Munich, Cod. Suec. 6; 'Hortus Uplandicus IV' (1731), Uppsala University Library, Linnésamlingen, D67b; 'Adonis Uplandicus' (1731), Uppsala University Library, Leufsta MS.
6. Hans-Jörg Rheinberger, 'Aspekte des Bedeutungswandels im Begriff organismischer Ähnlichkeit vom 18. zum 19. Jahrhundert', *History and philosophy of the life sciences* 8:2 (1986), p.237-50; Staffan Müller-Wille, 'Collection and collation: theory and practice of Linnaean botany', *Studies in history and philosophy of science*, part C, *Studies in history and philosophy of biological and biomedical sciences* 38:3 (2007), p.541-62.

This visual display of information, the alliance of diagram and table, prefigured the way Linnaeus displayed botanical information in his *Systema naturae* in 1735: with a dichotomous diagram illustrating the main distinctions of his sexual system (entitled 'Clavis systematis sexualis' ['Key to the sexual system']), and a table listing some 800 genera, enabling all of the information to be displayed in one go.

From notebooks to files

The garden catalogue manuscripts were primarily written for Linnaeus's students. By early 1731, Linnaeus's sexual system was more or less established and had been endorsed by the Royal Society of Sciences at Uppsala.[7] By late 1733, Linnaeus had conceived his ambitious and lifelong project of a universal catalogue of plants species, which he was only able to complete twenty years later, in 1753, with the publication of *Species plantarum*.[8] He began collecting material in a series of octavo bound notebooks, which he entitled 'Fundamenta botanica', recording the plants' 'specific differences', having pre-determined the space needed for each plant genus.[9] The 'Fundamenta botanica' manuscripts constitute a striking departure from the other paper technologies used earlier. They are much more open to revision and expansion than the systems and tables Linnaeus had used in 'Manuscripta medica', for example. As such, the 'Fundamenta botanica' manuscripts are unfinished works in progress, especially compared with the earlier five garden catalogue manuscripts. Rather than using paper space for displaying an existing body of knowledge in the most expedient way, the 'Fundamenta botanica' volumes exemplify a paper technology designed to accommodate an ever growing body of particulars. However, the fact that they were bound meant that Linnaeus could not control effectively the amount of space needed for each genus of plant.

Linnaeus left Sweden for the continent in 1735, heading first to Germany, and then to Holland, where he stayed until 1738. In his luggage, he carried the 'Fundamenta botanica' manuscripts, and the publications that were based on it would launch his career as a botanist.[10]

7. See *The Linnaean correspondence*, an electronic edition prepared by the Swedish Linnaeus Society, Uppsala, and published by the Centre international d'étude du XVIIIe siècle, Ferney-Voltaire. Carl Linnaeus to Kilian Stobaeus, late 1730 or early 1731, *The Linnaean correspondence*, http://linnaeus.c18.net/, letter L0008 (consulted 20 October 2014).
8. Carl Linnaeus to Gabriel Gyllengrip, *c.*5 October 1733, *The Linnaean correspondence*, http://linnaeus.c18.net/, letter L0027 (consulted 20 October 2014).
9. Carl Linnaeus, 'Fundamenta botanica', volumes VII and VIII, Library of the Linnean Society of London, LM/LP/BOT/3/1/13 and LM/LP/BOT/3/1/14.
10. Linnaeus has left an account of these manuscripts, written early in 1735, which today is preserved in a public library in the town of Linköping (Linköpings stiftsbibliotek). The

In Holland, Linnaeus published in quick succession his *Systema naturae* (1735), *Fundamenta botanica* and *Musa Cliffortiana* (both 1736), *Hortus Cliffortianus, Flora Lapponica, Genera plantarum* and *Critica botanica* (all 1737), as well as *Classes plantarum* (1738). It was only in 1746 – after returning from Holland to Sweden, getting married, working as a physician in Stockholm, and finally becoming professor of medicine and botany at Uppsala University – that Linnaeus returned to his project of a universal flora.

The two sets of preparatory manuscripts for *Species plantarum* that have survived from this period show a new approach to recording information. The first one consists of loose folio sheets that Linnaeus folded up to form fascicles.[11] Each fascicle was dedicated to a genus and listed its species and their numerous synonyms in the extant botanical literature. This was a great improvement on the previous notebooks. Unhindered by the constraints of covers and binding, Linnaeus could expand each fascicle at will, simply adding loose folio sheets within the fascicle. Moreover, since the fascicles were kept loose, genera could be shuffled around. The manuscript therefore resembles a filing system, much like Linnaeus's own herbarium, kept on loose sheets, which were stored in a purpose-built cabinet, described in detail in the 1751 *Philosophia botanica*.[12] Linnaeus filled his herbarium as he filled his 'Species plantarum' manuscript: on a day-to-day basis, as he encountered relevant information either through his reading, through his correspondence, or through the specimens he received.[13]

Linnaeus aborted this manuscript in the autumn of 1749, as he wrote to his friend Abraham Bäck (1713-1795).[14] It took another two years before he began to work on this project again and produced another manuscript that was bound at some stage.[15] The binding, however, proved counterproductive: each page was filled with numerous de-

---

numbering of the volumes in this account corresponds to the three volumes surviving in the collection of the Linnean Society; see Carl-Otto von Sydow, 'Den unge Linnés författarskap. En nyfunnen Linnéskrift', *Svenska Linnésällskapets Årsskrift* 45 (1962), p.3-19.

11. Carl Linnaeus, 'Species plantarum', Library of the Linnean Society of London, LM/LP/BOT/3/4/1. For a discussion on the dating of this manuscript, see Johan Markus Hulth, *Linnés första utkast till Species plantarum: några anteckningar* (Stockholm, 1912).
12. Carl Linnaeus, *Philosophia botanica in qua explicantur Fundamenta Botanica cum definitionibus partium, exemplis terminorum, observationibus rariorum* (Stockholm, Kiesewetter, 1751), p.291, 309.
13. Staffan Müller-Wille, 'Linnaeus's herbarium cabinet: a piece of furniture and its function, *Endeavour* 30:2 (2006), p.60-64.
14. Carl Linnaeus to Abraham Bäck, 17 October 1749, *The Linnaean correspondence*, http://linnaeus.c18.net/, letter L1060 (consulted 23 January 2017).
15. Carl Linnaeus, 'Species plantarum', Library of the Linnaean Society of London, LM/LP/BOT/3/4/2. Carl Linnaeus to Abraham Bäck, 9 July 1751, *The Linnaean correspondence*, http://linnaeus.c18.net/, letter L1285 (consulted 23 January 2017).

letions, insertions and crossings-out, all indicating that once the information was contained within a bound manuscript, Linnaeus experienced great difficulties in inserting new material, which seems to have come in ever greater quantities. Moreover, it was now impossible for Linnaeus to experiment in any way with the classification of new material, because the sheets could no longer be moved around.

Throughout the late 1740s and early 1750s, then, Linnaeus was gradually adopting a loose method of keeping information on species, both as specimens (in his herbarium) and as descriptions. However, this was more straightforward for a herbarium than for manuscript notes designed to form a book. What one can observe in Linnaeus's repeated attempt to produce a manuscript for *Species plantarum*, is a tension between the use of a flexible, and an, in principle, infinitely expandable filing system, and the bringing together of the information assembled into the linear and delimited space of a bound book.

Interleaves and multiple editions

Once a work was published, however, the same delimited, book-bound space proved to be the ideal working platform for its next edition. From the first edition of *Genera plantarum* in 1737, Linnaeus routinely used new editions, interleaved with blank pages, as templates for the next ones, adapting a long tradition of using interleaved botanical books for annotation purposes.[16] Linnaeus usually had several copies of the same edition of one of his works interleaved, and worked simultaneously on all of them, correcting the printed information, as well as adding other comments. It is interesting to note that such annotations did not get fewer with each new, revised edition of a work (as one would expect if the annotations' purpose was to contribute to its completion) but, on the contrary, they became more and more numerous, reflecting a profusion of information which only increased with Linnaeus's fame and the number of his correspondents and students.[17]

One of Linnaeus's personal copies of the first edition of *Genera plantarum* (1737), is equipped with a manuscript title page headed 'Species plantarum', and is dated 15 November 1752.[18] This copy used the space on the interleaved page facing the printed description of a genus to compile a handwritten list of species of that genus. The result

16. Alix Cooper, *Inventing the indigenous: local knowledge and natural history in early modern Europe* (Cambridge, 2007), p.73-75.
17. Staffan Müller-Wille, 'Vom Sexualsystem zur Karteikarte: Carl von Linnés Papiertechnologien', in *Nicht Fisch, nicht fleisch: Ordnungssysteme und ihre Störfälle*, ed. T. Bäumler, B. Bühler and S. Rieger (Zurich, 2011), p.33-50 (43-44).
18. Carl Linnaeus, *Genera plantarum* (Leiden, C. and G. J. Wishoff, 1742), Linnaeus's personal copy, Library of the Linnean Society of London, BL 51.

was yet another handwritten precursor to the *Species plantarum*, which Linnaeus dedicated to his son Carl, then twelve years old. Here the annotations neither served to amend the facing printed text, nor acted as a repository for comments and observations on the printed text, but rather to facilitate the preparation of an entirely different publication. The layout of the printed text thus functioned as a template against which information of a different kind could be collected – the collection of species within a particular genus.

Linnaeus was one of the first naturalists who opted for a serial publication of his main works – what Bettina Dietz calls an '"iterative" publication strategy'.[19] In the preface to the first edition of the *Genera plantarum*, Linnaeus apologised for coming out with a publication at such a young age.[20] Quick publication at an early stage of research was clearly an advantageous strategy: Linnaeus's fame spread immediately, and elicited feedback in the form of corrections and additions communicated by correspondents. Once this new information threatened to overwhelm Linnaeus's capacities for information storage, he would simply proceed to issue a new edition. In one stroke, a new, updated platform for annotation had been created which could accommodate the next cycle of feedback without having to return to the notes that predated the new edition. Linnaeus profited not only scientifically, but also financially from this 'cycle of accumulation'.[21] Surviving copies from private libraries of eighteenth-century botanists demonstrate that Linnaeus's contemporaries interleaved and annotated his publications in exactly the same way as he did – when they managed to procure themselves a copy.[22] In order to contribute to the 'cycle of accumulation' Linnaeus had initiated with his taxonomic publications, naturalists had to make sure to be up-to-date with the latest editions, and this is almost certainly a reason for Linnaeus's astounding success as an author. His major works were not made for reading, but provided a shared platform for collective annotation. They were designed and used for taking note

19. Bettina Dietz, 'Contribution and co-production: the collaborative culture of Linnaean botany', *Annals of science* 69:4 (2012), p.551-69 (552).
20. Staffan Müller-Wille and Karen Reeds, 'A translation of Carl Linnaeus's introduction to *Genera plantarum* (1737)', *Studies in history and philosophy of science*, part C, *Studies in history and philosophy of biological and biomedical sciences* 38:3 (2007), p.563-72 (570).
21. Bruno Latour, *Science in action: how to follow scientists and engineers through society* (Cambridge, MA, 1987), p.220.
22. See Petra Feuerstein-Herz, *'Die große Kette der Wesen': Naturgeschichte und Ordnung in der Frühen Neuzeit'* (Wiesbaden, 2007), p.163. Bettina Dietz discusses the difficulty of acquiring scientific literature in chapter 2 of this volume, 'What is a botanical author? Pehr Osbeck's travelogue and the culture of collaborative publishing in Linnaean botany', and in Bettina Dietz, 'Making natural history: doing the Enlightenment', *Central European history* 43:1 (2010), p.25-46 (39-40).

of and accumulating new observations, whether in the field, or in the cabinet.[23]

Paper slips

With the exception of a slim supplement volume to the first edition of *Genera plantarum* entitled *Corollarium generum plantarum* (1737), Linnaeus thus always published full, amended editions of his major taxonomic works. Yet, from 1767, Linnaeus resorted to publishing supplements: *Mantissa plantarum* (1767) and its successor *Mantissa plantarum altera*, published in October 1771. A third *Mantissa* was in preparation at the time of Linnaeus's death: the preparatory manuscript was copied out by two amanuenses, and contains corrections from both Linnaeus and his son Carl.[24] It was eventually published by Linnaeus the Younger, not as a further *Mantissa*, but as *Supplementum plantarum* (1781).[25]

The *Mantissae* volumes were meant to supplement the latest editions of *Species plantarum* (1762-1763) and *Genera plantarum* (1764). The first *Mantissa* alone contained twenty-five new genera and – as Linnaeus explained himself in an allegedly anonymous review in *Lärda Tidningar* – 'more than 400 of the rarest species, which partly have never been referred to a particular Genus, and partly are completely new'.[26] The reason why Linnaeus switched to a new publication strategy late in life simply seems to have been that he saw himself overwhelmed by the sheer amount of material that was returned to him due to the success of his prior publications, and by information on new plants in works such as *Descriptiones plantarum* (1767) by Peter Jonas Bergius (1730-1790), and *Flora Indica* (1768) by Nicolaas Burman (1734-1793), information that needed to be integrated into Linnaeus's universal catalogue. Linnaeus was acutely confronted by the problem of what historians have variously called 'information overload' or 'the challenge of plenitude'.[27] Since the early 1750s, Linnaeus had routinely complained about feeling overworked in letters to his good friend Abraham Bäck, and to numerous other correspondents. Illness also plagued him, and forced him to rely on

23. Müller-Wille and Charmantier, 'Natural history and information overload', p.11; Dietz, 'Contribution and co-production', p.566.
24. Carl Linnaeus fil., 'Supplementum plantarum', preparatory manuscript, Library of the Linnean Society of London, LM/LF/BOT/2/2/2/1.
25. Carl Linnaeus fil., *Supplementum plantarum* (Braunschweig, Impensis Orphanotrophei, 1781).
26. William T. Stearn, 'Introductory notes on Linnaeus's *Mantissa plantarum*', in Carl Linnaeus, *'Mantissa plantarum' 1767 & 1771: facsimile with an introduction by William T. Stearn* (Lehre, 1961), p.vi.
27. Ann Blair, *Too much to know: managing scholarly information before the modern age* (New Haven, CT, 2010); Gunnar Broberg, 'The broken circle', in *The Quantifying spirit in the eighteenth century*, ed. Tore Frängsmyr, J. L. Heilbron and Robin E. Rider (Berkeley, CA, 1998).

students to continue his work. Pehr Löfling, for instance, acted as Linnaeus's secretary when the latter was suffering from arthritis in 1750, and wrote several of his mentor's manuscripts, including the press manuscript for *Philosophia botanica*.[28] The feeling of having very little time left was acutely expressed in the preface to *Mantissa plantarum altera*: 'There have come to notice here and there while packing my luggage, as it were, in the evening of my life, scattered memoranda with observations not known to all; I have raked these together, to add to the first Mantissa.'[29] The word which William Stearn here translated as 'memoranda', is actually *adversaria*, which referred to a miscellaneous collection of excerpts from literature, 'generally presented in the order of the text from which they were produced'.[30] It was around 1769, for the preparation of *Mantissa plantarum altera*, that Linnaeus began to use paper slips, and the quoted passage from the preface to this book is the only instance where Linnaeus makes explicit reference to this new working method. He clearly associated the new tool with the feeling of having reached the limit of his working capacity, and having to resort to a change in publication strategy. The paper slips, of which there are 1030 in total (mostly botanical), were specifically designed to remain loose and expandable, yet could always be brought into a linear order. Linnaeus initially seems to have used the paper slips much like he had used loose sheets for his *Species plantarum* in the late 1740s, that is, as a preparatory manuscript. Later on, however, the slips seem to have become an integral part of his working method that functioned independently of particular publication projects, an information technology that he thereafter relied upon for the rest of his working life. Instead of discarding the slips and striking out information, Linnaeus kept them and added to them periodically, as variations in ink colours and handwriting indicate. Using a single paper technology enabled Linnaeus to assemble together information borne out of specimens and letters sent from his 'apostles' and other correspondents, or out of newly published books onto different paper slips which could be kept together according to their unifying category, the genus.

The paper slips took on another dimension when Linnaeus became too ill and fragile to work on his own after a stroke in the summer of 1774. The preparatory manuscript for the third *Mantissa* demonstrates that towards the mid-1770s Linnaeus was working in close collaboration with his son and two amanuenses. While the amanuenses neatly copied out Linnaeus's paper slips in a notebook, Linnaeus father and son

28. Carl Linnaeus and Pehr Löfling (scribe), 'Philosophia botanica', Library of the Linnean Society of London, LM/LP/BOT/3/3/1.

29. Stearn, 'Introductory notes on Linnaeus's *Mantissa plantarum*', p.xvii.

30. Ann Blair, 'Note-taking as an art of transmission', *Critical inquiry* 31:1 (2004), p.85-107 (87).

corrected some of the text. At the same time, his son continued the practice of writing new paper slips for new species of plants. The manuscript is a mixture of the amanuenses' tidy handwritings, and paper slips from both father and son inserted, glued or pinned on interleaved pages (see Fig.1, p.36). A few of these slips are Linnaeus's original paper slips, while others are more roughly cut paper slips, and were clearly written up while someone else was working on the manuscript, to be inserted later. Working from, and on, loose paper slips allowed up to four people to work on one piece of work simultaneously, without impinging upon the copying process. The manuscript was finally published as *Supplementum plantarum* in 1781, three years after Linnaeus's death.

Beyond Linnaeus

Daniel Solander

Beyond the Linnaeus household, the collaborative and information retrieval advantages of the paper slips were readily adopted. Back in 1752, Linnaeus had been asked to catalogue the royal natural history collections, and he called on his student Daniel Solander for assistance.[31] Queen Louisa Ulrika's collection of butterflies was catalogued using small paper sheets of roughly the same size (10.5 × 17cm approximately).[32] As in the case of Linnaeus's later paper slips, it is impossible to establish the original order of the manuscript.[33] Because the slips were used vertically, they are more akin to the pages of a small octavo volume than to customary index cards. The presence of a title page confirms this impression. What looks like a collection of cards, ready to be extended and reshuffled at will, is therefore rather an unbound manuscript put together for a particular book project.

31. Daniel Solander, *Collected correspondence 1753-1782*, ed. Edward Duyker and Per Tingbrand (Melbourne, 1995), p.15-16. Drottningholm and Ulriksdal were both royal palaces. For more on Daniel Solander, see Hanna Hodacs's contribution to this volume, 'The price of Linnaean natural history: materiality, commerce and change' (chapter 3).
32. Carl Linnaeus, 'Museum reginae', Carl von Linné, MS 1, Royal Swedish Academy of Sciences, Stockholm. The cards are dated *c.*1760 but can more accurately be dated to 1752 according to Solander's letter. By 1760, Solander was already in England, and it is therefore likely that the dating refers to the use of the manuscript for the publication of the catalogue in 1764.
33. It is now organised by bundles wrapped in paper inscribed with the names of insect orders in the hand of the Swedish entomologist Per Olof Christopher Aurivilius (1843-1928), who wrote a taxonomic review of Linnaeus's work on the queen's collection in 1882; see Aurivilius, 'Recensio critica lepidopterorum Musei Ludovicae Ulricae quae descripsit Carolus A Linné', *Kongl. Svenska Vetenskaps Academiens handlingar*, vol.19, part 5 (1882), p.1-189.

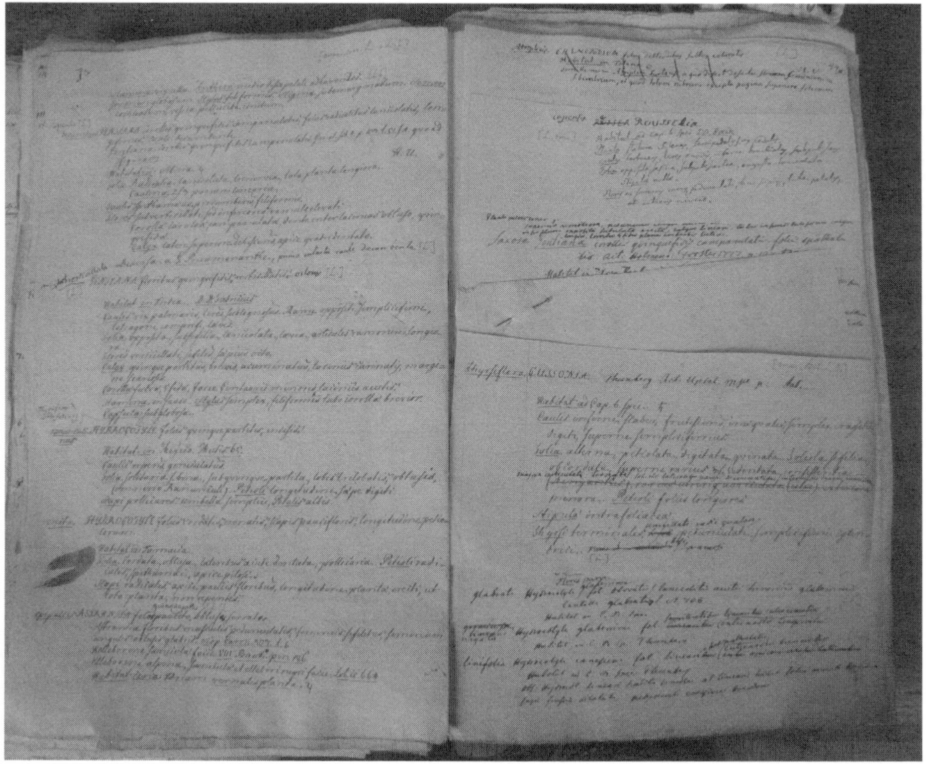

Figure 1: Pages 44 and 44a of 'Supplementum plantarum' preparatory manuscript, with the amanuensis's clean copy of Linnaeus's paper slips on the left page, and additional paper slips from both Linnaeus father and son on the facing interleaf. By permission of the Linnean Society of London.

One of the most distinctive features of this manuscript, if compared with other manuscripts left by Linnaeus, is that the slips cataloguing the queen's collection bear both Linnaeus and Solander's handwritings. Only nine slips are actually in Solander's very regular hand, most of which concern the genus *Phalaena*, now an obsolete genus in which Linnaeus included moths in general. All the other cards are in Linnaeus's distinctive scrawl (see Fig.2, p.38). The division of labour between the two was certainly an important motivation to work with loose sheets that could later be assembled to form a complete manuscript. That there was a division of labour between the professor and his student is revealed by a closer look at the structure and content of the slips. Solander was meticulously following the model provided by Linnaeus's earlier taxonomic publications, and also the prescriptions about how to lay out a proper species description put forward in Linnaeus's 1751 *Philosophia botanica*. By contrast, Linnaeus's slips are more cluttered and harder to decipher. They look more like private research notes than a manuscript carefully prepared for the typesetter. And indeed, in the preface to the catalogue of Queen Louisa Ulrika's collection that was published a good ten years later, Linnaeus recalled how he had originally collected some of his 'modest observations [*observatiunculas*] on paper sheets [*schedulas*], primarily devoted to my own use, so that I would retain some kind of idea of [the specimens in the collection].'[34]

Solander moved to England in 1760, and from 1763 he was employed at the British Museum to catalogue Sir Hans Sloane's collection, as Hanna Hodacs describes in chapter 3 of this volume. There, he started using paper slips similar to the ones cataloguing the queen's collection, although made from sturdier paper, and used horizontally (16 × 10cm approximately).[35] As Solander's letter offering his services to the Trustees of the British Museum indicates, the information contained on his slips continued to uphold the Linnaean model of species description: 'A catalogue such as I should think proper and fit for this purpose, should consist of: the true generical name, with a *differentia specifica* and a trivial name; a good Synonyme – the native country – the use, if any – and

34. Carl Linnaeus, *Museum [...] Ludovicæ Ulricæ reginæ Suecorum, in quo animalia rariora, imprimis insecta & conchilia describuntur* (Stockholm, L. Salvius, 1764), preface, unpaginated.

35. Solander's slips from this period are annotated with 'MB HS'. It is disputed whether this refers to Sloane's collection or another collection of plants present in the British Museum; see Judith A. Diment and Alwyne Wheeler, 'Catalogue of the natural history manuscripts and letters by Daniel Solander (1733-1782), or attributed to him, in British collections', *Archives of natural history* 11:3 (1984), p.457-88 (470); and John Braybrooke Marshall, *The Handwriting of Joseph Banks, his scientific staff and amanuenses* (London, 1978), p.2-3.

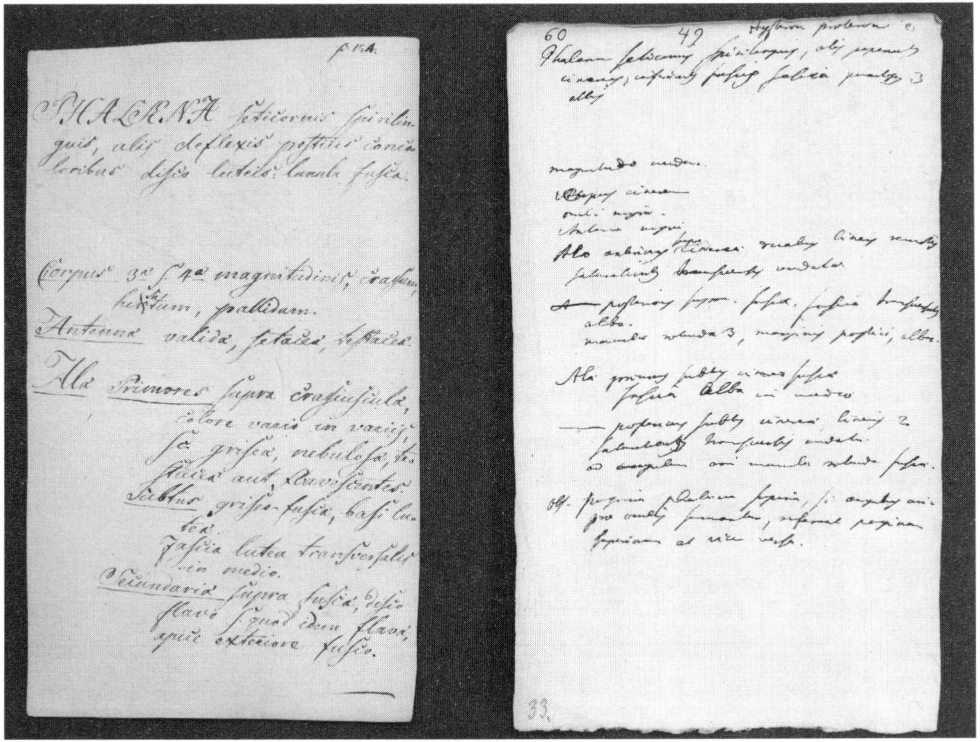

Figure 2: Catalogue of the Queen's insect collection, two paper slips for the genus *Phalaena*. On the left is a card drawn up by Daniel Solander, on the right one drawn up by Carl Linnaeus. Carl Linnaeus, 'Museum Reginae', von Linné, Carl, MS 1, Royal Swedish Academy of Sciences, Stockholm. Courtesy of The Centre for History of Science.

in case it was a new subject, then to add a short description.'[36] Solander continued to use this slip catalogue throughout his career, to take notes on species collected when he accompanied Joseph Banks (1743-1820) on Captain Cook's first circumnavigation from 1768 to 1771, and when working on herbarium specimens as curator and librarian to Banks after his return from this voyage.[37] These slips were initially kept loose in fifty-seven Solander boxes – boxes to contain manuscripts that Solander had introduced to Bank's library and that bear his name to this day – but were later bound into twenty-four volumes.[38] Solander also probably used slips when, for four years preceding his death, he catalogued and named new species from the duchess of Portland's shell collection.[39] Solander's slips bear a striking resemblance to Linnaeus's later slips in the way the name and description of the plants are laid out (see Fig.5a, p.48).

It is tempting to conclude that, while cataloguing Queen Louisa Ulrika's collection, the first seeds were planted for the later independent 'invention' of paper slips by Solander and Linnaeus. Both men could rely on a common previous experience of working with loose paper sheets of a standard size when cataloguing an especially rich and complex set of natural history specimens. What distinguishes this earlier use of loose paper sheets from the later use of smaller paper slips arranged in series is that the earlier sheets were still compiled to form the manuscript in the context of one particular publication project. Linnaeus persisted in thinking in terms of individual publications right up to the publication of his 1771 *Mantissa*, as has been shown above. By contrast, Solander seems to have realised the potential of paper slips much earlier, employing it in his cataloguing work eight years before his former teacher and mentor.

Solander's paper slips became an integral part of cataloguing specimens within the British Museum. Following Solander's death in 1782, Jonas Carlsson Dryander (1748-1810), another of Linnaeus's students, succeeded him as curator and librarian of the British Museum. Nearly as many cards as were written by Solander in the Museum's slip catalogue are in Dryander's hand. Upon his own death in 1810, Dryander was

36. Daniel Solander to William Watson, 26 January 1763, in Solander, *Collected correspondence*, p.254.
37. Marshall, *The Handwriting of Joseph Banks, p.2-4*; Diment and Wheeler, 'Catalogue', p.459, 470-71; John Braybrooke Marshall, 'Daniel Carl Solander, friend, librarian and assistant to Sir Joseph Banks', *Archives of natural history* 11:3 (1984), p.451-56.
38. Solander, 'Manuscript descriptions of plants', MS BANKS COLL SOL, Library and Archives, Natural History Museum, London.
39. Beth Fowkes Tobin, *The Duchess's shells: natural history collecting in the age of Cook's voyages* (New Haven, CT, 2014), p.209-11.

succeeded by Robert Brown (1773-1858), who had previously taken part in the voyage of the *Investigator* to Australia as the on-board naturalist, an expedition during which he consigned his notes to slips fashioned on the model of Solander's manuscript slips. He continued to do so as Joseph Banks's librarian, and these slips are in seventy-seven Solander cases of the Botanical Library (Natural History Museum), often written in pencil. On Banks's death in 1820, Brown inherited his library and herbarium, which were transferred to the British Museum in 1827. Brown then became Keeper of Sir Joseph Banks's Botanical Collection – or what would become the Department of Botany of the British Museum.[40] Solander's implementation of the Linnaean standardised method to describe plant species on loose paper slips for both Joseph Banks and the British Museum's collections was crucial. These slips, when united by the coming together of both collections, became a standardised catalogue that allowed naturalists to easily access the Museum's large amount of data and specimens.

Carl Linnaeus the Younger

Carl Linnaeus the Younger was born in 1741, and from an early age attended his father's lectures in mineralogy, zoology and botany.[41] Historians have not paid much attention to Linnaeus's son, partly because his father's pre-eminence cast him in the shadows, and partly because he died young, in 1783, before having really achieved anything of consequence. Moreover, in 1763, Linnaeus was granted the right to appoint his own successor, and he chose his son. The appointment of Carl Linnaeus the Younger as head of Practical Medicine at Uppsala University, and his taking over some of his father's teaching, was controversial and not widely approved. Posterity has also remembered Carl Peter Thunberg's frank and unfavourable opinion, expressed after the death of Linnaeus the Younger. Thunberg, who worked as demonstrator of botany under Linnaeus the Younger's professorship, thought him 'a very useless man and full of pride, even ignorant'.[42]

40. On Dryander and Brown, see Marshall, *The Handwriting of Joseph Banks*, p.3, 5, 16-18.
41. No standalone biography of Carl Linnaeus the Younger exists, but see Dietrich Heinrich Stoever, *The Life of Sir Charles Linnaeus [...] to which is added, a [...] list of his works, and a biographical sketch of the life of his son [...] translated from the [...] German by Joseph Trapp* (London, White, 1794). Many of Carl Linnaeus the Younger's notes from his father's lectures survive and are at the Linnean Society of London. The earliest set of lecture notes date from the early 1750s, before the publication of *Species plantarum*, when Linnaeus the Younger would have been ten years old.
42. 'Un home tres inutile et plein d'orgeuil, meme ignorant'. Carl Peter Thunberg to André Thouin, undated, received by Thouin 4 August 1785, Museum d'histoire naturelle, Paris, Thouin MS 1989/950-990.

However, manuscripts from both father and son reveal that Linnaeus the Younger's handwriting appears on his father's notes from the early 1760s, suggesting that he assisted his father, both with teaching and with wider scientific endeavours. As we have seen, his 1781 *Supplementum plantarum* was based to a large extent on his father's genus and species descriptions on paper slips. The preparatory manuscript was started probably shortly after the publication of the last 1771 *Mantissa*, possibly in 1775, and with Linnaeus father and son both actively collaborating on it, with the additional help of two amanuenses. Some slips that have been inserted in the manuscript are in Linnaeus the Younger's hand, indicating that by the mid-1770s he had readily adopted his father's paper technology, copying the rigorous order in which information was presented on the slip.

In April 1781, with the manuscript of 'Supplementum plantarum' ready for the printer, Linnaeus the Younger left for England, where he spent several months. Due to his connections with his compatriots Daniel Solander and Jonas Dryander, Linnaeus the Younger was fully integrated into the circle surrounding Joseph Banks. There he met Henry Smeathman (1742-1786) and Pierre Broussonnet (1761-1807), who accompanied him on excursions to botanical gardens and other attractions. Linnaeus the Younger was at Banks's Soho Square house when, at breakfast one morning in early May 1782, Solander suffered a severe brain haemorrhage which led to his death a week later. In England, Linnaeus the Younger visited botanical gardens such as Kew, James Lee's (1715-1795) garden in Hammersmith, Dr William Pitcairn's (1712-1791) garden in Islington, the marquess of Rockingham's (1730-1782) garden in Wimbledon, and parks further afield, such as Painshill Park, a landscape garden created by Charles Hamilton (1704-1786).[43] Linnaeus the Younger left England for the Continent and was in Paris by mid-November 1782, where he met André Thouin (1746-1824), visited the Jardin du Roi and Joseph de Jussieu's (1709-1779) herbarium. He then travelled through Holland, visiting Leiden, and was back in Uppsala by early 1783. The manuscripts kept at the Linnean Society show that the Continental trip was extremely productive and formative, especially the English stay. Numerous manuscripts on botany and zoology, begun in England and continued on the other side of the Channel, have survived, including notes indicating that Linnaeus the Younger had access to Banks's library and specimens. Most interesting for our purpose is a series of notes on paper slips, consisting of three sets on plants (412

43. See Carl Linnaeus fil., 'Notes from England', Library of the Linnean Society of London, LM/LF/PP/6/5.

slips),[44] another on mammals (602 slips), and a few stray slips on fish and insects (twelve slips).[45]

These slips, used to record information on individual species of organisms, provide invaluable information on Linnaeus the Younger's time in Europe and on the collections he visited. These included the Leverian and Hunterian museums, Banks's collections, the British Museum, the Royal Society collections in London; the menagerie ('vivario') of Versailles and the 'museo Parisii' which might refer to the Jardin du Roi in Paris; the collections amassed by Jean-Nicolas-Sébastien Allamand (1713 or 1716-1787) at the University of Leiden, as well as Vesalius's museum in Holland. The slips occasionally include vivid snippets that bring these collections to life: a malnourished and desultory marmot roamed Banks's collections, while in the breakfast room ('in Breakfest rum apud Banks'), hung the picture of an Arctic fox.[46]

Information that Linnaeus the Younger recorded in London was often corrected and expanded upon as he encountered other examples of specimens in Paris. For instance, Linnaeus the Younger encountered the specimen of *Hystrix macroura* (brush-tailed porcupine) at the British Museum in London ('specim: in Brit: mus:'). He wrote for it a formal species description, as well as a description on a slip of the specimen seen at the Museum.[47] In Paris, Linnaeus the Younger had the opportunity to see a live specimen of the animal ('vivam vidi Parisiis'), whereupon he corrected the description of the crown of its head and added that its eyes were small, globular and black. Linnaeus the Younger also frequently commented on the validity of the descriptions provided by the authors he quotes, or the quality of their illustrations – often noting 'fig. pulchra' or 'fig. mal.' next to a reference. For the species *Ursus mephitis* (striped skunk, now *Mephitis mephitis* Schreb.), for example, he notes that George Louis Leclerc de Buffon's (1707-1788) illustration of the animal in his *Histoire naturelle* is 'not perfectly expressed'.[48] The slips, then, were meant to help Linnaeus the Younger list references, provide a diagnosis, and pinpoint the most accurate descriptions – either textual or visual. Occasionally, he drew part of or the whole specimen seen during a garden or museum

44. Carl Linnaeus fil., Library of the Linnean Society of London: 'De Liliaceis et affinibus plantis', LM/LF/BOT/3/9/1; 'Palmarum descriptiones', LM/LF/BOT/3/9/2; 'Plant diagnoses', LM/LF/BOT/3/9/3.

45. Carl Linnaeus fil., 'Mammalia paper slips', Library of the Linnean Society of London, LM/LF/ZOO/3/6.

46. Linnaeus fil., 'Mammalia paper slips', f.400, 132.

47. Linnaeus fil., 'Mammalia paper slips', f.310. *Hystrix macroura* L. 1758 is the type species of the genus *Atherurus* (Cuvier 1829).

48. Linnaeus fil., 'Mammalia paper slips', f.178; George Louis Leclerc de Buffon, *Histoire naturelle générale et particulière, avec la description du Cabinet du Roi*, 15 vols (Paris, Imprimerie royale, 1749-1767), vol. 13 (Paris, Imprimerie royale, 1765), plate 41.

visit, such as the *Didelphis tridactylus*, seen in Banks's museum (see Fig.3a and 3b, p.44-45),[49] or the dried specimen of *Pancratium ornithogaloides* from Joseph de Jussieu's herbarium (see Fig.4, p.46).[50]

What is most interesting about these slips, however, is that Linnaeus the Younger's slips correspond exactly in their size (16 × 10cm) and in their paper to the slips that Solander and Dryander used to catalogue Sloane and Banks's collections. The watermarks and countermarks (Britannia/J. Whaterman and Britannia/C. Taylor) are identical. We can therefore conclude that the paper Linnaeus the Younger used for his slips came directly from Solander and Dryander's reserve of paper. There are still forty-five blank slips, ready to be used, an indication of a work in progress. In addition, the same way of grouping the Mammalia slips by genera was used by Linnaeus the Younger and Solander and Dryander: each group of slips, belonging to the same genus, is kept together by a longer piece of paper acting as a folder, and containing the genus description.[51] Within the folder are double or individual slips containing corresponding species diagnoses. Whilst Linnaeus the Younger borrowed paper and possibly ink from his hosts, careful comparison between his paper slips and those of Solander and Dryander show that Linnaeus the Younger did not simply copy their work, as attested also by the annotations 'vidi museo Dr Banks' on a number of his paper slips.[52]

Rather, it reveals that Dryander copied Linnaeus the Younger's slips verbatim, and acknowledged this unpublished work undertaken during the young Swede's English stay. Whereas Dryander referred to Linnaeus the Younger's *Supplementum plantarum* by 'Linn fil sup', the slips that were directly copied from Linnaeus the Younger's slip refer simply to 'Linn fil'. Dryander especially copied Linnaeus the Younger's work on genera from the families of Liliaceae or Palmae, as many of Dryander's slips refer to 'Linn fil' and correspond word for word to Linnaeus the Younger's own slips. Linnaeus the Younger's slip on *Narcissus pseudo-narcissus*, for example, contains numerous corrections and crossings-out which do not appear on Dryander's copy, indicating that Dryander copied Linnaeus the Younger's slips, rather than the other way around.[53] The

49. Linnaeus fil., 'Mammalia paper slips', f.296-97. The species is either a species of opossum (*Didelphis*) or a long-nose potoroo (*Potorous tridactylus*).

50. Linnaeus fil., 'De Liliaceis et affinibus plantis', f.73.

51. Daniel Solander, 'Manuscript descriptions of animals, written on slips and systematically arranged in accordance with Linné's "Systema naturae"', MS BANKS COLL SOL, Library and Archives, Natural History Museum, London, 27 volumes.

52. 'I saw [it] at Dr Banks's museum' (author's translation).

53. Linnaeus fil., '*Narcissus pseudo-narcissus*', in 'De Liliaceis et affinibus plantis', f.20; Daniel Solander and Jonas Dryander, '*Narcissus pseudo-narcissus*', in 'Manuscript descriptions of plants', MS BANKS COLL SOL, Library and Archives, Natural History Museum, London, vol.VIII, p.1602. Similarly, Dryander's slips on *Narcissus bicolor* and *N. minor* are also copied from Linnaeus the Younger's slips.

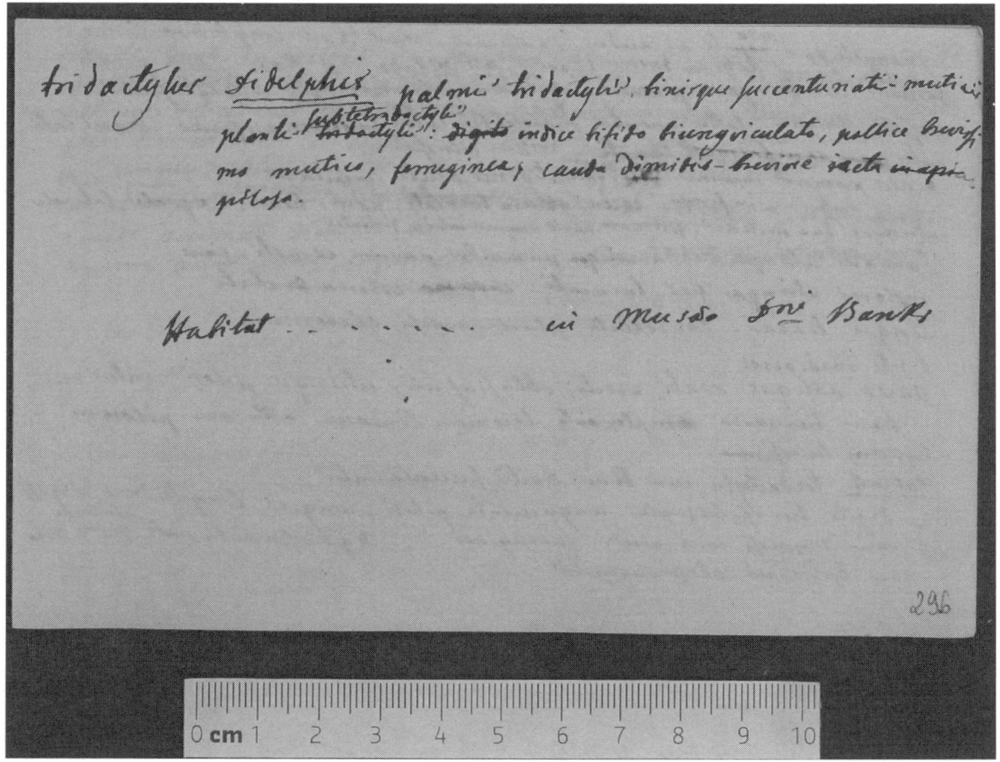

Figures 3a and 3b: Two paper slips for *Didelphis tridactylus*, with its textual description and a drawing by Linnaeus the Younger. Linnaeus the Younger, 'Mammalia paper slips', f.296-97. By permission of the Linnean Society of London.

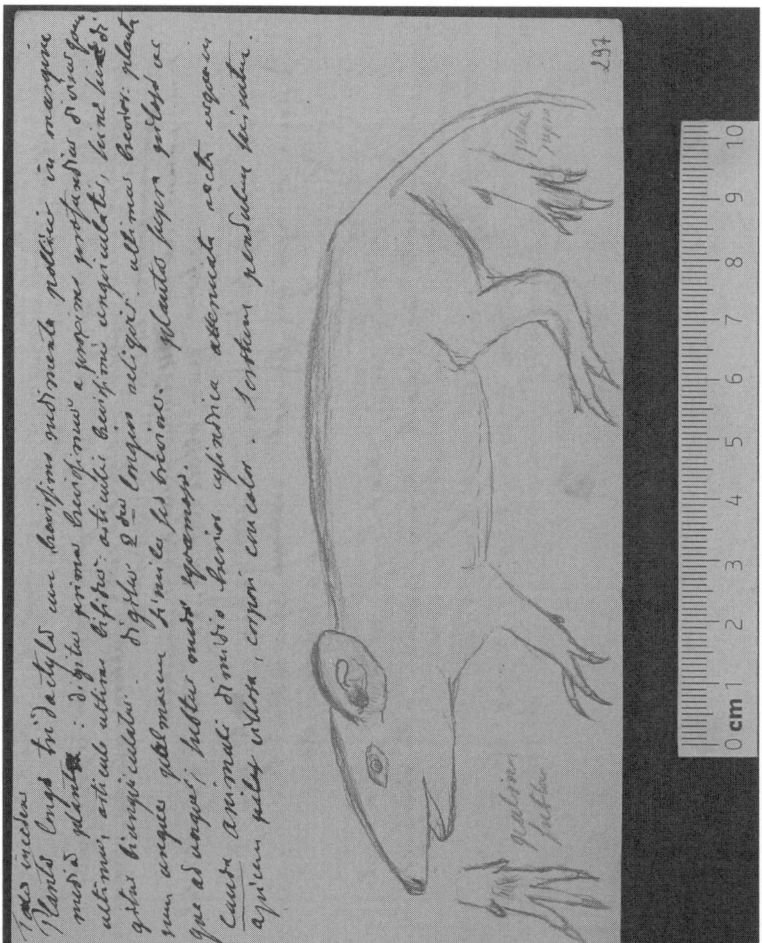

Figure 3b.

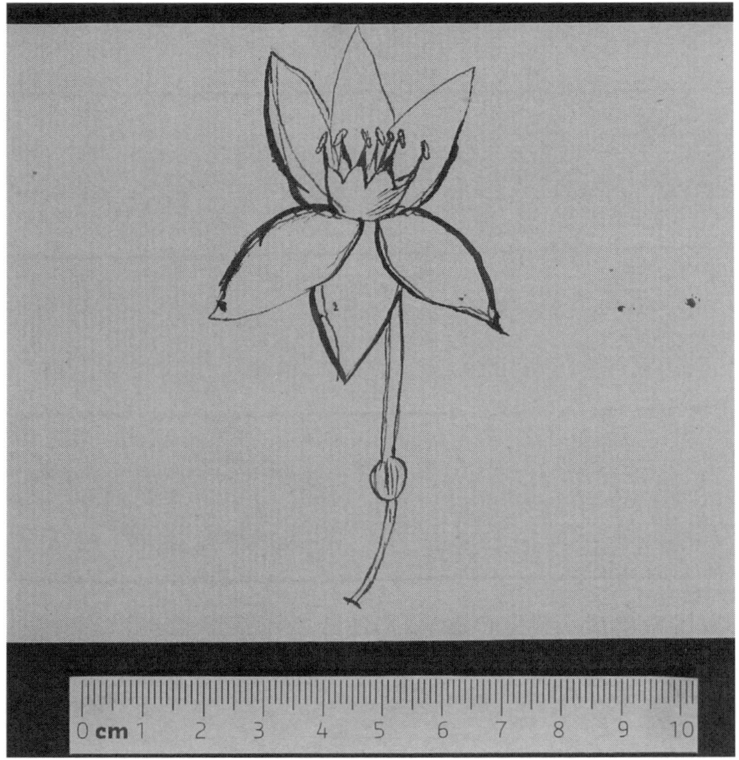

Figure 4: Ink drawing of the dried specimen of *Pancratium ornithogaloides* from Joseph de Jussieu's herbarium in Paris. Linnaeus the Younger, 'De Liliaceis et affinibus plantis', f.73. By permission of the Linnean Society of London.

case of *Cycas revoluta* shows that at the time of Linnaeus the Younger's visit, Solander's slip, probably written up when Solander was cataloguing Sloane's herbarium in the 1760s, simply bore a binomial name as well as a descriptive phrase (see Fig.5a, p.48).[54] On his own slip (see Fig.5b, p.49), Linnaeus the Younger seems to have reworked the descriptive phrase based on his examination of a second specimen, that collected by Banks in 'Nova Hollandia' (Australia). He also added references to botanical literature: Engelbert Kaempfer's *Amoenitates exoticae* (1712) and the first volume of Georgius Everhardus Rumphius's *Herbarium Amboinense* (1741).[55] Linnaeus the Younger's descriptive phrase and his additional references were then copied onto Solander's slip by Dryander.

Linnaeus the Younger's slips show that he took an active part in the scientific life of Banks's circle, and played a collaborative role in identifying, naming and describing new species. Linnaeus the Younger's work was fully integrated in the reassessment of information on new species, based on the examination of new specimens from the *Endeavour* voyage. As with the writing of the *Supplementum plantarum*, this collaborative research was facilitated by the technology of paper slips. This diagnostic work was undertaken on exactly the same paper slips, which could be exchanged, copied and given back individually. The assembling of knowledge for a foreign species such as *Cycas revoluta*, rarely seen as a live specimen, involved the reworking of previous published information and the attentive comparison of rare dried herbarium specimens. Working collaboratively and sharing the technology, as well as the data, facilitated communication exponentially. Interestingly, the name *Cycas revoluta* is now associated with Carl Peter Thunberg, who had seen a live specimen of the plant in Japan, collected its seeds and sent them back to his various patrons.[56] Thunberg was the first to publish information on *Cycas revoluta* in various publications of the early 1780s.[57] His descriptive phrase is directly lifted from Solander's slip, which Thunberg must have seen during his English stay in late 1778, two years before Linnaeus the Younger's stay in London. Thunberg's relations with Linnaeus the Younger were never easy, and he was probably unaware of Linnaeus the Younger's annotations concerning

54. Daniel Solander and Jonas Dryander, '*Cycas revolutus*', in 'Manuscript descriptions of plants', vol.XX, p.6076.
55. Linnaeus fil., 'Palmarum descriptiones', f.57.
56. Marie-Christine Skuncke, *Carl Peter Thunberg, botanist and physician: career-building across the oceans in the eighteenth century* (Uppsala, 2014), p.174.
57. Carl Peter Thunberg, 'Beschryving van twee nieuwe soorten van palmboomachtige gewassen, Uti Japan en va de Kaap des Goede Hope', *Verhandelingen, uitgegeeven door de Hollandsche Maatschappye der Weetenschappen, te Haarlem* 20:2 (Haarlem, J. Bosch, 1782), p.419-34; Carl Peter Thunberg, *Nova Acta Regiae Societatis Scientiarum Upsaliensis* 4:40 (1783); Carl Peter Thunberg, *Flora Japonica* (Leipsig, I. G. Müller, 1784), p.229.

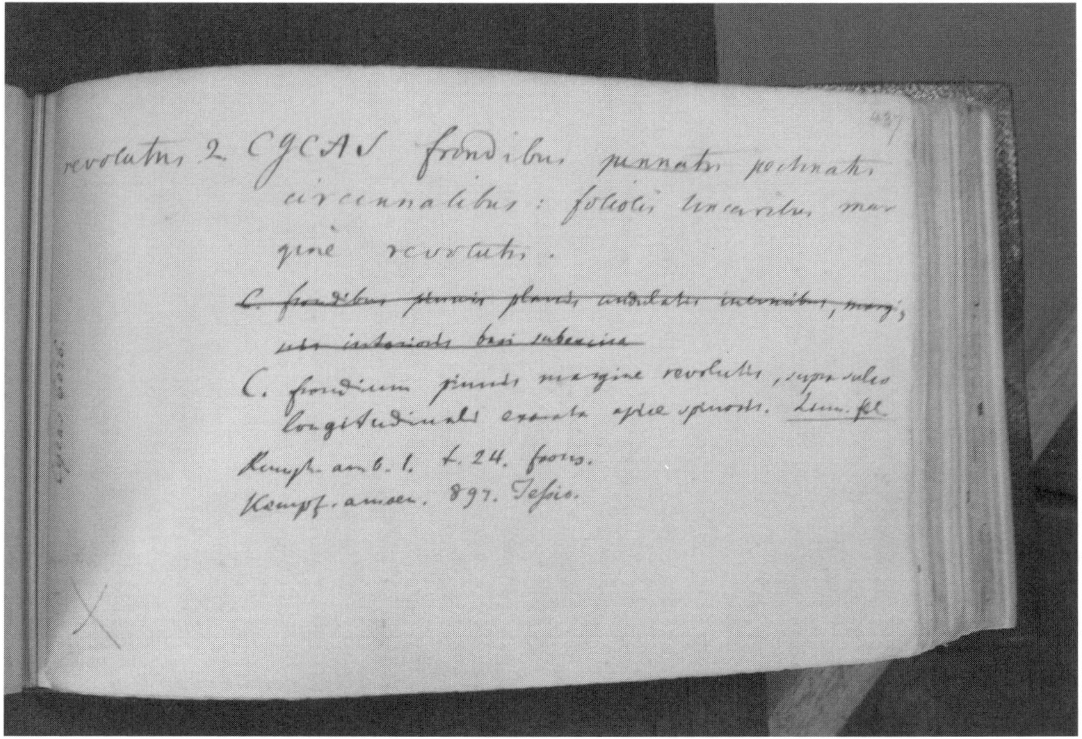

Figure 5a: Paper slip for *Cycas revolutus*, initially written by Daniel Solander, and with later additions, copied from Linnaeus the Younger's slip, by Jonas Dryander. Manuscript descriptions of plants, Library and Archives, Natural History Museum, MS BANKS COLL SOL, vol.XX, 6076, f.437. Reproduced by permission of the Trustees of the Natural History Museum, London.

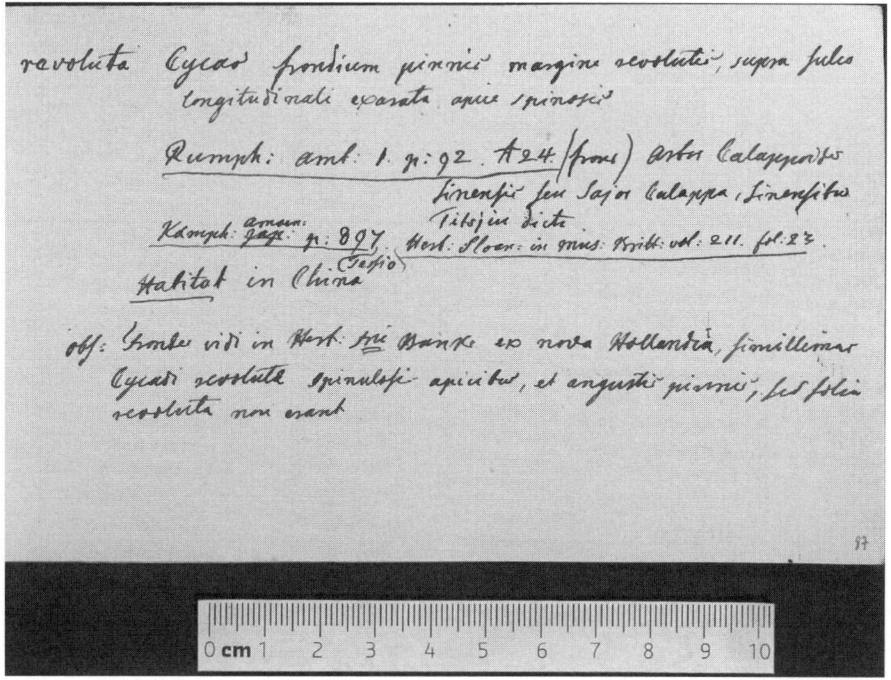

Figure 5b: Linnaeus the Younger's slip on *Cycas revoluta*, 'Palmarum descriptiones', f.57. By permission of the Linnean Society of London.

Cycas revoluta, or if he was, he probably did not think much of their quality. He would not have known that Dryander had copied these annotations onto the Solander slip during Linnaeus the Younger's 1781 visit. As a result, Linnaeus the Younger's work on *Cycas revoluta*, and on many other species of plants and animals, never made it into any published description. Paper slips were an efficient way of sharing information in manuscript form, but for a wider dissemination of this knowledge it needed to be converted into publications. Without the medium of print, the audience remained limited and, ultimately, the information risked being lost or overlooked.

Linnaeus the Younger died a few months after his return to Uppsala, on 1 November 1783. Apart from the *Supplementum plantarum* and a handful of other articles, Linnaeus the Younger's botanical legacy is virtually non-existent, yet these slips indicate that he did much more than visit and socialise during his English stay. He took an active part in the scientific life of prominent naturalists. The fact that he had used and published from his father's slips means that the Solander slips he encountered in England were not incompatible with his own ways of working. It is a system he seems to have readily adopted and continued not only during his French stay, but also upon his return to Uppsala.[58]

Carl Peter Thunberg

Carl Peter Thunberg, like Solander, was another of Linnaeus's students who travelled abroad. But while Solander became entrenched in England, and never returned to Sweden, Thunberg ultimately returned to settle in Uppsala, having greatly profited from Linnaeus's teaching, from various patronages, and from his travels. Thunberg was Linnaeus's student in the late 1760s, and gained his medical degree in 1767. He left Sweden for Paris in 1771, to continue his studies in medicine and natural history. From Paris, he moved to Holland, where he was commissioned by Johannes Burman (1707-1780) and his son Nicolaas (1734-1793) to visit the Dutch colonies and Japan to collect specimens for Dutch botanical gardens. En route to Japan, Thunberg stopped at the Cape of Good Hope where he remained until 1775, before moving on to Japan via Java and Batavia.[59]

58. Spencer Savage dated slips related to the *Amarylllidaceae* (Linnaeus fil., 'De Liliaceis et affinibus plantis') between September 1782 and November 1783. See Spencer Savage, 'Description and phylogeny: on a manuscript by the Younger Linnaeus dealing with certain genera now included in the *Amaryllidaceae*', *Herbertia* 4 (1937), p.85-86, 91-94, 97-100.

59. See Lars Hansen (ed.), *The Linnaeus apostles: global science and adventure*, vol.6, *Europe, southern Africa, East, southern & South East Asia: Carl Peter Thunberg* (London, 2007); Skuncke, *Carl Peter Thunberg*, p.39-145.

At the Cape, Thunberg collected many plant specimens for himself, his numerous funders, and for his former teacher, Linnaeus. It is important to stress that, while Thunberg turned to Linnaeus for scientific advice, Linnaeus was not one of Thunberg's patrons. Thunberg's wealthy and rival patrons were both Dutch and Swedish. In Holland they included the Burmans, VOC director Egbert de Vrij Temminck (1700-1785), Temminck's relative David ten Hove (1724-1787), the collector and botanical enthusiast Jean Deutz (1743-1784), as well as the institutions of the Hortus Medicus in Amsterdam. In Sweden his patrons included Peter Jonas Bergius (1730-1790), Abraham Bäck, Lars Montin (1723-1785) and the Royal Academy of Sciences in Stockholm. These patrons either paid for part of the trip, or they had promised to help further Thunberg's career on his return to Sweden. Thunberg was therefore obliged to send both dried and live plant specimens, accompanied with plant descriptions, to these various patrons, many of whom disliked each other, and from whom the extent of Thunberg's services needed to be concealed.[60] Indeed, his Dutch patrons had stipulated that they should have priority over Thunberg's specimens, a fact that most of his Swedish patrons (including Linnaeus) did not know.[61] Thunberg needed Linnaeus's long-standing knowledge and reputation, and as a result sent many specimens and plant descriptions to Linnaeus: eighteen letters from Thunberg to Linnaeus survive, written between 1771 and 1775. Yet, with many patrons supporting his endeavours, Thunberg was much less in need of Linnaeus's support than Osbeck had been twenty years earlier.[62]

Thunberg often collected plant specimens in multiple numbers to satisfy all of his patrons as well as members of his extensive network, and for himself. Marie-Christine Skuncke enumerated the number of correspondents for whom Thunberg would have needed multiple specimens while he was botanising at the Cape, and it amounts to fifteen individuals.[63] In order to collect plants, describe them, and deal with the logistics of shipping these letters and specimens to various destinations in Europe on Dutch VOC vessels, but also ships from the Swedish, British and French East India companies, Thunberg had to develop effective information management strategies.[64] As a result, Thunberg seems to have readily adopted Linnaeus's technology of the paper slip, perhaps to efficiently produce copies of the same plant description. On 27 May

60. Skuncke, *Carl Peter Thunberg*, p.26, 56-65.
61. Skuncke, *Carl Peter Thunberg*, p.61.
62. See Bettina Dietz's chapter in this volume, 'What is a botanical author?' (chapter 2).
63. Skuncke, *Carl Peter Thunberg*, p.78-79.
64. On the logistics of shipping mailings back to Sweden, see Skuncke, *Carl Peter Thunberg*, p.30-32.

1772, for example, Thunberg sent a letter, as well as some plants from the Cape of Good Hope, to Linnaeus. Thunberg wrote that he was unable to define the genera of some of these plants, and that he was looking forward to Linnaeus's opinion. Aside from discussing these plants within the letter, he also sent their formal description in the form of small slips of paper, headed by the genus name when known. Three slips, describing three species of *Oxalis* (*tormentosa*, *bulbosa* and *capensis*), were most probably sent with the letter (see Fig.6, p.53).[65] Given the number of people Thunberg had to supply specimens to, it is possible that he used slips in order to duplicate information quickly, collecting a number of specimens of the same species, and sending them on to various correspondents who then received the same information on slips.

Like Linnaeus before him, Thunberg often carefully numbered his specimens to allow Linnaeus to allocate each specimen to its corresponding description, especially if the plant in question was left unnamed. This seems to have been the case with *Thunbergia*. Thunberg sent a lengthy and detailed species description on a paper slip, lacking a genus name. The space left by Thunberg for the name, a series of full stops, was later filled in by Linnaeus, with 'Thunbergia', probably in 1773 or early 1774 when Linnaeus seems to have approved the name given by Lars Montin.[66]

It has been noted that most of Thunberg's manuscripts and publications 'lack in detail' and that 'most of them consisted of checklists giving binomial names to all species recorded'.[67] The slips, however, show that Thunberg provided a great deal of botanical information for plants newly discovered, and followed the Linnaean template for describing new genera and species.

In many ways, Thunberg's networks of patrons and correspondents were much more complex than Linnaeus's had ever been, and they grew increasingly complex as Thunberg travelled east and forged new professional relationships. Indeed, Thunberg's correspondents numbered '1,385 [...]; that is more than twice the number of Linnaeus' correspondents in the online edition of the Linnaean Correspondence'.[68] Much of his early career was spent walking a tightrope between his various

65. Carl Peter Thunberg, 'Diagnoses of *Oxalis* and *Septas*', Library of the Linnean Society of London, LM/MA/THU/1.
66. See Carl Peter Thunberg to Carl Linnaeus, 1 May 1774, *The Linnaean correspondence*, http://linnaeus.c18.net/, letter L5004 (consulted 27 October 2014). Lars Montin's *Thunbergia* was later discovered to be a genus of *Gardenia*, and the name *Thunbergia* was finally reattributed to a new genus in 1781 by Anders Jahan Retzius; see Skuncke, *Carl Peter Thunberg*, p.85.
67. L. C. Rookmaaker, *The Zoological exploration of southern Africa 1650-1790* (Rotterdam, 1989), p.149.
68. Skuncke, *Carl Peter Thunberg*, p.18.

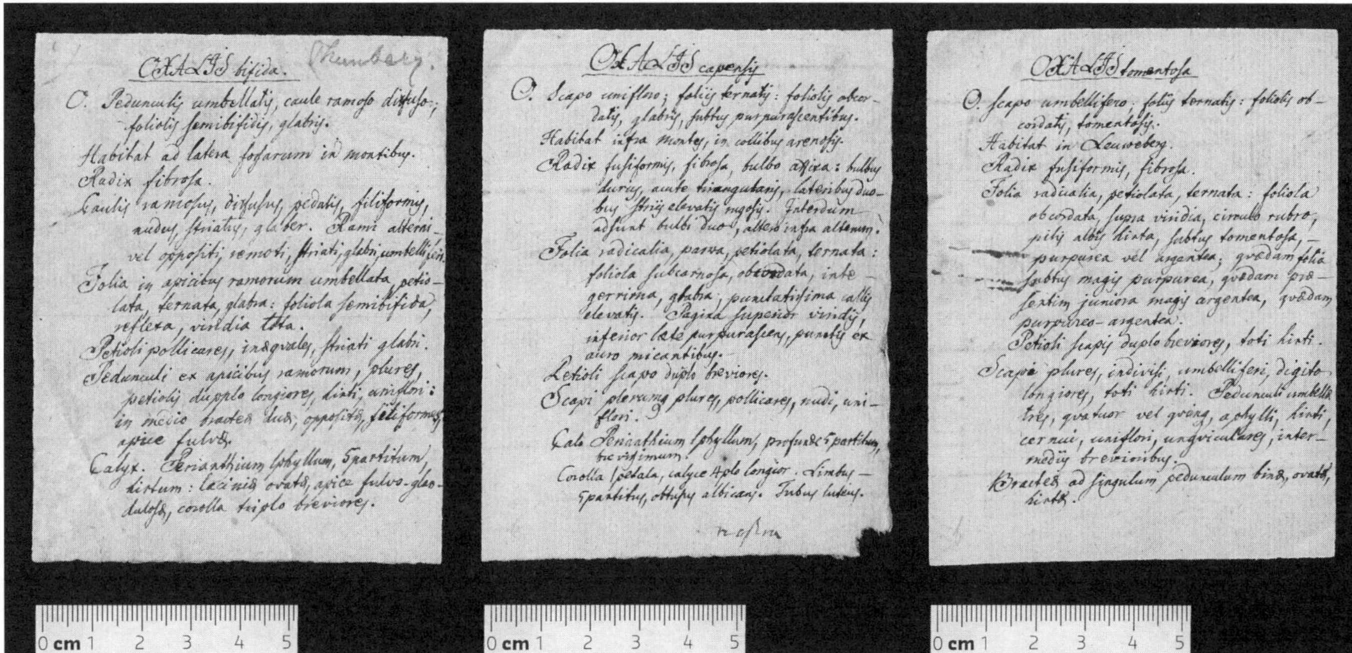
OXALIS bifida. Thunberg.

O. Pedunculis umbellatis, caule ramoso diffuso; foliolis semibifidis, glabris.
Habitat ad latera fossarum in montibus.
Radix fibrosa.
Caulis ramosus, diffusus, pedalis, filiformis, nudus, striatus, glaber. Rami alterni vel oppositi, remoti, striati, glabri, umbelliferi.
Folia in apicibus ramorum umbellata, petiolata, ternata, glabra: foliola semibifida, reflexa, viridia tota.
Petioli pollicares, inaequales, striati glabri.
Pedunculi ex apicibus ramorum, plures, petiolis duplo longiores, hirti, uniflori: in medio bracteis duo, oppositis, setiformes, apice fulvis.
Calyx. Perianthium 5phyllum, 5partitum, hirtum: laciniis ovatis, apice fulvo-glandulosis, corolla triplo breviores.

OXALIS capensis

O. Scapo unifloro; foliis ternatis: foliolis obcordatis, glabris, subtus purpurascentibus.
Habitat infra montes, in collibus arenosis.
Radix fusiformis, fibrosa, bulbo affixa: bulbus durus, acute triangularis, lateribus duobus striis elevatis rugosis. Interdum adsunt bulbi duo, altero infra alterum.
Folia radicalia, parva, petiolata, ternata: foliola subcarnosa, obcordata, integerrima, glabra, punctatissima albis elevatis. Pagina superior viridis, inferior lacte purpurascens, punctis ex auro micantibus.
Petioli scapo duplo breviores.
Scapi plerumque plures, pollicares, nudi, uniflori.
Cal. Perianthium 5phyllum, profunde 5partitum, brevissimum.
Corolla 5petala, calyce 4plo longior. Limbus 5partitus, obtusus albicans. Tubus luteus.
nostra

OXALIS tomentosa

O. scapo umbellifero, foliis ternatis: foliolis obcordatis, tomentosis.
Habitat in Leuweberg.
Radix fusiformis, fibrosa.
Folia radicalia, petiolata, ternata: foliola obcordata, supra viridia, circulo rubro, pilis albis hirta, subtus tomentosa, purpurea vel argentea; quaedam folia subtus magis purpurea, quaedam praesertim juniora magis argentea, quaedam purpureo-argentea.
Petioli scapis duplo breviores, toti hirti.
Scapi plures, indivisi, umbelliferi, digito longiores, toti hirti. Pedunculi umbellae tres, quatuor vel quinque, aphylli, hirti, cernui, uniflori, unguiculares, intermedii brevioribus.
Bracteae ad singulum pedunculum binae, ovatae, hirtae.

Figure 6: Slips for *Oxalis bifida*, *O. capensis* and *O. tomentosa*, written by Carl Peter Thunberg and sent to Linnaeus with his letter dated 27 May 1772. By permission of the Linnean Society of London.

patrons and interested correspondents, all vying for precious new plants arriving from the Cape, Java or Japan. While Thunberg may have appropriated some of his former teacher's paper technologies, he used them for very different ends during his travels. Paper slips not only enabled him to record information on new plants which could be shuffled at will to find their place in the Linnaean classificatory system, they also enabled him to duplicate the descriptions in order to distribute them more efficiently. For Thunberg, paper slips were not only a way of recording information on plants, but also a way of communicating this information efficiently from one part of the globe to another.

Before returning to Sweden, Thunberg spent six weeks in London, between December 1778 and January 1779. Having completed his significant voyage to Japan, he was welcomed into Banks's circle and given access to Banks's collections and library. Like Linnaeus the Younger two years later, he visited the famous gardens and nurseries of London, and forged links which he would maintain and nurture over the course of the next ten years.[69] Upon his return to Uppsala, he was appointed demonstrator of botany at the University of Uppsala in May 1779, with Professor Linnaeus the Younger as his superior. Thunberg contributed a large number of species from his recent travels to the Cape of Good Hope, Japan, Java and Ceylon to Linnaeus the Younger's *Supplementum plantarum*. In July 1779, Thunberg wrote to Pehr Wargentin (1717-1783), then secretary of the Royal Academy of Sciences in Stockholm, that he was working up to six hours a day with Linnaeus the Younger on the book. This is interesting, as the preparatory manuscript in the Linnean Society bears no trace of Thunberg's handwriting, and apart from the slips sent from Thunberg to Linnaeus, there is no trace of this collaborative work. The new species from Thunberg, especially those from the Cape of Good Hope (nearly 500 of them), are all listed in Linnaeus the Younger's hand on loose folio leaves which have been inserted in the preparatory manuscript.[70] Nevertheless, from his English stay, and from working on the 'Supplementum plantarum' manuscript, Thunberg would have been familiar with the technology of paper slips. It would be interesting to know whether Thunberg adopted this technology fully in his later work, as professor of botany and materia medica at the University of Uppsala.[71]

Following Carl Linnaeus the Younger's death and the loss of the

69. Skuncke, *Carl Peter Thunberg*, p.176-85.
70. See for example the list of species of *Anthericum* and *Thunbergia*, Linnaeus fil., 'Supplementum plantarum', p.50a, 50b.
71. Thunberg's manuscripts are for the most part kept at Uppsala universitets arkiv and Uppsala universitetsbibliotek. Unfortunately, the author has not been able to consult them.

Linnaean collections to James Edward Smith in England in 1783, Thunberg donated his natural history collections to the University of Uppsala in 1787. The upkeep and increase of the collections became the financial responsibility of the University, and no longer the private property of the professor of botany.[72] A new botanical garden was founded, and on 25 May 1807 a new botany building built in the neo-classical style, the Linneanum or Orangery Building as it came to be called, was opened. As in England, a national institution's collection was born out of the private collection of a prominent individual.

Linnaean science: a collaborative practice

This chapter has shown that Linnaean science was much more collaborative in practice than has previously been thought, that collaboration between naturalists who practised Linnaean science was underpinned by common information technologies, and that in time these information technologies played a crucial role in the institutionalisation of Linnaean science.

The practice of Linnaean science, like much of early modern science, has in the past been conceived of as solitary and independent work undertaken by a single man, yet current historiography is highlighting the sociability and networks of exchange that characterised the practices of collecting, exchanging, cataloguing and describing natural history specimens in early modern Europe.[73] The study of Linnaeus's manuscripts suggests that, despite the cultivated image of the hard-working naturalist labouring on his own, he quite often collaborated and depended upon the work of students (such as Löfling), amanuenses and his own son, all of whom helped him deal with the 'information overload' from which he suffered from the 1750s onwards. In particular, Linnaeus the Younger's role in the practice of natural history in the Linnaean household should be more closely investigated. The manuscripts indicate that by the 1760s some botanical, zoological and pharmaceutical work was undertaken in a collaborative manner between father and son.

A great part of Linnaeus's academic life was spent teaching botany, zoology, materia medica, and medical subjects such as pathology, nosology and dietetics. Linnaeus was a charismatic teacher, who inspired his students to travel and devote their lives to natural history. In exchange for his counsel, advice and support, Linnaeus received specimens and

72. Skuncke, *Carl Peter Thunberg*, p.216.
73. See for example Dietz, 'Contribution and co-production'; Mary Terrall, *Catching nature in the act: Réaumur and the practice of natural history in the eighteenth century* (Chicago, IL, 2014); Tobin, *The Duchess's shells*.

descriptions of new species from his 'apostles' abroad, as Bettina Dietz describes in her chapter concerning Osbeck, and as this chapter has shown in the case of Thunberg. Being surrounded by young, sometimes gifted, students had an undeniable impact on Linnaeus's work. Much like how the binomial nomenclature was devised in the 1740s to enable Linnaeus and his students to work together during a trip to Öland and Gotland, as well as on dissertations *Gemmae arborum* and *Pan svecicus*, the paper slip technology emerged from the close collaboration between Linnaeus and a gifted student.[74] Cataloguing the queen's insect collection collaboratively with Daniel Solander was most probably the impetus for the idea of individual paper slips for each species of insect. Solander seems to have recognised the technology as an efficient method for storing and retrieving information much earlier than his former mentor. Linnaean science in practice was never the product of one single man, even at the peak of Linnaeus's career.

Solander and Thunberg both in their own way contributed to the propagation of Linnaean science in England and in Sweden. Their practice of botany never strayed from the Linnaean precepts of nomenclature, description of species and classification. Due to their format and their flexibility, the paper slips lent themselves to the propagation of these Linnaean precepts. The examples of Solander and Linnaeus the Younger show that the slips were ideal for collaborating and allowing several naturalists to work at once on the same specimen or publication, while the example of Thunberg reveals how the slips allowed for efficient distribution of the same information to different people. The sustained use of the Solander slips by his successors Dryander and Brown is a testament to their efficiency and success as an information technology. In both England and Sweden, the first two decades of the nineteenth century saw an institutionalisation of natural history, with the establishment of the Linneanum and the transfer of Banks's collections to the British Museum. The technology used to record and communicate information on species of plants and animals was part and parcel of that institutionalisation. Through Solander, Dryander and Brown, the card catalogue had become one of the centrepieces of a natural history that was no longer dominated by individuals, such as Linnaeus, but by institutions whose workings reached beyond the time span of an individual's life. The paper slips had fully realised their collaborative and trans-generational potential, and became one of the earliest examples of a proper card catalogue.

74. William T. Stearn, 'The background of Linnaeus's contributions to the nomenclature and methods of systematic biology', *Systematic zoology* 8:1 (1959), p.4-22.

2. What is a botanical author? Pehr Osbeck's travelogue and the culture of collaborative publishing in Linnaean botany

BETTINA DIETZ

Introduction

This essay explores the culture of collaborative publishing in Linnaean botany by tracing the publication trajectory of Pehr Osbeck's report of his journey to Canton in southern China during the period of 1750 to 1752. Osbeck (1723-1805), a Swedish naturalist and student of Linnaeus, divided his material between a number of publications that appeared successively, building on and complementing each other. The aim of his project was to make a first version of his new material available to interested readers as quickly as possible, while leaving the process of completing and correcting it to be undertaken in stages at a later date. This long-term project was conceived as a collaboration from the start, and involved a number of authors. It began with most of Osbeck's newly discovered plants being published first not in his own book but, by mutual agreement, in Linnaeus's *Species plantarum* (1753). After Osbeck's travel report was published in Swedish (1757), translations into German and English gave additional authors a chance to expand and correct the text. This layering of information, along with the documentation of the provenance of each individual addition, shaped not only the appearance of the published text, but also a concept of scientific authorship. The various editions of Osbeck's travel report were published under a single name – his. But, as a botanical author, Osbeck was a 'composite', and his text was an assemblage of contributions by many others.[1]

1. The physicist Alan Thorndike used this expression in 1967 to aptly describe the transformation of experimental practice in particle physics: 'Who is the experimenter whose activities we have been discussing? Rarely, if ever, is he a single individual [...] *The experimenter, then, is not one person, but a composite* [emphasis added]. He might be three, more likely five or eight, possibly as many as ten, twenty, or more. [...] One thing, however, he certainly is not. He is not the traditional image of a cloistered scientist working in isolation at his laboratory bench'; Peter Galison, 'The collective author', in *Scientific authorship: credit and intellectual property in science*, ed. Mario Biagioli and Peter Galison (New York and London, 2003), p.325-55 (328-29).

An interest in scientific collaboration and the emergence of corresponding forms of scientific authorship was sparked off by studying how contemporary natural scientists worked. In the late 1970s and 1980s, ethnographically inspired descriptions and analyses of the working processes of modern laboratory science, of science in the making, by Bruno Latour, Karin Knorr-Cetina and others, demonstrated the extent to which scientific results depend on co-operation between a large number of visible and invisible actors – scientists, technicians, programmers and assistants – without whom laboratory work simply could not be done.[2] Under the name Actor-Network-Theory (ANT), non-human actants such as equipment, materials and laboratory animals were increasingly included in the analysis of factors on which the success or failure of research depends.[3]

The experimental procedures of modern particle physics in particular allow not only the escalating scale of scientific collaboration to be studied, but also the impact of a quintessentially collective research process on the publications it produced, something which is of special interest here. Starting from the question of the epistemic subject in collaborations of first a few dozen, then several hundred, and finally about 5000 scientists, Peter Galison analyses how these sorts of collectives publish the results of their research.[4] He shows that collaborations in particle physics have developed their own authorship protocols, which prescribe under what circumstances and in what order the names of authors are to be listed. And, as in the case of the Stanford Linear Detector (SLD) guidelines, which Galison cites as an example, they specify that the achievements of the group in all cases take precedence over the contributions of individuals.[5] In order to reflect the collaborative nature of experimental work, the writing process, too, is subject to strict regulations, especially concerning the publication of the results of the collaborative project as a whole. Every publication of this sort originates in a draft drawn up by a small group of authors, which is

2. A selection: Bruno Latour and Steve Woolgar, *Laboratory life: the construction of scientific facts* (Beverly Hills, CA, 1979); Sharon Traweek, *Beamtimes and lifetimes: the world of high energy physicists* (Cambridge, MA, 1988); Steven Shapin, 'The invisible technician', *American scientist* 77:6 (1989), p.554-63; Karin Knorr-Cetina, *Epistemic cultures: how the sciences make knowledge* (Cambridge, MA, 1999).
3. For an overview see Bruno Latour, *Reassembling the social: an introduction to actor-network-theory* (Oxford, 2005); Bruno Latour, *Science in action: how to follow scientists and engineers through society* (Cambridge, MA, 1987).
4. See Galison, 'Collective author'.
5. 'There should be no exceptions to the above, such as placing the student's name first if the paper originated as a thesis, as our first priority should be the coherence of the group and the de facto recognition that contributions to a piece of physics are made by all collaborators in different ways.' SLD Policy on Publications and Conference Presentations of 1 July 1988, quoted from Galison, 'Collective author', p.332.

presented to the whole group for criticism, after several further cycles of additions and revisions, before it is officially published.[6]

While collaboration and, consequently, collective authorship of such spectacular dimensions grow out of the conditions governing present-day laboratory work, scientific collaboration as such is nothing new. The question arises to what extent earlier forms of scientific practice, and not only those of an experimental nature, were organised collaboratively, and what impact this had both on the making of scientific publications and on the concept of scientific authorship. The few works which deal with these questions mostly concentrate on the early modern period, a time when the institutionalisation of science progressed with the founding of academies, learned societies and museums. While this was an important development, in the medium term it did not change the fact that many researchers were still dependent on private means. This explains why at that time the home or household of a scholar was a central location of research and experimentation, and why domestic protagonists – wife, children and servants – were often involved. Because of the demands imposed by the need for observations and the effort required in collecting, astronomy and natural history were quite often undertaken as a family project.[7] But there is also evidence of more far-reaching forms of scientific collaboration, for example, in the way Galileo Galilei worked. We now know that a large number of people – patrons, colleagues at the Lincean Academy, students and printers – were actively involved in the publication of his works as advisers.[8]

Yet, as a rule, these forms of early modern scientific teamwork were invisible to readers; on the publication of their results, they disappeared behind the name of the scientist who featured alone on the title page as the author. We do not yet have a history of the scientific author, but the concept of individualised authorship seems to have established itself as standard earlier in science than in literature. Research on the function of literary authorship, encouraged by Michel Foucault's pioneering 'Qu'est-ce qu'un auteur' (1969),[9] has shown that only in the late eighteenth and

6. See SLD Policy on Publications and Conference Presentations of 1 July 1988, quoted from Galison, 'Collective author', p.332.

7. On this see Alix Cooper, 'Homes and households', in *The Cambridge history of science*, ed. David C. Lindberg and Ronald L. Numbers, 8 vols (Cambridge, 2003-2013), vol.3, *Early modern science*, ed. Katharine Park and Lorraine Daston (Cambridge, 2006), p.224-37 (232-33.); from a different perspective, Gadi Algazi, 'Scholars in households: refiguring the learned habitus, 1480-1550', *Science in context* 16:1-2 (2003), p.9-42.

8. See David Freedberg, *The Eye of the lynx: Galileo, his friends and the beginning of modern natural history* (Chicage, IL, 2002), ch.5; Renée J. Raphael, 'Printing Galileo's *Discorsi*: a collaborative affair', *Annals of science* 69:4 (2012), p.483-513; Domenico Bertoloni Meli, 'Authorship and teamwork around the Cimento Academy: mathematics, anatomy and experimental philosophy', *Early science and medicine* 6:2 (2001), p.65-95.

9. Michel Foucault, 'Qu'est-ce qu'un auteur', *Bulletin de la société française de philosophie* 63:3

early nineteenth centuries was the literary author stylised into a solitary individual with the unique ability to create something genuinely new and original out of himself alone.[10] Remarkably, when it emerged, this concept was already at odds with the actual creative process of writers who played a large part in promoting it. Martha Woodmansee has shown how the English poet William Wordsworth eloquently presented himself as a solitary author, although his creative process was to a large extent organised co-operatively and involved his sister and his friend, the poet Samuel Taylor Coleridge.[11]

There are unmistakable similarities between this sort of literary self-description and the way in which Linnaeus represented his own significance for the progress of botany, never hesitant in the use of superlatives, as is well known. His *Genera plantarum* had 'smoothed the path for the science of the age' before he embarked on the *Species plantarum*, 'the greatest work in the discipline'.[12] Or in perhaps the most grandiose exaggeration of his own significance: 'No one had been a greater botanist or zoologist, no one had written a larger number of works, no one had become more renowned throughout the world.'[13] A more moderate version of this image, but one that still focuses on Linnaeus as supremely dominating his subject, has long persisted in the research literature, as William Stearn's Introduction in Wilfrid Blunt's widely read biography of Linnaeus, *Linnaeus: the compleat naturalist*, demonstrates. There we read: 'Linnaeus' essential task was to provide the means of identifying and naming all the organisms then known. He lived at a time, possibly the very last time, *when one person could accomplish this single-handed* – provided

(1969); for a correction of Foucault's chronology of the emergence of modern authorship from a historical perspective see Roger Chartier, 'Foucault's Chiasmus', in *Scientific authorship: credit and intellectual property in science*, ed. Mario Biagioli and Peter Galison (New York and London, 2003), p.15-31.

10. There is a huge body of literature on this topic. A selection: Paul Bénichou, *Le Sacre de l'écrivain* (Paris, 1973); Martha Woodmansee and Peter Jaszi (ed.), *The Construction of authorship: textual appropriation in law and literature* (Durham, 1994), especially Woodmansee and Jaszi, 'Introduction', p.1-13; and Woodmansee, 'On the author effect: recovering collectivity', p.15-28 (16).
11. See Woodmansee and Jaszi, 'Introduction', p.2; also Roger Chartier, 'Publishing Cervantes', in Roger Chartier, *The Author's hand and the printer's mind: transformations of the written word in early modern Europe* (Cambridge, 2014), p.150-57.
12. 'dem Zeitalter die Wissenschaft geebnet'; 'ein Werk, welches das größte in der Wissenschaft ist'. *Linné's eigenhändige Anzeichnungen über sich selbst, mit Anmerkungen und Zusätzen von Afzelius*, translated from the Swedish by Karl Lappe (Stockholm, 1823; Berlin, 1826), p.53.
13. Sten Lindroth, 'The two faces of Linnaeus', in *Linnaeus: the man and his work*, ed. Tore Frängsmyr (Berkeley, CA, 1983), p.1-62 (57). Lindroth here condenses a longer passage from *Vita Caroli Linnaei: Carl von Linnés självbiografier*, ed. E. Malmeström and A. Hj. Uggla (Stockholm, 1957), p.145-46.

he was willing to pay the price. The price for Linnaeus was a heavy one: for much of his life he was financially insecure and his health suffered as a result of overwork.'[14]

This image has only recently begun to be undermined, mainly as the result of increasing interest in the practices of information-gathering and data-processing that characterised Linnaeus's method of working. Staffan Müller-Wille and Isabelle Charmentier have investigated the paper technologies – the lists, boxes of paper, files, index cards, annotated books and especially paper slips – with whose aid Linnaeus organised the constantly growing stream of information about hitherto unknown plants, animals and minerals for his publication projects.[15] I myself have demonstrated how Linnaeus's ambition to record, name and classify all plants worldwide forced him to adopt a collaborative working mode, which, in turn, had a fundamental impact on his iterative style of publication. Behind the self-image of a lonely genius striding on ahead of everybody else that Linnaeus promoted in his autobiographical writings lay a collaborative practice and epistemology, which he described openly in his scientific publications and correspondences.[16]

In the following, the significance of this collaborative approach for Linnaeus's own working process will first be outlined, from the systematic gathering of information by means of correspondence and the techniques employed, to the iterative cycles of supplementing and correcting primarily the *Systema naturae* and *Species plantarum*, and the participatory architecture of an evolvable system. It will then be argued that this collaborative model shaped not only Linnaeus's publications but also the working methods of other botanists around him, so that we can speak of a culture of collaborative publishing in eighteenth-century botany. The essay will then turn to Osbeck's report on his trip to China and show that from the start its publication was planned not as a one-off act but as a long-term process that involved several authors, produced a number of different types of publication and required iterative cycles of updates.

14. Emphasis added. William Stearn, 'Introduction', in Wilfrid Blunt, *Linnaeus: the compleat naturalist* (Princeton, NJ, 2001), first published in 1971 as *The Compleat naturalist: a life of Linnaeus*, p.6-9 (6).
15. Isabelle Charmantier and Staffan Müller-Wille 'Carl Linnaeus's botanical paper slips (1767-1773)', *Intellectual history review* 24:2 (2014), p.215-38; Staffan Müller-Wille and Isabelle Charmantier, 'Natural history and information overload: the case of Linnaeus', *Studies in history and philosophy of science*, part C, *Studies in history and philosophy of biological and biomedical sciences* 43:1 (2012), p.4-15; see also Staffan Müller-Wille and Sara Scharf, *Indexing nature: Carl Linnaeus (1707-1778) and his fact-gathering strategies, Working papers on the nature of evidence: how well do 'facts' travel?* 36/08, http://www.lse.ac.uk/economichistory/pdf/factspdf/3909muellerwillescharf.pdf (last accessed 10 December 2015).
16. See Bettina Dietz, 'Contribution and co-production: the collaborative culture of Linnaean botany', *Annals of science* 69:4 (2012), p.551-69.

The focus here will be on the collaborative process of *writing* botanical works. The main protagonists are the European botanists of the eighteenth century, who not only shared information to supplement and correct their own publications among themselves, but also updated and expanded the works of others, drawing on the material available to them. In this context, the question of whether Osbeck had help from indigenous or local informants when *collecting* natural history specimens in China, and if so, how, will not be discussed here. The sources known to me contain no direct information on this. But it needs to be stressed that no botanists alone could accumulate the material for a publication covering a larger region, not even those who worked on regional European flora. Albrecht von Haller, for example, employed a whole host of assistants for his work on a Swiss flora. They systematically combed regions and mountain ranges on his behalf in order to gather the necessary plants.[17]

This essay concentrates less on the collection of material – an aspect that has long occupied praxis-oriented historians of science – than on the collaborative process of writing and publishing Osbeck's travel report, and the decentralisation of the botanical author associated with this.[18] The practice of superimposing layers of information, taken to an extreme by Linnaeus but also used by other botanists, had the effect of fragmenting the botanical author, although the general rule of featuring only one name on the title page of publications meant that

17. On this see Bettina Dietz, *Das System der Natur: die kollaborative Wissenskultur der Botanik im 18. Jahrhundert*, ch.1, 'Regionale Botanik als kollaboratives Projekt' (Cologne and Vienna, 2017); Luc Lienhard, '"La machine botanique": zur Entstehung von Hallers Flora der Schweiz', in *Hallers Netz: ein europäischer Gelehrtenbriefwechsel zur Zeit der Aufklärung*, ed. Martin Stuber, Stefan Hächler and Luc Lienhard (Basel, 2005), p.371-410; see also Alix Cooper, *Inventing the indigenous: local knowledge and natural history in early modern Europe* (Cambridge, 2007).

18. A selection: Jim Endersby, *Imperial nature: Joseph Hooker and the practices of Victorian science* (Chicago, IL, 2008); Emma C. Spary, *Utopia's garden: French natural history from old regime to Revolution* (Chicago, IL, 2000); Anke te Heesen and Emma C. Spary (ed.), *Sammeln als Wissen: das Sammeln und seine wissenschaftsgeschichtliche Bedeutung* (Göttingen, 2001). On the significance of local informants for the identification and cultivation of economically useful plants in colonial settings see Deepak Kumar, 'Botanical explorations and the East India Company: revisiting "plant colonialism"', in *The East India Company and the natural world*, ed. Anna Winterbottom, Vinita Damadoran and Alan Lester (Basingstoke, 2014), p.16-34 (Kumar here also discusses the precariousness of usable sources); Anna Winterbottom, 'Medicine and botany in the making of Madras, 1680-1720', in *The East India Company and the natural world*, ed. Anna Winterbottom, Vinita Damadoran and Alan Lester (Basingstoke, 2014), p.35-57; on China see the publications by Carla Nappi, 'Surface tension: objectifying ginseng in Chinese early modernity', in *Early modern things: objects and their histories 1500-1800*, ed. Paula Findlen (New York, 2013), p.31-52; Paula Findlen, *The Monkey and the inkpot: natural history and its transformations in early modern China* (Cambridge, MA, 2009).

the author continued to appear as a single subject.[19] But the input of various contributors involved, in the case of Osbeck's travel report especially the two translators, was minutely documented, demonstrating to readers that this text had not been produced by one individual, but had grown out of a culture of collaborative publishing.

The collaborative culture of Linnaean botany

Linnaeus's systematic works, such as the *Systema naturae* and the *Species plantarum*, emerged out of a collaborative working process in which modules of information and knowledge contributed by many people were put together in such a way that the end product was recognisably the result of a co-operative effort.[20] For all involved, correspondence was an indispensable instrument of information-gathering and thus provided the basis of a contribution-based epistemology. An arsenal of techniques – reference catalogues, lists, and illustrations, asking and answering questions, and reporting mistakes to each other – was used to enable efficient information-gathering, that is, to save time, prevent the loss of information, allow questions and answers to be precisely correlated, and the countless details communicated to be located and used efficiently. The huge input, both quantitative and qualitative, of information generated via correspondence had a direct impact on the form of botanical publications. The more ambitious a project's size, the more of an aggregate it became.[21]

Reference must be made to an essential information technique that this process of aggregating relied on, namely, reporting errors. Scholars regularly pointed out errors that they had identified either in the work of their correspondence partner or in the publications of other natural historians, to each other by letter. Even botanists such as Linnaeus, who had abundant resources, could find themselves in the position of not having to hand a living plant that they urgently needed for purposes of comparison. Pressed and dried specimens were the necessary alternative, but the processes of drying and pressing could deform details and make

19. On Linnaeus in this context see Dietz, 'Contribution and co-production'.
20. This section is compiled from Dietz, 'Contribution and co-production', with the kind permission of the journal's editors. Central aspects that can only briefly be touched on here – correspondence as natural history's information system and the techniques of efficient information-gathering (especially the reporting of errors), the character of Linnaeus's publications as aggregates and the principle of iterative publication – are discussed in detail there. See also Dietz, *Das System der Natur*.
21. On accumulation from a global perspective see Lissa Roberts, 'Accumulation and management in global perspective: an introduction', *History of science* 52:3 (2014), p.227-46; Lissa Roberts, '"Le centre de toutes les choses": constructing and managing centralisation on the Isle de France', *History of science* 52:3 (2014), p.319-42.

precise observation difficult. In cases where neither a living nor a dried specimen was available, recourse had to be taken to illustrations and descriptions in the botanical literature. As mistakes could and did happen at every stage of this process – more mistakes than one author alone could identify and correct – error reports became an essential component of the correspondence routine. The correct identification of a plant was therefore, as a rule, the result not only of an individual act of observation, but to a large extent also of a collective one. The more eyes were involved in examining the features of a plant that had not yet been definitively identified, the more quickly was accuracy likely to be achieved.

This frequently used form of feedback, sent either solicited or unsolicited via the botanical correspondence, makes two things clear: first, in Linnaean botany, precision was often only possible as the result of a long and co-operative process of correction; and second, it shows that published work was fundamentally provisional, although to varying degrees. New material was constantly turning up, new observations were being made, and new observers were entering the discussion, so that correcting descriptions, illustrations and classifications was not only a necessity, but also accepted as such. Publications, especially the largest ones, were therefore, as a rule, iterative in nature. This means that they were not intended to be definitive, but were from the start designed to be published in several editions.

Nobody knew how to instrumentalise this principle as systematically as Linnaeus, and the consecutive editions of his *Systema naturae* and *Species plantarum* demonstrate how he aggregated information from his correspondents to publish constantly augmented and corrected versions – twelve of the *Systema naturae* and two of the *Species plantarum.* Hardly had a new edition of one of his works been published before it needed to be revised and supplemented. New books had appeared in the interim, and new species and genera had to be extracted from them, checked, and incorporated at the relevant place. His travelling students and numerous correspondents constantly sent him dried plants and seeds, which had to be identified, before also being incorporated.

Given this situation, Linnaeus developed a publication strategy which took account of the insight that botanical projects with large regional let alone global aspirations by their nature have to be works in progress, and require an appropriate publication mode in order to be of maximum scientific use. In the rapidly growing field of natural history, especially botany, projects of this magnitude could not be accomplished by an individual scholar, nor in a single, one-off publication. This would have meant putting a text on the market, if at all, only after a long process of correction and enlargement, and it would have become out of date quickly all the same. Publication would happen late, rarely, or, in the worst case, never. Linnaeus chose a different path. Instead of late and rarely, he published early and often. Completeness and precision could be attained only in the long term, through collaboration and iteration.

Linnaeus's works were long term and evolvable projects, whose success was based on their structural capacity to grow. While his genera and species were established in a way that largely guaranteed the stability of already familiar taxa despite the constant discovery of as yet unidentified specimens, the strategy of iterative publication made it possible to incorporate the influx of new specimens and to absorb the stream of corrections received from correspondents. This *perpetuum mobile* did not stop with Linnaeus's death, but continued to produce updated editions of his works for decades under the direction of various scholars.

The role of Linnaeus's systematic works as central registers of a globally conceived botany gave them a special status among eighteenth-century botanical publications. Their topicality and reliability were essential for the collaborative practice of Linnean botany: the regular publication of newly identified, classified and named plants at a central location made it possible to limit, if not exclude, multiple namings and competing classifications, allowing the work of many scientists to be efficiently co-ordinated. As a result – and this should be emphasised here – for many botanists who found themselves in the possession of rare and sought-after material, sending plants and information to Linnaeus so that he could supplement and correct his books took priority over working on their own publications. This also applied to Pehr Osbeck, whose strategy of publishing the natural history yield of his journey to China in stages, and of distributing it between various authors and publications, will be discussed in the following section.

The mutual transfer of information between Osbeck and Linnaeus

As a theology student at the University of Uppsala, Pehr Osbeck had attended Linnaeus's lectures on natural history for several years before leaving for Canton (today Guangzhou) in southern China in 1750 as chaplain on board a Swedish East India Company ship. Stops on this journey, which Osbeck used to collect plants and other natural history objects, were made at Cadiz, Java and Ascension Island in the South Atlantic. The first recipient of his botanical discoveries was Linnaeus, and the first medium in which he announced them was correspondence. Immediately on his return, in June 1752, Osbeck began sending plants, either dried or preserved in alcohol, and seeds, to an impatient Linnaeus. Linnaeus was not satisfied with this, but also asked for Osbeck's observations of living plants – an indication of the dimensions of the information transfer expected.[22]

When Osbeck returned from China, Linnaeus was in the final stages of his work on the *Species plantarum*, a critical overview, organised in line with his own classification system, of all plant species known to date which he had, wherever possible, seen and investigated in person.[23] While it is known that a considerable number of the species contained in the *Species plantarum* had been collected and sent to Linnaeus by his travelling students, known as the 'apostles', their role as informants and the mutuality of the information exchange will be examined further here.[24] Informants in this sense should not be seen as subordinate suppliers of information who merely followed instructions, but, as we will see, as partners, advisers and correctors, whose indispensability made them into co-authors and gave them the right to expect a quid pro quo.

In the letter cited above, in which Linnaeus asked Osbeck for all the details he had observed on living plants, he listed those he was most interested in and asked a number of detailed questions. Osbeck replied immediately and answered as precisely as he could. For example, in response to Linnaeus's question of whether one of the plants he had received from Osbeck was a *Melastoma*, Osbeck compared his Chinese material with Linnaeus's description of *Melastoma*, and came to the conclusion that the plant in question must belong to a new species.[25] Just one week later, Osbeck thanked Linnaeus for another letter and, on the basis of the botanical descriptions in his travel notes, replied, sometimes in great detail, to many of the one hundred plus questions

22. Letter of Carl Linnaeus to Pehr Osbeck, 21 August 1752, in Carlo Hansen and Anne Fox Maule (ed.), 'Linnés korrespondance med Pehr Osbeck 1750-1753: med inledning og anmærkninger', *Svenska Linnésällskapets årsskrift* (1972-1974), p.75-145 (97); facsimile of the original and an English summary at http://linnaeus.c18.net/ Letter/L1468 (consulted 14 Jan. 2015). Osbeck replied that he had to revise the notes he had made during the journey to make them legible and useful for another person. See letter of Osbeck to Linnaeus, 23 August 1752, in Hansen and Fox Maule, 'Linnés korrespondance', p.98-99 (98); facsimile of the original and an English summary at http://linnaeus.c18.net/ Letter/L1467 (consulted 14 January 2015).

23. William Thomas Stearn, 'An introduction to the *Species plantarum* and cognate botanical works of Carl Linnaeus', in '*Species plantarum*': *a facsimile of the first edition 1753*, vol.1 (London, 1957), p.1-176 (97-109).

24. Mariette Manktelow and Kenneth Nyberg, 'Linnaeus's apostles and the development of the *Species plantarum*', in *'Species plantarum' 250 years: proceedings of the 'Species plantarum' symposium*, held in Uppsala 22-24 August 2003, ed. Inga Hedberg (Uppsala, 2005), p.73-80; Carlo Hansen and Anne Fox Maule, 'Pehr Osbeck's collections and Linnaeus's *Species plantarum* (1753)', *Botanical journal of the Linnaean Society* 67:3 (1973), p.189-212.

25. See letter of Linnaeus to Osbeck, 21 August 1752, in Hansen and Fox Maule, 'Linnés korrespondance', p.97.

that Linnaeus had asked about numbered plants. For example, Linnaeus asked how many stamens plant number 4 had (the name of the plant is not mentioned) and what its fruit was like, and whether number 18 was the same plant as number 12?[26] Osbeck also reported that by comparing some of the preserved fish, roots and fruits that he had sent Linnaeus with descriptions and illustrations in the natural history literature, he had managed to identify some of them as already known, including the fruit of a plant that he thought was *Crinum* from Paul Hermann's *Flora Zeylanica* (1747), edited by Linnaeus. But as he was not quite sure, he asked Linnaeus to examine the specimens he had, and to tell him of his findings.[27]

Osbeck sent Linnaeus not only plants, seeds and observations, but also plant names that he had coined himself, following the rules of Linnaean nomenclature. The project of cataloguing all existing plants worldwide required not only countless contributors, but also clear rules and standards to co-ordinate the many contributions. Due to the growing international success of Linnaeus's sexual system, a simple method for the identification and classification of plants in combination with his standardised binomial nomenclature, a structure and a language were created that allowed this accumulative dynamic to be efficiently regulated. Only then did it become possible to overcome the old problem that, without a binding nomenclature, the same plant was inevitably given different names by different authors, making it difficult if not impossible to find relevant information in the botanical literature.

Another factor in the growing success of this procedure for realising a global project, in addition to the standardisation of modules to be integrated into the system, was that Linnaeus was not alone in acting as a nomenclator. He was joined by increasing numbers of other botanists, including those who described many hitherto unknown plants of non-European, particularly American and Asian, provenance. Osbeck was one of the first to use Linnaean nomenclature, initially in his travel notes and later in his 1757 travel report *Dagbok öfwer en ostindisk resa åren 1750, 1751, 1752* [Diary of a trip to East India], to name the plants he collected on his travels. In this way, the plants he discovered and named became compatible modules which Linnaeus no longer had to identify from scratch. After more or less extensive checking and, as a rule,

26. Letter of Osbeck to Linnaeus, 30 August 1752, in Hansen and Fox Maule, 'Linnés korrespondance', p.102-105; facsimile of the original and an English summary at http://linnaeus.c18.net/Letter/L1464 (consulted 14 January 2015). On the use of lists of questions and answers in the context of the correspondence-based information system of Linnaean botany, see Dietz, 'Contribution and co-production', p.556-57.

27. See letter of Osbeck to Linnaeus, 11 October 1752, in Hansen and Fox Maule, 'Linnés korrespondance', p.108-109 (109).

retaining their names, he was able to integrate them into his system.[28] Both sides, Osbeck as well as Linnaeus, commented on the dimensions and efficiency of the information transfer that resulted from the use of a common language. In the foreword to his travel report, Osbeck pointed to the compatibility of his names with the *Species plantarum* and to their incorporation into it: 'from this [journal] I gave him [Linnaeus] some descriptions of new plants found in Spain, China, and other places, which were immediately incorporated into that capital book then printing under the title of species plantarum, and with which my names of plants agree.'[29]

Linnaeus praised Osbeck's procedure for naming plants when he wrote to congratulate him on the publication of the *Dagbok* [Diary]. He stressed that, unlike most earlier travellers, Osbeck gave his discoveries 'precise' names, with the aid of binomial nomenclature, so that the learned world could understand him:

> most of the voyages hitherto published, by imposing barbarous names on their discoveries, have rather sharpened our desire after knowledge, than afforded any real instruction. You, Sir, have everywhere travelled with the light of science: you have named every thing so precisely, that it may be comprehended by the learned world; and have discovered and settled both the genera and species. For this reason, I seem myself to have travelled with you, and to have examined every object you saw with my own eyes.[30]

The urgency of the mutual need for information and the way in which this was taken for granted speaks out of Linnaeus and Osbeck's letters, drawing them into an intensive exchange of objects, names and observations that made them into indispensable points of reference for each other. Each was dependent on the other's knowledge for their own publication project. In order to identify the new species from China correctly in his *Species plantarum*, Linnaeus needed not only Osbeck's plants, but also the observations he had made of the living plants during his journey. Osbeck, in turn, had a substantial interest in drawing on Linnaeus's knowledge of plants for the identification, naming and classification of unknown species in the planned publication of his own extensive material.

Although Osbeck's status as a scholar and career as a botanist

28. See Elmer Drew Merrill, 'Osbeck's dagbok öfwer en ostindisk resa', *American journal of botany* 3:10 (1916), p.571-88 (571).

29. Pehr Osbeck, *A Voyage to China and the East Indies, by Peter Osbeck, Rector of Hasloef and Woxtorp, Member of the Academy of Stockholm, and of the Society of Upsal;* [...] *translated from the German by John Reinhold Forster; to which are added a Faunula and Flora Sinensis*, 2 vols (London, Benjamin White, 1771), vol.1, author's preface, n.p.

30. Letter of Linnaeus to Osbeck, undated, reprinted in the English translation of his travel report *A Voyage to China and the East Indies*, vol.2, p.127-28.

depended on the publication of his material on Chinese flora, he allowed most of his sensational discoveries to be published first in Linnaeus's *Species plantarum*. This was neither the selfless act of a devout disciple, nor did it signify the end of his career chances. On the contrary, the imminent publication of Linnaeus's *Species plantarum* gave Osbeck and other botanists who had collected spectacular botanical material while travelling a unique opportunity to go public with their findings for the first time in one of the most high-profile and visible publications in the field. And it allowed them to publish their findings much more quickly than they could have done in their own books, whether a flora or a travel report, which could take years to prepare.[31] Linnaeus, in turn, who could only achieve the global aspiration of his works by systematically integrating the contributions of others, gave his informants the recognition they deserved by acknowledging them as the discoverers or observers of plants, or as the authors of published or forthcoming descriptions of plants. This mutually profitable principle of give and take was one of the pillars of the collaborative publication culture of Linnaean botany.[32] After about two months of intense correspondence, Linnaeus wrote to Osbeck in October 1752 saying that all new species in the *Species plantarum* had a reference to him, so that he would soon be famous;[33] and in December of the same year, when the first printed sheets of the *Species plantarum* had been published, he wrote to Osbeck that he had inserted his description of a tree called *Mangha* with the note 'haec [this] Osbeck'.[34]

To complete this account of an exchange of information that was profitable for both partners, it should be mentioned that a few months earlier, Linnaeus had promised Osbeck a copy of the *Species plantarum*. This was a gift of considerable value given the limited availability of botanical publications, which meant that many botanists did not own a copy of this key work.[35] Osbeck replied that he was certainly not alone in

31. This was also the case with Nikolaus Joseph Freiherr von Jacquin (1727-1817), for example, a botanist who worked in Vienna and also first published the findings of his expedition to the Caribbean in Linnaeus's *Species plantarum*. On this see Dietz, 'Contribution and co-production', p.561-65.
32. See Dietz, 'Contribution and co-production', p.554-59.
33. See letter of Linnaeus to Osbeck, 16 October 1752, in Hansen and Fox Maule, 'Linnés korrespondance', p.110-111 (111); facsimile of the original and an English summary at http://linnaeus.c18.net/ Letter/L1496 (consulted 14 January 2015).
34. See letter of Linnaeus to Osbeck, 4 December 1752, in Hansen and Fox Maule, 'Linnés korrespondance', p.128-29 (129); facsimile of the original and an English summary at http://linnaeus.c18.net/ Letter/L1503 (last accessed 14 January 2015).
35. On the difficulty or even impossibility of acquiring scientific literature, especially that published abroad, in book shops, even in a town such as Leipzig which hosted a book fair, see Bettina Dietz, 'Making natural history: doing the Enlightenment', *Central European history* 43:1 (2010), p.25-46 (39-41).

eagerly anticipating the publication of the *Species plantarum*, which indicates its central methodological as well as practical significance in the field of Linnaean botany.[36] As an up-to-date systematic catalogue of all plant species known at the time, the *Species plantarum* functioned as a register that made clear which plants were known and had been named and described. This, in turn, explains why a young scientist at the beginning of his career had a vital interest in having his capital of botanical objects published for the first time in this work.

Linnaeus's significance to the publication of Osbeck's Chinese plants was not limited to their incorporation in the *Species plantarum*. His standing in the Swedish scientific community gave Linnaeus the institutional authority to encourage the publication of Osbeck's findings and observations, something in which not only he, but most botanists working systematically at the end of the eighteenth century, had a vital interest. The more species were named, described and classified, the more precisely plants (and animals) that had not yet been unequivocally identified could be compared with known ones, advancing the global cataloguing project. In his letter to Osbeck of December 1752, already cited here a number of times, Linnaeus announced that he would ensure that Osbeck's descriptions of individual species were published in the proceedings of the Royal Swedish Academy of Sciences in Stockholm, the next stage in the long path by which Osbeck's Chinese material found its way into print.[37] It was also a further strategy for presenting at least a few of his species to international experts while he was working on a comprehensive monograph.

In December 1752 Linnaeus first encouraged Osbeck to start working on a report of his trip to China.[38] Osbeck replied that if the report were really to be published, he would have to meet Linnaeus to discuss the plants, insects and fish that were difficult to identify, and other things, with him in person.[39] Osbeck now claimed the same services that he had provided to assist in the completion of the *Species plantarum* – the supply of specimens, information, observations and corrections, in order to produce a work that was as complete and free of errors as possible.

36. Letter of Osbeck to Linnaeus, 11 October 1752, in Hansen and Fox Maule, 'Linnés korrespondance', p.108-109 (109); facsimile of the original and English summary at http:// linnaeus.c18.net/ Letter/L1475 (consulted 14 January 2015).

37. See letter of Linnaeus to Osbeck, 4 December 1752, in Hansen and Fox Maule, 'Linnés korrespondance', p.128-29.

38. Letter of Linnaeus to Osbeck, 17 December 1752, in Hansen and Fox Maule, 'Linnés korrespondance', p.131-32 (132); facsimile of the original and an English summary at http://linnaeus.c18.net/ Letter/L1523 (consulted 14 January 2015).

39. Letter of Osbeck to Linnaeus, 23 December 1752, in Hansen and Fox Maule, 'Linnés korrespondance', p.132-34 (132); facsimile of the original and an English summary at http://linnaeus.c18.net/ Letter/L1519 (consulted 14 Jan. 2015).

Linnaeus had, it seems, already promised his support, because Osbeck thanked him for it in the same letter in which he asked for it.[40]

Gained in translation: the translator as co-author

The first version of Osbeck's travel report was published in 1757, largely in Swedish, entitled *Dagbok öfwer en ostindisk resa åren 1750, 1751, 1752* [Diary of a trip to East India in 1750, 1751, 1752].[41] It contained information on everything that early modern apodemic literature considered worth knowing: the physical appearance, clothing and way of life of the people; agriculture and commerce; landscape; and above all, as was to be expected with Osbeck, the natural history, in great detail, of the places visited. Noteworthy in terms of Osbeck's publication strategy is the fact that while the text is mainly in Swedish, Latin passages containing botanically precise descriptions of hitherto unknown plants and other natural objects are scattered throughout it. The *Dagbok* contains, in two languages, information of varying complexity, so that one work could satisfy two groups of readers: one that could read Swedish and was interested in a general travel narrative; and a group of international botanical experts who were unversed in Swedish, but could concentrate on the plant descriptions in Latin.[42] In order to make his findings public as quickly as possible, Osbeck chose a form for his monograph that did not require too much work. The *Dagbok*, a revised version of his travel diary supplemented with additional observations, was published relatively quickly, whereas to prepare a Chinese Flora for publication would have taken years. This decision also meant that early in his career Osbeck had a significant publication to his name, one that showed him to be a serious author and qualified him for scientific awards. In 1757 when Linnaeus proposed Osbeck as a member to the Royal Swedish Academy of Sciences in Stockholm, he could point to the sensational natural history yield of his travels, and to the fact that the report would shortly be published, something that he had repeatedly urged on Osbeck.[43]

40. See letter of Osbeck to Linnaeus, 23 December 1752, in Hansen and Fox Maule, 'Linnés korrespondance', p.132-34 (132).
41. Osbeck, Pehr, *Dagbok öfwer en ostindisk resa åren 1750, 1751, 1752* (Stockholm, Lor. Ludv. Grefing, 1757).
42. Linnaeus selected a similar format for the publication of the notes made by his student Fredrik Hasselquist, who travelled through Palestine and Egypt between 1749 and 1752, and died in Smyrna in 1752. The general travel narrative was published in Swedish, while the descriptions of animals and plants were in Latin, although, unlike in Osbeck's book, they were not dispersed throughout the text, but placed in a systematically organised second part. See Fredrik Hasselquist, *Iter Palaestinum eller resa til Heliga Landet, 1749-1752* (Stockholm, Lars Salvius, 1757).
43. See Linnaeus to the Secretary of the Royal Swedish Academy of Sciences in Stockholm, 15

The fact that Osbeck's travel report, so eagerly awaited by the learned world, was published in a language that practically nobody outside Sweden could read, meant that within a relatively short time it was translated into two other languages: in 1765 it was translated from Swedish into German, and in 1771 the German version was translated into English.[44] These translations, however, represented far more than the mere transfer of the original text from one language into another. They corrected errors of fact and added relevant information, thus playing a crucial role in a publication process that must be seen as a long-term and collaborative undertaking.

The initiator and editor of the German translation, *Reise nach Ostindien und China* (1765), was Johann Christian Daniel Schreber (1739-1810), a German medic and botanist who held two professorships, of medicine and of cameralism, at the University of Erlangen.[45] He had studied with Linnaeus in Uppsala, and as a champion of his classification system had translated a number of Linnaeus's works into German himself, and produced updated editions of others.[46] He justified the translation of Osbeck's travel report, which was done not by himself, but by the German scholar Johann Gottlieb Georgi (1729-1802), by arguing that the German translation was superior to the excellent Swedish original because it contained more information.[47] In fact, in addition to Osbeck's

April 1757; see facsimile of the original and an English summary at http://linnaeus.c18.net/Letter/L2186 (last accessed 14 January 2015); see also letter Linnaeus to Osbeck, 25 January 1754, in which he advises Osbeck to make the publication of his travel report a priority; see facsimile of the original and an English summary at http://linnaeus.c18.net/Letter/L1696 (consulted 14 January 2015).

44. The translation of a report about his travels through North America (1747-1751) by Pehr Kalm, another student of Linnaeus, followed a similar pattern. First published in Swedish, it was then translated into German, and then from the German into English. On translations of Carl Peter Thunberg's journey to Japan see Marie-Christine Skuncke, *Carl Peter Thunberg, botanist and physician: career-building across the oceans in the eighteenth century* (Uppsala, 2014), p.252-70; on translations of Linnaeus's works see Bettina Dietz, 'Linnaeus's restless system'.

45. Pehr Osbeck, *Herrn Peter Osbeck, Pastors zu Haßlöf und Woxtorp, der Königl. Schwedischen Akademie zu Stockholm und der Kön. Gesellschaft zu Upsala Mitgliedes, Reise nach Ostindien und China, nebst O. Toreens Reise nach Suratte und C. G. Ekebergs Nachricht von der Landwirthschaft der Chineser*, translated from the Swedish by J. G. Georgi (Rostock, Johann Christian Koppe, 1765).

46. Schreber translated *Des Herrn Archiaters und Ritters von Linné Reisen durch einige schwedische Provinzen*, 2 vols (Halle, Johann Jakob Curts, 1764-1765), vol.1, *Reisen durch Oeland und Gothland*, vol.2, *Reisen durch Westgothland*. He supervised the publication of, among others, extended editions of Caroli A. Linné *Genera Plantarum [...], editio octava post Reichardianam secunda [...] auctior atque emendatior, curante D. Jo. Christiano Dan. Schreber* (Frankfurt am Main, Varrentrapp and Wenner, 1789-1791); *Caroli Linnæi Materia Medica; editio secunda auctior, curante Jo. Chr. Dan. Schrebero* (Leipzig and Erlangen, Wolfgang Walther, 1772).

47. Georgi had run a pharmacy in Stendal (Saxony-Anhalt), before he moved to St Petersburg in 1769, from where he accompanied Peter Simon Pallas's expeditions through Russia in the 1770s. His later publications were based on these travels. In his Foreword to Osbeck's

travel report, it also contained the following texts: a description of Olof Torén's (1718-1753) journey to Surat (in northern India) and China, also translated from the Swedish, which had been included in the Swedish version of Osbeck's report;[48] the inaugural lecture that Osbeck had delivered when he became a member of the Royal Swedish Academy of Sciences in 1758;[49] and a report about agriculture in China by Carl Gustav Ekeberg (1716-1784), a captain with the Swedish East India Company.[50] The most interesting additions from the point of view of publication strategy, however, were made by Osbeck and Schreber, and these will be discussed in greater detail below.

Osbeck was actively involved in the translation process. He sent Schreber new, as yet unpublished material, which Schreber then inserted at the appropriate place and, where necessary, supplemented further and annotated.[51] Osbeck seems to have welcomed if not expected the opportunity to publish additional material, or to correct observations that had already been published. Schreber put any observations by Osbeck that arrived after the printing process had begun into his Foreword, where he supplemented and commented on them further. There, for example, Schreber pointed to the considerable differences that he had noticed between the generic character of *Cassytha*, a genus of semi-parasitic climbing plant, as described by Osbeck on the one hand and by Nikolaus Joseph von Jacquin on the other.[52] As both authors had proved themselves to be knowledgeable and reliable observers through their journeys and the publications based on them, it was not to be assumed, Schreber wrote, that one or the other had made a blunder in this matter. He

---

travel report, Schreber mentions that Georgi had studied in Uppsala for a number of years, and had already demonstrated his knowledge of Swedish acquired there in other translations. See Osbeck, *Reise nach Ostindien und China*, preface, p.xvi.

48. The Swedish original of Torén's text was appended to the Swedish version of Osbeck's travel report as an appendix under the title *En Ostindisk Resa Til Suratte, China &c. Från 1750 April 1. til 1752. Jun. 26 Förrättad af Olof Torén*; in the German translation, the title is *Eine Ostindische Reise nach Suratte, China etc., von 1750. den 1 April bis 1752. den 26 Jun. Verrichtet von Olof Toreen, Schiffsprediger der ostindischen Compagnie.*

49. *Anleitungen zu einer nützlichen Aufmerksamkeit bey chinesischen Reisen, in einer Rede bey der Aufnahme in die Königl. Schwed. Akademie der Wissenschaften zu Stockholm, den 25 Hornung 1758. ertheilet von Peter Osbeck.*

50. *Kurzer Bericht von der chinesischen Landwirtschaft, von dem Herrn Schiffscapitain Carl Gustav Eckeberg.*

51. Unfortunately this process is not documented in Schreber's correspondence. His papers, held by the University Library Erlangen, contain only a single letter from Osbeck, dating from the 1770s.

52. See Nikolaus Joseph von Jacquin, *Selectarum stirpium Americanarum historia, in qua ad Linnaeanum systema determinatae descriptaeque sistuntur plantae illae, quas in insulis Martinica, Jamaica, Domingo, aliisque, et in vicinae continentis parte, observavit rariores* (Vienna, Kraus, 1763), p.115; Osbeck, *Reise nach Ostindien und China*, preface, p.xiii.

therefore invited the informed reader to decide: 'whether [...] the plant that he [Jacquin] saw, the *Cuscuta baccifera barbadensium Pluk. Alm. 126. t. 172. f. 2.*, was of a different genus from the Chinese *Cassutha Rumph. Herb. Amb. 5. p. 491. t. 184 f. 4* observed by Pastor Osbeck [...] especially as there seems to be a marked difference in the structure of the two.'[53]

Schreber's achievement, therefore, was to have identified a classificatory problem, which he then asked others to solve. Noting inconsistencies in the main text or footnotes of publications, and asking questions to which the author himself was unable to provide definitive answers, was part of the arsenal of strategies used by Linnaeus and others to generate knowledge in a collaborative publication culture.[54] Osbeck had already inserted footnotes that served this purpose into the Swedish original version of his travel report. For example, with reference to the description of a Chinese fruit called by its local name of *Lang-ann* in the text, a footnote asks whether this might be the same plant that Georg Eberhard Rumpf had called *Cussambium* in his *Herbarium amboinense* (Amsterdam, 1741-1750)?[55] Elsewhere Osbeck expresses his belief that a plant he described was Linnaeus's *Convolvulus batatas*, which corresponded to Hans Sloane's *Convolvulus radice tuberosa esculenta minore purpurea*.[56]

The frequency of such questions and conjectures in botanical publications shows that authors and their readers saw these works, especially those with an ambitious scope, as provisional. It was crucial for the accumulative process of knowledge acquisition in botany to make new information public as quickly as possible, so that those who needed it for their own work could use it. The iterative model of publication described above took account of this need. Few large botanical publications could claim long-term relevance unless they were to some degree based on this principle of constantly collecting and incorporating information. In this sense, to publish not only knowledge that was reliably established, or considered to be secure, but also questions and conjectures to be answered, confirmed, or refuted in later editions on the basis of readers' reactions, made a constructive contribution to the long-term completeness of a work.

53. 'ob nicht [...] diejenige Pflanze, welche er [Jacquin] gesehen hat, die *Cuscuta baccifera barbadensium Pluk. Alm. 126. t. 172. f. 2.*, von der chinesischen vom Herrn Pastor Osbeck beobachteten *Cassutha Rumph. Herb. Amb. 5. p. 491. t. 184 f. 4* [...] dem Geschlecht nach unterschieden sey, zumal da sich in der Structur von beyden eine merkliche Differenz zu finden scheint.' Osbeck, *Reise nach Ostindien und China*, preface, p.xv (author's own translation).

54. See Dietz, 'Contribution and co-production', p.565.

55. See Osbeck, *Dagbok*, p.193 ('An Cussambium? Rumpf. libr. 1. p.154. t. 56'). Schreber retained the footnote unchanged, in Latin, in the German translation. See Osbeck, *Reise*, p.251.

56. See Osbeck, *Dagbok*, p.253.

Schreber continued this process initiated by Osbeck. He retained Osbeck's questions which were still unresolved, and added notes that gave the readers of his translation more material for solving a problem, signing his additions with his initials 'D. S.' (Daniel Schreber). To Osbeck's detailed description of a tree named *Dracaena ferrea*, for example, he added a footnote in which he identified this tree as a plant that had already been described in Linnaeus's *Species plantarum* and Rumpf's *Herbarium amboinense*: 'The tree in question here is called *Asparagus terminalis* LINN. Spec. pl. p. 450. n. 13. *Terminalis alba* RUMPH. amb. t. 4. p. 79. t. 34. D.S.'[57] Given Linnaean botany's enormous need for information and the constant need to modify and correct already published observations, names, classifications and so on, a translation offered a vital opportunity to update the original.

Layers of information: multiple authors

Six years after the publication of the German translation, Osbeck's travel report went through another cycle of updating, in the form of a second translation. In 1771 the German natural historian Johann Reinhold Forster (1729-1798) published an English version, which continued the project begun by Osbeck and Schreber, and further decentralised it.[58] It was not that English natural historians could not benefit from the Swedish-Latin original or the German translation but, as the English merchant and amateur botanist Peter Collinson explained, travel reports in English were generally bestsellers on the English book market.[59]

57. 'Der Baum von welchem hier die Rede ist, heißt *Asparagus terminalis* LINN. Spec. pl. p. 450. n. 13. *Terminalis alba* RUMPH. amb. t. 4. p. 79. t. 34. D.S.' See Osbeck, *Reise*, p.328 (translation by the author); on the translation of botanical texts as a process of textual engineering, and on the growing complexity of information layering see Dietz, 'Linnaeus's restless system'.
58. On Johann Reinhold Forster as a translator of travel literature see Vladimir Kapor, 'Translating the great maritime explorations: on Johann Reinhold Forster's translation of Bougainville's *Voyage autour du monde*', in *Travel narratives in translation, 1750-1830: nationalism, ideology, gender*, ed. Alison Martin and Susan Pickford (Abingdon, 2012), p.93-109; on Georg Forster's translations of travel literature see Alison Martin, 'Übersetzung und die Entdeckung der Welt: Georg Forster und die Reiseliteratur', in *Übersetzung / Translation / Traduction: ein internationales Handbuch zur Übersetzungsforschung / an international encyclopedia of translation studies / une encyclopédie internationale de la recherche sur la traduction*, ed. Harald Kittel, Armin Paul Frank *et al.*, 3 vols (Berlin, 2008-2011), vol.2 (2008), p.1634-41; Christine Haug, '"Diese Arbeit unterhält mich, ohne mich zu ermüden": Georg Forsters Übersetzungsmanufaktur in Mainz in den 1790er Jahren', *Georg-Forster-Studien* 13 (2008), p.99-128.
59. See letter of Peter Collinson to Linnaeus, 20 April 1754, in Carl Linnaeus, *A Selection of the correspondence of Linnaeus and other naturalists, from the original manuscripts*, ed. James Edward Smith, 2 vols (London, 1821), vol.1, p.31-32 (32). In this letter Collinson commented on the publication in Swedish of Pehr Kalm's *En Resa til Norra America, på Kongl. Swenska Wetenskaps Academiens befattning, och publici kostnad*, 3 vols (Stockholm, Lars Salvius, 1753-1761): 'I hope

Forster translated the German version of Osbeck's travel report, that is, one that had already been expanded, and now enriched it by adding more information. Readers could reconstruct the provenance of the observations inserted into the original by stages because Forster also worked with footnotes which he signed with the initial 'F' (for 'Forster'). In this way, the text went through two cycles of revision by three authors – Osbeck himself, Schreber and Forster – during which it was updated or revised where relevant new information had come to light.

Thus Forster, for example, commented on the description of *Dracaena ferrea* quoted above, which had already been expanded by Schreber, by adding another footnote on botanical nomenclature. Forster referred to the entry on *Dracaena* in the most recent version of Linnaeus's *Systema naturae* available at the time when his translation was published, that is, the twelfth, and the classificatory changes that Linnaeus had made there, repeating the synonyms that Schreber had used in his German translation: '(*Dracaena ferrea, Linn. Syst. Nat., ed. 12. p. 246*). D.S. remarks that it is called *Asparagus terminalis* in *Species Plantar* and *Terminalis alba* by *Rumph. Am. vol. IV. p. 79. tab. 34*, but Linnaeus, in his new system, on the aforementioned page, calls that species of *Asparagus, Dracaena terminalis.* F.'[60]

Or to take another example, in a long footnote on the tea shrub, Schreber had reported a claim by the English botanist John Hill that, contrary to general assumption, black tea and green tea were not just the result of different methods of picking and preparation, but were made from the leaves of different tea plants. Forster retained the footnote, but added a final sentence, in which he cast doubt on the validity of this claim:

> It has been universally believed that all teas came from the same shrub, and are only distinguished by different age, gathering, and preparation. But Dr. Hill has lately discovered that the brown tea comes from the tea shrub with six petals, or flower leaves, which Kaempfer has described, and represented: but that the green tea is taken from the tea shrub with nine leaved flowers. The former in *Linnaeus's* Spec. Plantarum, Ed. ii. p. 734 is *Thea bohea*, and the latter *Thea viridis*. Linnaeus distinguishes it, besides the flowers, by the longer and narrower leaves. *Dr. Schreber.* It is, notwithstanding, very doubtful whether the plant of the green tea is really different from that of the bohea tea. F.[61]

---

some ingenious man will translate it into either Latin, English, or French. All books of voyages and travels, printed in English, sell the most of any books in England.'

60. Osbeck, *Voyage to China*, vol.2, p.14. References to Linnaeus's nomenclature also play a central part in Forster's translation of Bougainville's *Voyage autour du monde*. See Kapor, 'Translating', p.99.

61. Osbeck, *Voyage to China*, vol.1, p. 246.

In addition to these footnotes with comments and corrections, Forster offered the readers of his translation of Osbeck a further addition. Under the title *Faunula sinensis* and *Flora sinensis*, he put together an overview, organised by Linnaean principles of classification, of all the Chinese animals and plants known so far, compiled from Linnaeus's *Species plantarum*, Osbeck's travel report, and other botanical publications. His aim was to record the status quo of European knowledge of Chinese flora and fauna at the time, to draw attention to classificatory discrepancies or difficulties, to identify cases where different authors had given the same plant different botanical names, and to refer to the best descriptions in botanical literature.

Under the generic name of *Panicum*, for example, Forster's *Flora sinensis* presents a list of known Chinese species of millet as follows: '1. *Alopecurodeum*, 2. *Glaucum*, 3. *Crus galli*, 4. *Brevifolium*, 5. *Arborestens*, 6. *Patens*, 7. *dissectum*. Osbeck. Perhaps the *P. dimidiatum* Linn.'[62] Forster here pointed out that the seventh species might have been listed both by Osbeck and Linnaeus but under different names. He could not settle this question himself, but left it to be cleared up by others. Yet the reference to this potential duplication alone was of great benefit, as it localised a problem that others now merely had to solve, without first having to find or identify it.

Compiling a Chinese Flora was, like Linnaean botany in general, a long-term project that depended on countless contributions, observations and corrections. Osbeck's expedition and his travel report provided the core that others built on. This sort of collaboration required a central authority, someone to record newly identified and named species in a central register and to update and correct these entries when necessary. After Linnaeus had served as this centre of global botany for several decades, publishing expanded editions of his systematic writings at intervals – the twelfth and last edition of the *Systema naturae* that he edited himself was published between 1766 and 1768 – the publication of Osbeck's travel report pointed to a structural change. Immediately after Osbeck's return from China, it was still taken for granted that newly discovered species would be incorporated as fully as possible into the global list of species, Linnaeus's *Species plantarum*. But Forster's *Faunula* and *Flora sinensis* were symptoms of a process of differentiation, from a central, global register to several regional registers that were no longer maintained by a single scientist but by several, in order to keep up with the explosive growth in botanical and zoological information.

62. Osbeck, *Voyage to China*, vol.2, p.343.

What was a botanical author in the eighteenth century?

There is, as yet, no comprehensive answer to this question, but it will be closely connected to the answer to a second question, namely, what is a botanical publication, or how did it come about? What has been shown here is that far-reaching forms of collective authorship were not a product just of the twentieth century, but had already emerged much earlier, at a time when the knowledge claim of a scientific project or discipline so far exceeded the capacity of an individual scientist that collaboration, and thus the aggregation of contributions by many scientists, became a necessity. The global project of Linnaean botany and the modular and participatory architecture of Linnaeus's systematic works as well as of the publications of other botanists – and thus of a *Systema naturae* in a wider sense – is probably the most spectacular pre-modern example of this sort of dynamic. The process of layering information, however, driven to extremes by Linnaeus but also practised by other botanists, fragmented the botanical author, who still appeared to be an individual because, as a rule, only one name was printed on the title page of publications. Yet the input by all the contributors involved, in the case of Osbeck's travel report, the two translators in particular, was minutely documented. This demonstrated to every reader that this text had not been produced by one individual, but had grown out of a culture of collaborative publication.

As the result of collaborations of various dimensions, which had to reflect this fact, botanical publications of the eighteenth century faced challenges not entirely different from those of present-day laboratory research. The long-term accumulation of a *Systema naturae* (in the wider sense described above) created a publication culture that anticipated what, in the biomedical community of the 1990s, was proposed as a radical solution to the increasingly complex problem of authorship: replacing the concept of 'author' by that of 'contributor' in order to take account of the fundamentally collaborative nature of the knowledge-production process. A proposal published in the *Journal of the American Medical Association* in 1997 suggests that '[whoever] has added usefully to the work' should be listed as a 'contributor', and should briefly describe his or her own contribution in a list of contributors. This would give readers a clearer view of the variety and diversity of the contributions than conventional selective author by-lines permits.[63] This step towards

63. See Mario Biagioli, 'Rights or rewards? Changing frameworks of scientific authorship', in *Scientific authorship: credit and intellectual property in science*, ed. Mario Biagioli and Peter Galison (New York and London, 2003), p.253-79 (264-66); Drummond Rennie, Veronica Yank and Linda Emmanuel, 'When authorship fails: a proposal to make contributors accountable', *Journal of the American Medical Association* 278:7 (1997), p.579-85. The three

a concept of modularised authorship that adequately reflects that a publication is an aggregate had already been taken, in its way, by Linnaean botany.

authors' response to the question of who, in the context of modularised scientific authorship, is responsible for the content of a publication is to suggest that 'special contributors' should be designated.

3. The price of Linnaean natural history: materiality, commerce and change

HANNA HODACS

Tens of thousands of plants, fish, insect specimens, shells and minerals crossed the North Sea in 1784. Together with numerous manuscripts, books and letters they made up the bulk of Carl Linnaeus's collection. Purchased by the budding English naturalist James Edward Smith (1759-1828), the collection left Uppsala to be housed by its new owner in London.[1] Today the specimens, manuscripts, letters and books are the property of the Linnean Society of London, and kept at Burlington House, Piccadilly. Having been carefully conserved and cared for, many of Linnaeus's specimens are today defined as *holotypes*, representing some of the earliest primary material that allows scientists to apportion names and identities to some of the many millions of plant and animal species populating our world.[2] The significance of the collection reflects Linnaeus's invention of the binomial nomenclature, first applied to plants and introduced to an international audience in *Species plantarum* (1753). In this respect, the Linnaean collection forms part of a pre-history to today's globally applied taxonomic protocol.[3]

While we know the price Smith paid for the collection, namely £1088 and 5 shillings, including £50 for the transport, there has been no attempt to study the relocation of the Linnaean collection in the light of markets for natural history specimens in late eighteenth-century Europe.[4] Neither have questions been asked about how the purchase of Linnaean specimens informed the taxonomic dimension of collection-building in the eighteenth century. By focusing on trade, materiality

The work for this paper was conducted within the research project 'Westward science between 1760-1810: on social mobility and the mobility of science', funded by the Swedish Research Council (Vetenskapsrådet). I would particularly like to thank Staffan Müller-Wille, Stéphane Van Damme, Jacob Orrje, Linda Andersson Burnett and Leos Müller for their helpful comments while preparing the final version of this paper.

1. For a recent biography of Smith see Tom Kennett, *The Lord Treasurer of botany: Sir James Edward Smith and the Linnaean collections* (London, 2016).
2. On the scientific significance of the different parts of the collections see Brian Gardiner and Mary Morris, *The Linnaean collections*, *The Linnean special issue* 7 (Oxford, 2007).
3. Lorraine Daston, 'Type specimens and scientific memory', *Critical inquiry* 31:1 (Autumn 2004), p.153-82 (158).
4. Theodor Magnus Fries, *Linné: lefnadsteckning*, part 2 (Stockholm, 1903) p.420.

and consumer culture the intention of this paper is to provide a new perspective on the pre-history to Linnaeus's central position in modern taxonomy. Assessments of the value in money of Linnaeus's conserved plants and animals before Smith's purchase will be mapped out in the first section of this paper. In the second section the focus will be on the material dimension, foremost on aspects to do with preservation, presentation and fashion. Questions regarding how, in late eighteenth-century London, evaluations of specimen collections Linnaeus used in his taxonomic work changed, and how this reflected Linnaeus's shifting status within the history of natural history, will be discussed in the third section. Fourthly, the market and the use of binary names when selling natural history specimens at auctions in eighteenth-century Europe, particularly Sweden and England, will be explored. In combination these sections will help to provide an economic history of Linnaean natural history which, I would argue, form an important but overlooked part in the development of taxonomic praxis in use today.

The increase in monetary value

There was sometimes money to be made in the pursuit of natural history, and Linnaeus was prepared, from early on, to cash in on the results of his hard scholarly labours. In 1746 Linnaeus declared himself willing to donate his 'herbariums, stones, *bibliotheca botanica*' and more specimens to the 'Academy' in return for an annual pension of 100 ducats. The proposed host was probably the Royal Swedish Academy of Sciences, which had been established in 1739.[5] Given that Linnaeus had another thirty-two years of life ahead of him, the offer – had it been accepted – would have cost the Academy 3200 ducats, a substantial sum in the mid-eighteenth century and the equivalent of more than £1421 (1746 exchange rates).[6] How serious Linnaeus's offer was is hard to judge; as the earliest evaluations, in money, of Linnaeus's collections, the author of this chapter has come across, the proposal provides an interesting backdrop to later calculations made towards the end of Linnaeus's life, as well as the transaction that followed on his death.

5. The proposal was made in a letter from Linnaeus written in 1746, the recipient was Sten C. Bielke (1709-1753), one of Linnaeus's patrons and a co-founder of the Academy; Carl Linnaeus to Bielke, 8 November 1746, *The Linnaean correspondence*, http://linnaeus.c18.net, letter L0737 (last accessed 20 October 2016).
6. When not otherwise specified all early modern Swedish and non-Swedish currencies in this paper have been converted into pounds sterling with the help of the web resource 'Prisomräknare från medeltiden till 2100' (http://www.historia.se/Jamforelsepris.htm), (last accessed 4 May 2016). The conversion rate to British sterling has been used drawing on Rodney Edvinsson, 'Foreign exchange rates in Sweden 1658-1803' in *Exchange rates, prices and wages, 1277-2008*, ed. Rodney Edvinsson, Tor Jacobson and Daniel Waldenström (Stockholm, 2010), p.238-290, especially Table A5.22, p.283-86, and Table A5.24, p.288.

Linnaeus's various testaments and the negotiations surrounding the division of the inheritance after his death in 1778 further illuminate the estimated value of the collections. Linnaeus's surviving family members included his wife Sara Elisabeth, née Moræa (1716-1806), and daughters Elisabeth Christina (1743-1782), Lovisa (1749-1839), Sara Christina (1751-1835), and Sofia (1757-1830), as well his son Carl (1741-1783). The value of Linnaeus's whole estate was estimated at 11,000 rix-dollar (*riksdaler banco*), or £2786 (1778 exchange rates), although this was an underestimation of its true value. The Hammarby estate, Linnaeus's summer residency, and his other land holdings represented a little bit less than half of the inheritance. Together with most of the chattels these properties were to remain in the sole possession of his wife, Sara Elisabeth, until her death.[7] Scholarly articles, books and specimens were, however, to be passed on to Linnaeus's children immediately upon his death. The process of distributing these things illuminates a range of issues to do with professional and social capital, family politics and the monetary value attached to different parts of the collection.

There were, for example, two different libraries, one located in Hammarby, and one located in his Uppsala residence. Linnaeus's son inherited one of the libraries and his daughters jointly inherited the other.[8] The specimens were also divided: the daughters inherited what was then referred to as the 'two herbaria', while the son was granted the rest of the specimen collections. The distribution reflected the existing schism between Linnaeus the Younger and the rest of the family. In a will written in 1769 the elder Linnaeus had explained that his son would inherit his professorship, which together with 'the library' (here not divided into two) equalled more than what his sisters would receive between them. Hence the daughters were to inherit the two herbaria, but not for their own use; they were 'to be sold to the highest bidder' although preferably to Uppsala University.[9] In 1776, in a document entitled 'Cries from the grave to my presently beloved wife', Linnaeus described his two herbaria as his 'greatest work', 'priceless in themselves which would in time become increasingly desirable. They are the largest the world has ever seen.' His advice to his family was *not* to sell them 'under 1,000 ducats specie', or £498 (1776 exchange rates).[10]

7. Carl Forsstrand, 'Linnés ekonomi och hans kvarlåtenskaps öden', *Svenska Linnésällskapets årsskrift* (1922), p.70-89.
8. Fries, *Linné*, part 2, p.415. There is some confusion as to who got which library; compare Fries, *Linné*, part 2, p.415 with Arvid Hj. Uggla 'Linné den yngres brev till Abraham Bäck, 1778', *Svenska Linnésällskapets årsskrift* (1957), p.138-65 (147, n.6).
9. Quoted from Fries, *Linné*, part 2, p.414 (author's translation): 'sälljes till den mästbjudande för Döttrarnes räkning. Dock såge jag hälst att Uppsala Academie det inlöste'.
10. Quoted from Fries, *Linné*, part 2, p.414-15 (author's translation): 'Rop ur grafwen till min i tiden kiära Hustru. I. Herbarier 2:ne främst i Museuo, mitt största arbete i tiden.'

What about the value of the other parts of the collection his children inherited from Linnaeus? In 1764/65 Linnaeus estimated the value of his various sub-collections of insects, amphibians and minerals at 10,000 copper dollars *each*, equalling £109 per collection (1764/65 exchange rates), while the herbarium (here singular) was estimated to be worth 50,000 copper dollars, or £543 (1764/65 exchange rates).[11] In 1776, in the 'Cries from the grave' document, Linnaeus estimated the value of one of his libraries at '3,000 dlr' (presumably copper dollars, the equivalent of £42 at 1776 exchange rates).[12] While he names the other collections, 'the Stone cabinet', 'the Insect cabinet' and 'the Shell cabinet', only the latter is given an estimated value of '12,000 dlr' (presumably copper dollars), the equivalent of £169 (1776 exchange rates).[13] The estate inventory after Linnaeus's death in 1778 tells us only about the value of one library (estimated to be worth 166 rix-dollar (*riksdaler banco*), or £42 at 1778 exchange rates); the specimen collections were found to be too hard to evaluate and were consequently not priced.[14]

What can the future destiny of the various parts of the collection say about their perceived value? Against the explicit will of his now dead father Linnaeus the Younger *did* take ownership of the two herbaria as well as his father's manuscripts in 1778 by swapping his share of the tenure of two of the farms (Hubby) granted to the Linnaeus family by Gustav III and estimated to be valued at 2000 copper dollars, *and* paying his sisters 6000 copper dollars (in total the equivalent of £113 at 1778 exchange rates).[15] Soon after the deal, Joseph Banks offered up to £1200 for the scholarly material Linnaeus left behind.[16] Lars Montin (1723-1785), a provincial physician and former student of Linnaeus, and also the uncle of Jonas Carlsson Dryander (1748-1810) who served as Bank's librarian, deliberately delayed passing on Banks's offer, thereby helping Linnaeus the Younger to secure the deal. In return Montin, who himself owned a

'Owärderlige i sig sielfwe, med tiden mer och mer begiärlige. Äro de största werlden någonsin sett. Sälg dem intet under 1 000 Ducater specie.'

11. Will by Linneaus ('Min yttersta wilja'), dated 1764/65, *The Linnaean correspondence* (no document number) http://linnaeus.c18.net/eds/bs/01/01/t-0269.pdf (last accessed 20 October 2016).
12. Fries, *Linné*, part 2, p.415.
13. Fries, *Linné*, part 2, p.415.
14. Sigurd Wallin, 'Urkunder rörande familjen Linnés ägodelar', *Svenska Linnésällskapets årsskrift* (1951), p.67-94 (84).
15. Sigurd Wallin, 'Urkunder rörande famijen Linnés ägodelar. IV', *Svenska Linnésällskapets årsskrift* (1962), p.54-102 (73). Note the price is quoted in both copper dollars and *riksdaler* (444 riksdaler, 21 sk. 4 r.) See also Fries, *Linné*, part 2, p.416.
16. Kennett, *The Lord Treasurer*, p.58.

great herbarium, was promised duplicates from Linnaeus's collections, a deal Linnaeus the Younger struggled to keep.[17]

After the purchase, Linnaeus the Younger wanted to move the whole collection from Hammarby into Uppsala as soon as possible. One reason for this was to prevent his family from retracting their agreement in response to Banks's 'cruel offer'; representing almost ten times the amount Linnaeus the Younger had paid his family, removal of the goods made sense.[18] It is likely Linnaeus the Younger even felt entitled to take over the collection. Households which contained libraries, specimens, and scientific tools were significant sites for reproducing scholarly work across generations and within families in early modern Europe.[19] The elder Linnaeus indicated several times that he was aware of this; while he might have been unwilling to pass on his herbaria to his son for personal reasons, as late as 1776 he was, for example, willing to let a botanically minded son-in-law have it.[20] Also, before the relationship between son and father broke down, the latter argued in favour of his son being awarded the position of Demonstrator at the Uppsala botanical garden, specifically referring to Linnaeus the Younger's access to his father's collections and library as a supporting factor.[21] Collections were pedagogical assets important to those who wanted to attract students, including private students who provided university professors with lucrative side incomes. Linnaeus's teacher in Lund, Kilian Stobæus (1690-1742), had, for example, given his students access to his collection in order for them to learn natural history.[22] Similarly, Linnaeus used his collection for education purposes, and the museum he had built at Hammarby contained a pulpit and three benches for students to sit on.[23]

In the light of this we should understand Linnaeus the Younger's purchase as a reflection of his need to secure resources central to his own future career, and income as a scholar. Linnaeus the Younger needed his father's collections in order to pursue scholarly work, teaching and writing books, including producing new editions of his father's work.[24]

17. Arvid, Hj. Uggla, 'Linné den yngres brev till Abraham Bäck, 2. 1779-1783', *Svenska Linnésällskapets årsskrift* (1958), p.61-100 (64-65).
18. Linnaeus fil. to A. Bäck, 21 July 1778, in Uggla, 'Linné den yngres brev till Abraham Bäck, 1778', p.158.
19. Alix Cooper, 'Homes and households', in *The Cambridge history of science*, ed. David C. Lindberg and Ronald L. Numbers, 8 vols (Cambridge, 2003-2013), vol.3, *Early modern science*, ed. Katharine Park and Lorraine Daston (Cambridge, 2006), p.224-37.
20. Fries, *Linné*, part 2, p.415.
21. Fries, *Linné*, part 2, p.413-14.
22. Yngve Löwegren, *Naturaliekabinett i Sverige under 1700-talet: ett bidrag till zoologiens historia* (Uppsala, 1952), p.84.
23. Löwegren, *Naturaliekabinett*, p.239.
24. Olle Franzén, 'Carl Linné, von', *Svenskt biografiskt lexikon*, http://sok.riksarkivet.se/sbl/artikel/10736, (last accessed 4 May 2016).

Even before Linnaeus the Younger bought out his sisters he had started to build up his own collections, indicating his intention to pursue a scholarly life. Once he purchased his father's herbarium, Linnaeus the Younger's plan was to sell his own herbarium; he hoped to receive 8000 dlr (presumably copper dollars) or £113 (1778 exchange rates) for it.[25] It is of course important to acknowledge that the specimens making up the collections of both father and son to no small extent were the result of a gift-exchange system stretching across Europe, the Atlantic and Asian world.[26] Gifts replaced cash in this system; as Linnaeus explained to his best friend Abraham Bäck (1713-1795), rare plant specimen were better than 'ducats' in the exchange between naturalists.[27] Linnaeus the Younger's plan to repay Montin with duplets from his father's collections, in return for his favour of delaying the news of Banks's offer, is another example of the gift economy in which naturalists also operated.

It was only after Linnaeus the Younger died, five years after his father, that the Linnaeus collection was sold to Smith and transported to London. Before then the estate inventory of Linnaeus the Younger listed the value of the library at 500 rix-dollar (*riksdaler specie*), or £131 and the value of the other collections at 666 rix-dollar (*riksdaler specie*), or £174 (both 1783 exchange rates).[28] Smith, however, paid what was then 4150 rix-dollar (*riksdaler specie*), which according to 1783 conversion rates equalled £1084 for the collections or more than three and a half times its estimated value as noted in the estate inventory after the death of Linnaeus the Younger.[29]

The history of how Smith got his hands on the Linnaean collection and the upheaval it caused has been retold many times. While the deal was swiftly closed, preventing some Swedish and other contenders from offering a higher bid, it is also true that those who were given the chance to match Smith's bid failed to match the Englishman's price.[30] By

25. Linnaeus fil. to A. Bäck, 21 April 1778, in Uggla, 'Linné den yngres brev till Abraham Bäck, 1778', p.151.
26. One exception was the herbarium of Patrick Browne consisting of plants from Jamaica which Linnaeus bought in 1758. Originally Uppsala University was to fund the purchase but once Linnaeus discovered the value of the herbarium he decided to buy it himself, and returned the 600 copper dollars which the university had allocated towards the purchase; see Löwegren, *Naturaliekabinett*, p.186.
27. C. Linné to A. Bäck, 30 September 1751, in *The Linnaean correspondence*, http://linnaeus.c18.net/Letter/L1310, letter L1310 (last accessed 20 October 2016); Staffan Müller-Wille, 'Nature as a marketplace: the political economy of Linnaean botany', in *Oeconomies in the age of Newton*, ed. Margaret Schabas and Neil De Marchi (Durham, NC, 2003), p.154-72.
28. Wallin, 'Urkunder rörande familjen Linnés ägodelar', p.93.
29. Löwegren, *Naturaliekabinett*, p.236. In the negotiation the price for the collections was reduced from 1000 guineas to 900 guineas as Linnaeus fil.'s own collection was excluded from the sale; Kennett, *The Lord Treasurer*, p.59-61.
30. Kennett, *The Lord Treasurer*, p.61-62. Claes Alströmer's offer of £462 and 10 shillings is listed here.

purchasing the collection, Smith gained a platform on which to establish himself as a scholar. Smith was the eldest son of a wealthy Unitarian wool merchant; he had studied medicine in Edinburgh, where he had become acquainted with Linnaean taxonomy. Smith never practised medicine, instead he came to devote his life to natural history. After purchasing the Linnaean collection he was elected a Fellow of the Royal Society; Smith also undertook a European grand tour, during which he obtained a medical degree in Leiden. Once back in England, Smith became the president of the Linnean Society, formed in London in 1788, a position he held for forty years until his death in 1828. Dryander became the society's librarian and Banks an honorary member. Natural history brought Smith into contact with high society. The early 1790s saw him organise the royal herbarium and teach Queen Charlotte and her daughters zoology and botany. In the second half of the 1790s, Smith moved to Norwich, taking the collection with him, and spending only a few months each year in London. His position as a member of high society was confirmed when Smith was knighted in 1814.[31]

Given that he was just in his mid-twenties when he obtained the Linnaean collection, Smith's future career was open-ended. Some of the social and financial implications of buying the collection were, however, clear to the Smith family. The letters to and from James Edward Smith and his father prior to the purchase reveal an awareness of just what was entailed in housing and displaying the contents of the collection. As Smith senior wrote: 'All this [...] convinces me that it will require no small nor inelegant house to place so capital a collection and library in a commodious manner.'[32] Many issues relating to the purchase were discussed, including the value of individual parts of the collection, foremost the extensive library and its most precious and rare volumes.[33]

It is also worth noticing that Smith sold off Linnaeus's minerals and the cabinet in which they were kept. Before the mineral collection left Sweden it comprised 2424 specimens. The sale brought in just over £71 to Smith.[34] Once Smith had died, what remained of Linnaeus's collections, the zoological and botanical material, the manuscripts and the

31. G. S. Boulger, 'Smith, Sir James Edward (1759-1828)', revised by Margot Walker, *Oxford dictionary of national biography* (Oxford, 2004), http://www.oxforddnb.com/index/25/101025825/ (last accessed 5 May 2016); Peter S. Dance, *A History of shell collecting*, revised edn (Leiden, 1986), p.83.

32. James Edward Smith, *Memoir and correspondence of the late Sir James Edward Smith*, 2 vols, (London, 1832), vol.1 p.96.

33. Smith, *Memoir*, p.102. Banks also offered Smith £109 for the ninety-seven wooden blocks in the Linnaean collection which originally had been used to print *Campi Elysii*; see C. G. Myrin, 'Om Linnés naturhistoriska samlingar och deras bortförande till England: ett bidrag till Sveriges litteraturhistoria', *Scandia* (1833), p.242-88 (267).

34. Arvid Hj., Uggla, 'Ur Linnésamlingarna i London: försäljningen av Linnés mineralkabinett', *Svenska Linnésällskapets årsskrift* (1953), p.125-34 (125-26).

library were sold to the Linnaean Society for a sum of £4488, almost five times the price Smith had paid Linnaeus's family in 1783.[35] The purchase also meant a change in status of the collection, as it went from being private to institutional property.

Table 1: Evaluations, bids and prices paid for the Linnaean collection (whole or parts of) between 1746 and 1835.

Year	Price	Currency	Converted to pounds sterling	Estimator/buyer/bidder
1746	100	ducats annually	£ 47	estimation by Linnaeus
1764/65	80,000	copper dollar	£ 870	estimation by Linnaeus
1776	15,000 1000	copper dollar and ducats	£ 709	estimation by Linnaeus
1778	444	rix-dollar (*riksdaler specie*) or 8000 copper dollar	£ 113	price paid by Linnaeus the Younger
1778	1200	pound sterling	£ 1200	bid by Joseph Banks
1783	4159	rix-dollar (*riksdaler specie*)	£ 1084	price paid by James Edward Smith
1829-1835	4488	pound sterling	£ 4488	price paid by Linnean Society of London

Sources: C. Linnaeus to Bielke, 8 November 1746, *The Linnaean correspondence*, http://linnaeus.c18.net/, letter L0737 (consulted 20 October 2016); Theodor Magnus Fries, *Linné: lefnadsteckning*, part 2 (Stockholm, 1903) p.415-16; Yngve Löwegren, *Naturaliekabinett i Sverige under 1700-talet: ett bidrag till zoologiens historia* (Uppsala, 1952), p.236-37. For currency conversions see 'Prisomräknare från medeltiden till 2100', http://www.historia.se/Jamforelsepris.htm (last accessed 20 October 2016), and Rodney Edvinsson, 'Foreign exchange rates in Sweden 1658-1803' in *Exchange rates, prices and wages, 1277-2008*, ed. Rodney Edvinsson, Tor Jacobson and Daniel Waldenström (Stockholm, 2010), Table A5.22, A5.24 (p.283-86, 288).

Table 1 summarises the fluctuating monetary value of the Linnaean collection, estimated as well as actual. While the prices refer to different parts of the collection and hence are not comparable, they do illustrate the discrepancy between Swedish and British evaluations of natural history collections in the late 1700s. London buyers were prepared to pay a price significantly higher than those Linnaeus had envisioned his collections could be sold for.

What about Swedish price developments before the rise of the British interests? Again Table 1 should be read carefully. According to Yngve Löwengren the value Linnaeus ascribed to the most important

35. Löwegren, *Naturaliekabinett*, p.237.

components of his collection in the mid-1760s did, for example, *not* reflect the domestic decline in prices of natural history specimens in Sweden.[36] Another aspect *not* illuminated by Table 1 is the value of different parts of the collection, and how they changed. While the library, the shells and the herbarium material are regularly priced, other parts of the collection were not always given a price tag. Below we shall discuss further the changing monetary value of natural history collections in Sweden as well as price differences between Sweden and Britain, and what it can tell us of the rise of London as a market for natural history specimens. First, however, we shall look closer at the aspects which can help explain the relationship between different types of natural history specimens and their perceived monetary value, but also the relationship between conservation, order and value, with a short-term and a long-term perspective.

The shifting state and status of the Linnaean collection

While natural history specimens could be converted into money, they could just as easily turn into dust. The histories of early modern collections are overwhelmingly histories of deterioration. This is also the case with different parts of the Linnaean collection. Aside from fear that his mother and sisters would withdraw from the deal following news of Banks's generous offer, Linnaeus the Younger was keen to bring his father's herbarium from Hammarby to Uppsala in 1778 in order to clear it of mould and moths.[37] Linnaeus senior had built a museum for all his collections at his Hammarby estate, at a safe distance from Uppsala where he feared fires might destroy it. The museum was purpose-built, but Linnaeus the Younger's letters also tell us that dampness threatened the specimens kept in the museum.[38] From Linnaeus senior's admissions we also know that moths very quickly became a threat to the insect collection, which possibly explains why he refrained from putting a price on his insects in 1776.[39] That pests were an issue also becomes apparent when Smith negotiated buying the collection. Olof Acrel (1717-1806), who was responsible for negotiating the deal with Smith on behalf of

36. Löwegren, *Naturaliekabinett*, p.234.

37. Linnaeus fil. to A. Bäck, 21 July 1778, in Uggla, 'Linné den yngres brev till Abraham Bäck, 1778', p.158.

38. Linnaeus fil. to A. Bäck, 16 September 1780, in Uggla, 'Linné den yngres brev till Abraham Bäck. 2. 1779-1783', p.77.

39. Fries, *Linné*, part 2, p.415. According to Olof Acrel the insects were a valuable part of Linnaeus's collection; he told Smith a Swedish purchaser had offered the family 170 ducats or £87 (1783 exchange rates) for the insects only (Myrin, 'Om Linnés', p.264-65).

Linnaeus's family, actually certified that the herbarium contained no moths.[40]

In terms of its role as a host for natural history collections, Uppsala University has a long history of neglect. Ynge Löwegren, in commenting on the deterioration of the collections of Carl Peter Thunberg (1743-1828) – Linnaeus's most successful student who returned to Uppsala after long journeys in Asia with a very large number of specimens – concludes that 'one cannot but be happy that the Linnaean collection were sold to England'.[41] Such arguments, however, hide the variation in natural history specimens with regard to their more *or* less delicate nature depending entirely on what they are made of. While moths threatened Linnaeus's insect collection, possibly preventing Linnaeus from estimating its value, he *did* put a price tag on his shell collection. As Peter Dance has pointed out, the 'durability' of molluscan shells together with their beauty helps to explain their prominent position in the market of natural history objects, something we shall discuss in more detail in the next section.[42] First, though, we shall consider the relationship between preservation, presentation, order and value in greater detail.

Preservation was an important aspect of natural history. We can trace its history as a parallel development to the history of collections.[43] Knowledge of how to preserve was shared between naturalists. Linnaeus is likely to have learned the art of, for example, stuffing birds from Olof Rudbeck the younger (1660-1749); a skill that came in handy when Linnaeus senior ventured on his trip to Lapland.[44] Preservation did not come cheap though. Zoological specimens were frequently conserved in costly glass jars and bottles filled with alcohol. Condensation meant these containers needed to be topped up regularly. As Löwegren has noted, Linnaeus donated to the university those of his specimens which demanded most looking after while he kept specimens which were easy to store and cheap to maintain.[45]

40. Myrin, 'Om Linnés', p.264.

41. Löwegren, *Naturaliekabinett*, p.213 (author's translation): 'man inte vara annat än glad över, att de Linnéanska försåldes till England'. On Thunberg's life see Marie-Christine Skuncke, *Carl Peter Thunberg, botanist and physician: career-building across the oceans in the eighteenth century* (Uppsala, 2014).

42. Peter S. Dance, 'Shells', in *Natural history auctions 1700-1972: a register of sales in the British Isles*, compiled by J. M. Chalmers-Hunt (London, 1976), p.45-51 (45).

43. Robert McCracken Peck, 'Preserving nature for study and display', and 'Alcohol and arsenic, pepper and pitch', in *Stuffing birds, pressing plants, shaping knowledge: natural history in North America 1730-1860*, ed. Sue Ann Prince (Philadelphia, PA, 2003), p.11-53; Stephen T. Asma, *Stuffed animals and pickled heads: the culture and evolution of natural history museums* (Oxford, 2001).

44. Löwegren, *Naturaliekabinett*, p.66.

45. Löwegren, *Naturaliekabinett*, p.186.

Another way to protect specimens was to store them in specially designed cases and furniture that safeguarded them from dust and insects. Of course, protection was hardly the only thing that mattered when it came to storing collections. Workmanship and expensive materials, exotic wood such as 'East India cypress wood', or fine woods like oak, walnut and inlays, were regularly used to construct cabinets to house collections, lined and padded on the inside by colourful satin silk with specimens arranged in decorative patterns.[46] As Bettina Dietz and Thomas Nutz have argued in relation to Parisian eighteenth-century natural history collections, aesthetic ideals were central. Artefacts such as paintings, sculptures and porcelain objects were typically combined with natural history specimens; a collection was judged on the total visual effect.[47] From a modern perspective the investments required to obtain natural history specimens did not always pay off though, at least not in the long run. Dance has pointed out that shells which cost their owners fortunes in the first half of the eighteenth century could be worthless today, while the art objects bought and displayed together with natural history specimens might be 'priceless'.[48]

In this light it is worth highlighting that no famous art pieces, or exotic woods and textiles, seem to have been involved in the Linnaeus collection. According to Acrel, Linnaeus was reluctant to show his collections to visitors *specifically* because the latter 'expected to see Walnut- and Mahogany-cabinets, gilded frames and splendour, and such he did not own to show'.[49] Instead, Acrel points to another dimension of Linnaeus's collections, they 'were to him like a book, which he read daily. Hence he could not keep it as new [fresh], as prodigious, as elegant, as others [collections] less used.'[50] The need for daily access probably stopped Linnaeus from enclosing his collections in too elaborate cases. More importantly, Linnaeus designed his cases so as to match his taxonomy. As Staffan Müller-Wille has pointed out in relation to Linnaeus's herbarium, the manner of its storage reflected Linnaeus's practices as a taxonomist. The herbarium was made up of loose sheets *not* bound

46. Dance, *A History*, p.40, 42, 51, 57. On the history of furnishing in the eighteenth century see Dena Goodman and Kathryn Norberg (ed.), *Furnishing the eighteenth century: what furniture can tell us about the European and American past* (New York, 2006).

47. Bettina Dietz and Thomas Nutz, '*Collections curieuses*: the aesthetics of curiosity and elite lifestyle in eighteenth-century Paris', *Eighteenth-century life* 29:3 (2005), p.44-75. See also Emma C. Spary, 'Scientific symmetries', *History of science* 42:1 (2004), p.1-46.

48. Dance, *A History*, p.32.

49. Quoted from Löwegren, *Naturaliekabinett*, p.233 (author's translation): 'få se Valnöte- och Maghogny-skåp, förgylde lister och prakt, och sådant ägde han icke att visa'.

50. Quoted from Löwegren, *Naturaliekabinett*, p.233 (author's translation): 'Hans Naturaliesamlingar voro honom en bok, den han dageligen läste, följakteligen ej kunde hålla så ny, så prunkande och zirlig, som andra mindre brukade'.

together in book volumes, as was the tradition. These sheets were placed in a specially designed cabinet with twenty-four shelves, corresponding to the classes in the sexual system. The flexible construction enabled Linnaeus to compare herbarium sheets when settling taxonomic issues. Moreover, the shelves in the cabinet were intentionally not fixed in place so as to allow for the expansion or reduction of shelf space to various classes.[51]

Another aspect worth remarking on is that while Linnaeus's herbarium was unique, the cabinet could be copied. In *Philosophica botanica* Linnaeus actually reproduced an image of his herbarium cabinet, its twenty-four shelves/classes marked with Roman numerals.[52] Linnaeus's floras offered further information on which plant belonged to which class, helping his followers to establish order and sometimes develop their own ordering systems. For example, Pehr Osbeck (1723-1805), one of Linnaeus's students also discussed by Bettina Dietz in the present volume, kept his duplets of dried plants in a book between sheets the number of which corresponded to the species number Linnaeus had given them in *Flora Svecica*.[53] Those of greater means could purchase order by employing Linnaeus or one of his students to organise their collection. Linnaeus, assisted by his student Daniel Solander (1733-1782), arranged the collections of the Swedish monarchs, the descriptions of which were published in 1754 (*Museum Regis Adolphi Frederici*) and 1764 (*Museum S. R. M. Ludovicae Ulricae, Reginae Suecorum*).[54] Meanwhile, top-rated Swedish architects and designers were engaged to supply the interior design of rooms in which the royal collections were housed, incorporating references to the natural history of fish, animals, shells and corals in the wall panels.[55]

The example of the Swedish royal collections also illustrates that Linnaeus's taxonomy was compatible with collection cultures aimed at demonstrating taste, style and wealth. This connection to a world of fashion was, however, neither permanent nor, perhaps, very strong. It is possible that the 1750s represented the high watermark of Linnaean natural history in Sweden, at least as regards royal support. By the late 1750s the interest of Louisa Ulrika and her husband Adolf Fredrik was in

51. Staffan Müller-Wille, 'Linnaeus's herbarium cabinet: a piece of furniture and its function', *Endeavour* 30:2 (2006), p.60-64.
52. Carl Linnaeus, *Philosophia botanica in qva explicantur Fundamenta botanica cum definitionibus partium, exemplis terminorum, observationibus rariorum* (Stockholm, Kiesewetter, 1751), p.309.
53. Travel diary written by Johan Lindwall (1743-1796), the private tutor of Carl Bäck (1760-1776), describing a visit to Osbeck's home on 1 August 1774; Ihre 187:1, Uppsala University Library.
54. Edward Duyker, *Nature's argonaut: Daniel Solander 1733-1782* (Melbourne, 1998), p.18-19.
55. Löwegren, *Naturaliekabinett*, p.299-300, 305-306.

decline.[56] The 1760s and the following decades saw changes both to the political system and the economic and ideological context in which Linnaeus had operated so successfully.[57] Meanwhile, in Britain, and particularly London, Linnaeus's systematic work continued to attract followers, a development that was to no small degree helped by the presence of Linnaeus's students in the British capital. By following Linnaean traces to London we can illuminate not only the reception the Linnaean collection got but also its changing status from if not fashionable at least modern, to historically significant, something which can provide us with a better understanding of why both Banks and Smith were prepared to cash out in order to secure ownership of Linnaeus's dried plants, insects, shells and minerals, among other things.

From modern to historically important: the changing status of the Linnaean collection

Daniel Solander arrived in London in 1760. Before becoming Banks's secretary and travel companion on Captain Cook's first expedition, Solander was employed by the British Museum. As Isabelle Charmantier discusses in this book, using loose-leafed paper slips Solander composed several catalogues indexing the newly opened museum's collections, drawing on Linnaeus's taxonomic principles.[58] Another possible material testimony to Linnaeus's impact is the shell collection in Banks's possession: the design and use of fitting metal containers to house the collection mirrors those used by Linnaeus.[59] Whether Solander or someone else influenced the layout of Banks's shell containers is hard to tell but we know that Solander was an expert on shells. He organised the natural history collection of Margaret Cavendish-Bentinck, duchess of Portland (1715-1785). From the beginning of 1778 to the middle of 1781, Solander spent most Tuesdays between 11 a.m. and 6 p.m. identifying and naming the duchess's shells.[60] Together with Banks the duchess was one of the biggest investors in natural history in Britain at the time. After Solander died in 1782, his role as a 'taxonomic oracle' seems, to some degree, to have been taken over by Dryander, who

56. Löwegren, *Naturaliekabinett*, p.302.

57. Karin Johannisson, 'Naturvetenskap på reträtt: en diskusssion om naturvetenskapens status under svenskt 1700-tal', *Lychnos* 29 (1979-1980), p.109-54.

58. See also Duyker, *Daniel Solander*, p.69. For a more general discussion of Linnaeus's Swedish scholars in London between 1760 and 1810 see Hanna Hodacs, 'Local, universal and embodied knowledge: Anglo-Swedish contacts and Linnaean natural history', in *Global scientific practice in an age of revolutions, 1750-1850*, ed. Patrick Manning and Daniel Rood (Pittsburgh, PA, 2016), p.90-104.

59. Dance, *A History*, p.80.

60. Dance, *A History*, p.74.

followed in Solander's footsteps by assisting London naturalists with taxonomic queries, while also holding down a position as Banks's librarian.[61] In this capacity he wrote a five-volume bibliography, *Catalogus bibliothecae historico-naturalis Joseph Banksi* (1798-1800), detailing the library's contents and using the sexual system as the organising principle for the botanical section.[62]

In other words, the ground was well prepared for the arrival and reception of the Linnaean collection. But while some of Linnaeus's taxonomic innovations, most notably his binary names, became universally accepted others, such as the sexual system, became perceived as outdated relatively quickly. Linnaean taxonomy was even associated with an old, conservative order in a political sense, a situation that came to influence the discussion around the arrangement and cataloguing of the growing collections of the British Museum well into the nineteenth century.[63] One can even argue that Linnaeus's legacy halted taxonomic development in the capital. When discussing conchology, the study of mollusc shells, and taxonomy, Peter Dance has suggested that the delayed introduction of Lamarckian taxonomy reflects a hesitation among the younger generation of naturalists; they wanted to 'avoid annoying' James Edward Smith and perhaps risking their membership of the Linnean Society of London.[64]

Simultaneously, though, and of greater significance here, a growing appreciation emerged for the *historical* value of specimen collections which Linnaeus had worked with, which we can trace in the trade in natural history specimens. While Banks declined to buy Linnaeus's collection the second time it was put up for sale, on the death of Linnaeus the Younger, he did not miss the opportunity to purchase three herbaria closely connected with Linnaeus's work in the following decade. By following the discussion about value and price of collections with a Linnaean provenance we can map how the historic interest in them evolved.

The first herbarium Banks bought had belonged to George Clifford III (1685-1760), a wealthy Dutch merchant banker who employed Linnaeus as his in-house naturalist between 1735 and 1737, at Clifford's estate Hartekamp, close to Haarlem in the Dutch Republic. Here Linnaeus

61. Quote from Duyker, *Daniel Solander*, p.276.

62. Arvid Hj. Uggla, 'Jonas Carlsson Dryander', *Svenskt biografiskt lexikon*, vol.11 (Stockholm, 1945), https://sok.riksarkivet.se/sbl/artikel/17665 (last accessed 12 June 2016).

63. Gordon McOuat, 'Cataloguing power: delineating "competent naturalists" and the meaning of species in the British Museum', *The British journal for the history of science* 34:1 (2001), p.1-28.

64. Dance, *A History*, p.83. For a critical discussion on the impact of Linnaeus's taxonomy on British natural history see David Elliston Allen, *The Naturalist in Britain: a social history* (Princeton, NJ, 1994) p.40-51.

wrote *Hortus Cliffortianus*, published in 1737. It was this connection that made the herbarium interesting to Banks in 1791. How valuable the Clifford herbarium was to Banks, though, depended on the extent to which it contained the examples of plants which Linnaeus had used as his primary material when he described some of the species listed in *Hortus Cliffortianus*.[65] Evaluating the herbarium without inspecting it *in situ* was not uncomplicated. Samples of handwritten material seem to have been sent over to London in order for Banks and Dryander to establish whether it truly was Linnaeus's handwriting. Banks thought it looked like notes taken by a younger person, possibly a 'more attentive & careful' Linnaeus who conformed to his junior status but also to Clifford's expectations; the latter was probably used to the 'Excellent hand writing' of the clerks working in his merchant and bank business.[66] Dryander was not happy to just make an educated guess but since Smith was out of town he could not access Smith's Linnaean material for some comparative palaeographic work. Optimally Dryander wanted James Edward Smith, himself, or another Swedish botanist living in London at the time, Adam Afzelius (1750-1837), to inspect the herbarium prior to Banks making a bid for it.[67] As it was, they had to settle for trusting the evaluation of Sebald Justinus Brugmans (1763-1819), later appointed professor in natural history in Leiden, who also secured the Clifford herbarium for Banks when it was finally auctioned off in Leiden in the winter of 1791/92.[68]

In the end, Banks paid 245 florins plus 5 florins tax (in total £22 and 10 shillings) for the collection, a reasonably cheap price, at least when compared to what Banks had been prepared to pay for Linnaeus's herbarium.[69] Of greater interest are the references to other prices for Dutch natural history collections which arose in the discussion regarding what to bid for Clifford's collection. In this context Dryander referred to the price of '1000 florins' paid by Lord Bute for a herbarium owned by the Gronovius family. Bute purchased the material in 1778 when, as

65. J. Dryander to J. Banks, 24 and 19 October 1791 (letters 1064, 1067); J. Banks to J. Dryander 22 October 1791 (letter 1065), and S. J. Brugmans to J. Banks, 16 April 1792 (letter 1105) in Joseph Banks, *Scientific correspondence of Sir Joseph Banks, 1765-1820*, ed. Neil Chambers, 6 vols (London, 2007), vol.4, *The Middle period 1785-1799* (London, 2007), p.75-78, 115-17.

66. J. Banks to J. Dryander 22 October 1791 (letter 1065), in Banks, vol.4, *The Middle period*, p.77.

67. J. Dryander to J. Banks, 24 October 1791 (letter 1067), in Banks, vol.4, *The Middle period*, p.78.

68. S. J. Brugmans to J. Banks, 16 April 1792 (letter 1105), in Banks, vol.4, *The Middle period*, p.115-17.

69. See the Marteau early eighteenth-century currency converter http://www.pierre-marteau.com/currency/converter/hol-eng.html (last accessed 10 October 2016).

Dryander put it, 'money was a more plentiful commodity in Holland'.[70] The reference to Gronovius's collection is interesting for two reasons.

The first of these reasons was because it too involved material with a Linnaean connection that moved from the Continent to Britain in the last third of the eighteenth century. The Gronovius family were scholars and naturalists of Dutch-German origin. Johan Frederik Gronovius (1668-1762) had been one of Linnaeus's patrons during Linnaeus's years in the Dutch Republic. He sponsored the publication of *Systema naturae* and employed Linnaeus to organise the American collections of John Clayton along Linnaeus's principles, a work that resulted in *Flora Virginica* published from 1739-1743.[71] Johan Frederik's son Laurens Theodorus Gronovius (1730-1777) continued in his father's footsteps but, after his death, the family herbarium was put up for auction in Leiden and Lord Bute's commissioner, the bookseller Bennet of Rotterdam, outbid the other competitors. Sixteen years later, in 1794, and two years after Banks had bought the Clifford herbarium, Banks likewise bought Gronovius's herbarium when Lord Bute's collections were put up for sale in London.[72] A year earlier he had also bought another herbarium with a Linnaean provenance, the herbarium assembled by Paul Hermann (1646-1696), a medical officer of the Dutch East India Company (VOC). Banks had paid £75 for that herbarium which had been consulted by Linnaeus when he wrote *Flora Zeylanica* (1747) and also *Species plantarum* (1753). At the time of Banks's purchase, Herman's herbarium was owned by August Günther, a Danish Apothecary Royal.[73]

The second reason concerns the case of the Gronovius collection

70. On the price Brugmans secured the herbarium for see S. J. Brugmans to J. Banks, 16 April 1792 (letter 1105), in Banks, vol.4, *The Middle period*, p.115-16. Compare what Dryander thought he remembered Bute paying for the Gronovius's herbarium (J. Dryander to J. Banks, 19 October 1791, letter 1064, in Banks, vol.4, *The Middle period*, p.76) with Alwyne Wheeler who records Bute's price at 650 guilders; see Alwyne Wheeler, 'The Gronovius fish collection: a catalogue and historical account', in *Bulletin of the British Museum (natural history), historical series* 1:5 (1953-1959), p.187-249 (196-97). The latter source says the bookseller Bennet of Rotterdam, who was Lord Bute's commissioner at the auction, had a commission of £250-300!
71. Karel Davids, 'The scholarly Atlantic circuits of knowledge between Britain, the Dutch Republic and the Americas in the eighteenth century', in *Dutch Atlantic connections, 1680-1800: linking empires, bridging borders*, ed. Gert Oostindie and J. Roitman (Brill, 2014), p.224-49.
72. Wheeler, 'The Gronovius', p.196-97. On the auction of Bute's collections in London in 1794 see G. L. E. Turner, 'The auction sales of the earl of Bute's instruments, 1793', *Annals of science* 23:3 (1967), p.213-42 (214, n.5).
73. Natural History Museum London, 'About Paul Hermann', http://www.nhm.ac.uk/research-curation/scientific-resources/collections/botanical-collections/hermann-herbarium/about-paul-hermann/index.html (last accessed 11 June 2016). On the purchase see J. Banks to Hermann Treschow, 14 June and 6 July 1793 (letters 1202, 1208), in Banks, vol.4, *The Middle period*, p.227-28, 233-34.

which, as Dryander's comment indicates, reflects the declining market for natural history specimens in the Dutch Republic. We can find further evidence of this elsewhere. Not long before he bought Grovonius's herbarium from Lord Bute's estate, Banks was offered the option to purchase another part of the Gronovius family's collection, around 500 fishes, dried and glued on to paper. This fish collection had failed to find a purchaser when it was put up for sale in 1778 and had instead remained in Leiden.[74] The offer came from Pieter Boddaert, who was commissioned by the surviving family to find a British buyer since, as Boddaert put it, 'here [in the Dutch Republic] natural history is dead, and France is too much occupied with revolutions and troubles'.[75] If Banks was not interested Boddaert thought perhaps James Edward Smith or 'the owner of the cabinet of M. Ashton Levers' would be interested in the specimens, including the descriptions and illustrations that went with the collection.[76] The latter reference refers to the collection of 27,000 objects belonging to Sir Ashton Lever and displayed to a paying audience at the Holosphusikon museum in Leicester Square from 1775 onwards.[77]

The above mentioned discussions indicate an awareness of the English market, even of specific potential purchasers, among those on the Continent with collections to sell. It is also worth noticing that while Banks seems to have declined the offer, Gronovius's fish collection did make it to London in the end, although we don't know where it spent the first half of the nineteenth century. In 1854 the collection was put up for sale in the auction rooms of Mr Phillips of Bond Street, in London. J. E. Gray, the keeper of the department of zoology in the British Museum, inspected the collection and studied the manuscript that accompanied it and recommended purchase. Only later on did members of the staff recognise the historical dimension of the collection which contained several Linnaean holotypes. Several of these fish specimens of the Gronovius collection had been consulted by Linnaeus, who specifically referred to them in the tenth and twelfth editions of *Systema naturae*.[78]

As Lorraine Daston recently argued, in contrast to scholars in arts and humanities who frequently engaged previous generations, dead or alive, we tend to think about scientists as profoundly disinterested in the past of their own subjects beyond the narrow scopes of the very latest publications. This does, however, not mean that scientists lack an

74. Wheeler, 'The Gronovius', p.194-95.
75. Wheeler, 'The Gronovius', p.195.
76. Wheeler, 'The Gronovius', p.195.
77. Clare Haynes, 'A "natural" exhibitioner: Sir Ashton Lever and his Holosphusikon', *Journal for eighteenth-century studies* 24:1 (2001), p.1-13.
78. Wheeler, 'The Gronovius', p.187-88.

understanding of history, and their own role in it, as they are trying to appreciate how their subject will evolve in the future. Daston argues that timeframes, as well as collective identities of scientists, can also be recreated by studying the histories of scientific archives, collections and libraries.[79] As the discussion above illustrates, the assembling, buying and selling of collections with a Linnaean provenance does offer just such an opportunity to study the historiography of natural history scholarship across Sweden and Britain in the second half of the eighteenth century. How collections were valued and how the evaluations changed over time reflect their perceived importance and function. As the case of the Swedish royal collections illustrates, Linnaean taxonomy was compatible with the aesthetic culture that perhaps reached a peak in mid-eighteenth century Paris, a culture in which natural objects and human artefacts, shells and paintings were displayed together. While Linnaean-styled collections had a longer life span than those dictated by Parisian fashion, they did become the victim of taxonomic reforms; the last decades of the eighteenth century and the beginning of the nineteenth century saw Linnaean taxonomy undergo a shift of status in London but also in Sweden. In 1848, for example, insects arranged according to Linnaeus's system became part of the 'ancient collection' in Sweden's second university, Lund. The display's purpose was mainly pedagogical, showing the visitor 'the systematic ancient school achieving its perfection and its end'.[80] While Linnaeus's systems became 'ancient' a growing appreciation of Linnaeus's significant role in the historical development, largely played out in London, meant that collections with a Linnaean provenance did not lose their value, as indicated by Lord Bute and Banks's purchases of material assembled by Clifford, Hermann and Gronovius. Combined with social and political circumstances local to London, this resulted in the conservation of the various collections of material linked to Linnaeus's work.

To sum up the last two sections, moths and dampness form a constant threat to natural history specimens, although some materials are more sensitive than others. Given Uppsala University's poor record of conserving collections, and the general decline of Swedish interest in natural history, it is unlikely that the more delicate insect- and plant-specimens that formed parts of the Linnaean collection would have survived had they not been sold to Smith. However, such an argument is somewhat anachronistic; it chiefly reflects our contemporary appreciation of the Linnaean collection, where the emphasis is on the holotypes it contains.

79. Lorraine Daston, 'The sciences of the archive', *Osiris* 27:1 (2012), p.156-87.

80. A. G. Dahlbom quoted in Löwegren, *Naturaliekabinett*, p.144-45 (n.41, author's translation): 'den systematiskt-antika skolan vunnit sin högsta fulländning och sitt slut.' Note that both Fabricius and Linnaeus's systems were referred to as 'ancient'.

What is lost is that the Linnaean collection materialised a new, modern way of arranging nature, possibly compatible with, but definitely not restricted to, eighteenth-century elite aesthetic ideals. From the mid-century and onwards this new modern way of organising natural history was diffused by Linnaeus himself and by his students. It reached Britain too. In 1760 Solander migrated to London and set to work in the collections of the newly opened British Museum, Banks's collections and the duchess of Portland's collections. Twenty-four years after Solander's arrival in London, the Linnaean collection followed. The relocation is crucial as London and its circle of conservative naturalists offered the Linnaeus collection a place where it could make the transition from modern to increasingly historically significant collection without having to suffer many years of neglect, moths and dampness in the meantime. This brings us back to the importance of London as a market for natural history collections as indicated in the prices both Smith and Banks were prepared to pay for Linnaeus's collections in Uppsala.

The market for natural history collections

The trade in collections with a Linnaean provenance, including Linnaeus's own but also those assembled by the likes of Clifford, Gronovius and Herman, is not only indicative of a growing awareness of the importance of the history of natural history. Their relocations *from* Scandinavia and the Dutch Republic *to* England are also symptomatic of a more general trend: London's emerging role as a central market place for natural history material.

Central market places, or 'staple markets' to use a historical term, attract both national and international buyers. The bigger the market, the better it functions, or at least so it is believed. Typically, goods are sold at auctions. The preference for auctions is particularly strong among sellers of perishable goods, such as fruit, or goods which are hard to price, such as antiques. As noted above, natural history specimens were particularly vulnerable to rapid perishing, but more importantly interest in them changed over time, creating a fluctuating market with hard to price goods. Auctions were, of course, also important meeting places for collectors and traders. In addition, they served, as Cynthia Wall has argued in relation to eighteenth-century London, as places where social boundaries were crossed and identities were invented.[81]

Different branches of natural history material sold more or less well in the British capital in the late eighteenth and nineteenth centuries. In his survey of entomological auctions in London J. M. Chalmers-Hunt points

81. Dance, *A History*, p.39; Cynthia Wall, 'The English auction: narratives of dismantlings', *Eighteenth-century studies* 31:1 (1997), p.1-25.

to the period between 1870 and 1930 as particularly flourishing, while the period between 1940 and 1963 was also strong.[82] The auctioning of other categories of animal specimens, including birds, shells and fossils, and plant specimens, but less so minerals, also saw phases of bust and boom, and there were various explanations for these trends.[83] The rise in shell auctions in London following the end of the Second World War can, as Peter Dance argues, be explained by the return to London of British military personnel from exotic locations across the globe.[84]

Another factor influencing the market and the auctioning off of natural history material in the two last centuries was the rise of institutional buyers such as museums. While London was an important market for the British Museum and, from 1881 onwards, the Natural History Museum, not all goods stayed in the capital. Take the example of the herbarium collections of British naturalist Aylmer Bourke Lambert (1761-1842). It was made up of 50,000 specimens originating from the over 130 collections Lambert had purchased over the course of his life time. Put up for sale in 1842, the collections were divided up and sold to numerous buyers, including the British Museum, but also many located on the continent of Europe.[85]

Of course, the establishment of national museums devoted to natural history was not unique to Britain. An important factor turning London into *a* or perhaps *the* main modern market place for natural history trade in Europe is largely related to political developments elsewhere. Political upheavals can be detrimental to collection building. As W. D. Ian Rolfe puts it, 'The breakdown of social barriers brought about by the French Revolution opened the doors to many palaces on the continent'. In contrast, London and 'Britain represented a stable bulwark'.[86] We recognise this sentiment in Pieter Boddaert's letter to Banks in 1791, seeking to determine the interest in Gronovius's collection in London. As he put it, 'France is too much occupied with revolutions and troubles' to be interested in dried fish glued to sheets of paper.[87]

82. J. M. Chalmers-Hunt, 'Entomological sales', in *Natural history auctions 1700-1972: a register of sales in the British Isles*, compiled by J. M. Chalmers-Hunt (London, 1976), p.3-14.
83. Alwyne Wheeler, 'Zoological specimens other than insects, molluscs and bird'; William T. Stearn, 'Botanical auction sales'; Clive Simson, 'Ornithology'; W. D. Ian Rolfe, 'Fossils'; Peter G. Embrey, 'Minerals'; Peter S. Dance, 'Shells', all in *Natural history auctions 1700-1972: a register of sales in the British Isles*, compiled by J. M. Chalmers-Hunt (London, 1976), p.15-51.
84. Dance, 'Shells', p.51.
85. Stearn, 'Botanical', p.20-22.
86. Rolfe, 'Fossils', p.36.
87. Note this was probably a cliché, as Pierr-Yves Lacour has recently shown. The Revolutionary era saw a great interest in and a change in status of natural history collection in France; see Pierre-Yves Lacour, *La République naturaliste: collections d'histoire naturelle et Révolution française, 1789-1804* (Paris, 2014).

Stretching the perspective into the nineteenth and even twentieth century we need also to consider the rise of the British Empire. As Gordon McQuat writes: 'The ability of the British metropolis to amass, name and classify the world's population of natural inhabitants would, in the eyes of its patrons [British Museum] exemplify British imperialism's brilliance in the post-Napoleonic period.'[88] In this the British Museum was helped by 'the unstable European climate and the impoverishment of continental collectors and museums'.[89] However, even before the French Revolution, never mind the rise of the second British Empire, London was already the largest city in Europe. London had outbid Amsterdam as the main early modern trading centre by the first half of the eighteenth century.[90] This shift is also reflected in Dryander's comment on how money no longer was as 'plentiful' in the Dutch Republic as it had been.[91] In London there was seemingly no lack of it, as the purchase of the Linnaeus collection also underlines.

Tracing the buying and selling of natural history collections in Sweden in the early eighteenth century we can see how the Dutch Republic had previously played a dominant role as a market place for natural history specimens. In general, the consumption of natural history increased across the whole of the Baltic area, and the Low Countries was the main market. Peter the Great, for example, bought several Dutch collections, including the first two assembled by Albertus Seba (1665-1736), for which the Russian tsar paid 15,000 guilders in 1717 (£1350).[92] This was not a trade restricted to the ruling elite though. Linnaeus's teacher in Lund, Stobaeus bought specimens from Seba too, although Stobaeus complained about the price tags.[93] The Swedish royals were keen consumers of Dutch natural history collections. Queen Louisa Ulrika spent, for example, 20,283 copper dollars, or the equivalent of £480 (1751 exchange rates), buying up two Dutch collections in the early 1750s. One had previously been owned by Jacob Elias Shot in Utrecht and the other belonged to Henrik ten Kate in Amsterdam; aside from natural history specimens, the former also included books and miniatures. When the late Seba's last collection was auctioned off in 1752, both the queen and

88. McQuat, 'Cataloguing', p.6.
89. McQuat, 'Cataloguing', p.6.
90. David Ormrod, *The Rise of commercial empires: England and the Netherlands in the age of mercantilism, 1650-1770* (Cambridge, 2003).
91. On the role of Paris and Amsterdam as key markets for shells in the early eighteenth century see Dance, *A History*, p.40.
92. Dance, *A History*, p.36; Löwegren, *Naturaliekabinett*, p.310-11. For conversion from guilders to pounds see http://www.pierre-marteau.com/currency/converter/hol-eng.htm (last accessed 15 May 2016).
93. Löwegren, *Naturaliekabinett*, p.87.

her husband Adolf Fredrik purchased items.[94] According to one eyewitness, Peter Ascanius (1723-1893), collectors in Amsterdam complained that the Swedish queen had made shells too expensive for everyone else. Linnaeus, too, found the prices paid by the Swedish royal house exorbitant. Conspicuous consumption was one problem; as Linnaeus put it, if the queen had let someone purchase it on her behalf 'privately' then a 'pile of goods could be had for reasonable money, however when someone royal is buying, he has to pay with a piece of gold the size of an egg'.[95]

Nonetheless, the potentially lucrative dimension to natural history was something Linnaeus communicated to his students. He lectured on *Turbo scalaris* to them; it was one of the most expensive shells he knew of, and typically sold for between 100 and 200 ducats, the equivalent of between £49 and £98 (1752 exchange rates), as he told his students. Linnaeus also discussed the prices of the *Hirundo* shell, 8-9 ducats, or between £3 and 18 shillings and £4 and 8 shillings (1752 exchange rates), and *Cunnus veneris*, of which he writes: 'When first discovered, [it] cost up to 30 *riksdaler* [silver dollars?], but now a beautiful couple sell for a couple of *riksdaler*.'[96] Linnaeus also warned his students that 'it is not for every man to assemble such a [shell] cabinet; because once you have gathered a sizeable amount you immediately want more, and it becomes easy to spend yourself into poverty in trying to complete a collection'.[97] For many of Linnaeus's students this information was perhaps less relevant as they tended to come from relatively poor backgrounds. As many as a fifth of Uppsala's students in the second half of the eighteenth century were the sons of farmers or people of lower social status.[98]

Rather than buying specimens, Linnaeus's students seem largely to have assembled their own collections of regional and local flora and fauna, using Linnaean order rather than gilded cabinets and artwork to frame their collections. However, several collections gathered by Linnaeus's so-called apostles, those who travelled outside Sweden, ended up in the royal household. Pehr Kalm (1716-1779), who travelled to North America, and Pehr Osbeck, who gathered material en route to and

94. Löwegren, *Naturaliekabinett*, p.310-11.

95. Löwegren, *Naturaliekabinett*, p.312 (author's translation): 'en hop gods fås för drägelige pgr, men när en Konung skall kiöpa, måste han betala ett lika stort gukdstycke för ett ägg.'

96. From a lecture given by Linnaeus in 1752, transcribed and printed in Carl von Linné, *Linnés föreläsningar öfver djurriket: utgivna och försedda med förklarande anm. af Einar Lönnberg* (Uppsala, 1913), p.325, 345, 351 (author's translation): 'När den först blef uptäckt, kostade den till 30 Riksdaler, men nu säljes ett wackert Par för ett par Riksdlr.'.

97. Linné, *Linnés föreläsningar*, p.310 (author's translation): 'det är ej heller hwar mans wilkor, at lägga sig tll ett sådant Cabinett; ty då man får en hoper, will man ändtligen hafwa mer, och kan man snart köpa sig fattig, innan man får samlingen complett.'

98. Sten Lindroth, *A History of Uppsala University 1477-1977* (Uppsala, 1977), p.145.

in China, both contributed to the royal collection. Most famous perhaps is the royal purchase of the collection of Fredric Hasselquist (1722-1752), which the queen bought for 14,000 copper dollars or £360 (1753 exchange rates). A former student of Linnaeus, Hasselquist had died at Smyrna after having spent more than two years investigating the natural history of the Middle East.[99]

Some auction catalogues from the first half of the eighteenth century can provide snapshots of what was for sale on the Swedish natural history market. Take Peter Hoffman's collection, for example, put up for sale in Stockholm in 1747; this collection showed no influences of Linnaean taxonomy in its organisation. According to the title of the catalogue it contained 'Rarities, natural history specimens, minerals, ore, coins and rifles'.[100] Rather than listing the objects in this order, however, it was the furniture in which the objects were contained that provided the structure in the catalogue. Each piece of furniture was given a letter. Under 'A' we find 'A Black Cabinet', and lists of lots contained within it noted all sorts of precious stones and corals. Aside from the items' colours, the wood used to construct the cabinets – walnut, pine and birch – are also detailed.

Unsurprisingly, the most elaborately designed cabinet, with gilded features, contained shells. Under 'G. A Large Green Painted Cabinet' we find lot 17: 'Many types of rare shells', including '*Concha Veneris*'. The latter is the only zoological specimen given a scientific name, perhaps a synonym to the *Cunnus veneris* Linnaeus mentioned in his lecture. This shell came with 'an elegantly lathed box'.[101] Towards the end of the catalogue we find loose items which today would be labelled ethnographical material, like Sami drums and 'Indian bow and 2 quivers', as well as singular larger zoological specimens, such as 'lion heads' and 'two crocodiles', listed.[102] One is left with the impression that Hoffman's cabinet conformed to ideals associated with the pre-Enlightenment *Wunderkammern*, or cabinets of curiosities.

Among the few Swedish auction catalogues that have survived we also find the remains of the collection aesthetic we associate with French elite culture in the eighteenth century. Take the example of Carl Gustaf Tessin (1695-1770): Swedish ambassador to Paris between 1739 and 1742,

99. Löwegren, *Naturaliekabinett*, p.314. Carl-Otto von Sydow, 'Fredric Hasselquist', *Svenskt biografiskt lexikon*, vol.18 (Stockholm, 1961-71), https://sok.riksarkivet.se/sbl/artikel/12645 (last accessed 1 June 2016).

100. *Förteckning, Uppå framledne stads-auctuariens Herr PETER HOFFMANS samlingar, af åtskilliga rarieteter, naturalier, mineralier, malmer, mynt och gewär, som igenom offentlig auction komma at försäljas den 9 juni 1747* (Stockholm, Peter Jör. Nyström, 1747). Sales catalogue, copy located at Royal Library (Kungliga biblioteket), Stockholm (vardagstryck, ekonomi).

101. *Förteckning, Uppå framledne*, p.18.

102. *Förteckning, Uppå framledne*, p.29-31.

Tessin bought art and coins but he was also interested in natural history and, like Louisa Ulrika, he purchased specimens from auctions in Amsterdam, some of which had belonged to Seba. Tessin also employed naturalists such as Anders Philip Tidström (1723-1779), Johan Haartman (1725-1787) and Peter Jonas Bergius (1730-1790) to collect minerals for him in Sweden. Further, Osbeck was employed to organise his collection and write a catalogue. Linnaeus had already published a catalogue (*Museum Tessinianum*) in 1753. Tessin himself specialised in shells and was, according to Linnaeus, quite knowledgeable. In 1757 Linnaeus helped Tessin arrange his shells according to Linnaeus's new system. Meanwhile, however, Tessin's finances deteriorated and in 1762 his mineral collections were packed up and sold to the Danish king. Two years later, Tessin's herbarium and insect collection were shipped over to Denmark, to the new owner Duke Moltke.[103] The fate of Tessin's shell collection is not known. An auction catalogue from 1771 gives some indication that what had once comprised its most precious parts had been sold. The section of the catalogue devoted to natural history starts off with a lot of five drawers of shells, described as 'duplets' and the 'remains of the large and costly natural history collection' arranged in compartments.[104]

As Löwegren points out, the 1750s was the high watermark of the trade in natural history in Sweden, particularly in terms of Swedes making purchases abroad. After that a decline took place. By 1770 even Osbeck, now dean, was in the process of selling his insect collection; financial pressure was forcing him to get rid of the collection he had spent twenty-four years building up. The collection was made up of specimens named according to Linnaeus's system and gathered in the East Indies, in and around Uppsala, Stockholm, and his deanery in the south west of Sweden. From Osbeck's letter to the secretary of the Royal Swedish Academy of Sciences we learn that the collection was kept in a 'well-made cabinet', painted to look like oak, with sixty-eight labelled drawers. Wax, cork and Bohemian glass helped seal and protect the insects. Built by the queen's carpenter it was 'supposedly similar to the royal insect cabinet' in design. The cabinet alone was worth 30 ducats, according to Osbeck; with the insects he priced it at 160 ducats or £65 (1770 exchange rates) to a 'foreign gentleman'.[105] As it was, Osbeck's collection only sold

103. P. Osbeck to P. Wargentin, 10 December 1770, quoted in Löwegren, *Naturaliekabinett*, p.359.

104. *Förteckning på hwad i synnerhet förtjenar utmärkas, blad de saker, som förekomma wid auction på Åkeröö den 4:de och följande dagar till och med den 16:e i febrarii månad, 1771.* Sales catalogue, copy located at Royal Library (Kungliga biblioteket), Stockholm (vardagstryck, ekonomi).

105. P. Osbeck to P. Wargentin, 10 December 1770, quoted in Löwegren, *Naturaliekabinett*, p.359 (author's translation) 'skall vara likt H:s Maj:ts insektsskåp.'

in 1788, to the Portuguese minister in Copenhagen albeit for a higher price than Osbeck had originally asked, namely 4000 silver dollars or £160 (1788 exchange rates).[106]

The Danish market seemed more buoyant than the Swedish in the last decades of the eighteenth century. In Stockholm the collection of the late Daniel Solander failed to find any buyers. Solander had crossed the North Sea in 1760 and, since his plan was to return to Sweden after he had visited London, he had left his natural history books and specimens behind. His library, minerals and cabinets containing shells, insects and a herbarium with more than 7500 plants, were put on the market in late 1786, four years after Solander died. In early 1787, the collection was advertised several times in Stockholm newspapers. No single buyer came forward; it was converted into four lots and put up for sale at an auction on 24 February 1787. Lots 1-3 were made up of cabinets: the first one contained shells, fossilised plants and mosses; the second 'a complete herbarium'; and the third insects, snakes and so on. The fourth lot included around 2000 minerals. The notes which accompanied the lot description suggested a minimum price of 200 rix-dollar (*riksdaler species*) or £49 for the herbarium (1787 exchange rates) and half that for the shells. Neither found a buyer willing to bid more than 50 and 15 rix-dollar, or £12 and 2 shillings and £3 and 15 shillings respectively (1787 exchange rate). There was no interest shown for the other collections either, and the seller, who was acting on behalf of Solander's family, withdrew all the lots from the sale. Who eventually purchased the material and for what price we do not know but some of Solander's material did survive and can today be found at the Natural History Museum (Naturhistoriska riksmuseet) in Stockholm.[107]

Natural history markets could change rapidly; fashion, for example, caused sharp increases in shell prices in the first half of the eighteenth century. By the late 1750s the *mode* changed drastically in Paris and Amsterdam, and prices started to decline.[108] One reason for the downturn was the lack of new shell material. Consequently, the market for fabricated new shells grew; as Dance puts it: 'less scrupulous collectors and dealers resorted to the creation of fakes and many man-made

106. Löwegren, *Naturaliekabinett*, p.360.

107. Olof H. Selling, 'Daniel Solanders naturalkabinett och dess öden', *Svenska Linnésällskapets årsskrift* (1962), p.128-37 (135). Note that the English traveller Edward Daniel Clarke (1769-1822) bought Solander's copy of *Flora Svecica* while visiting Sweden in 1799. Clarke also bought a herbarium from Daniel Erik Naezen, and he also got hold of a handful of dried plants glued onto papers with Linnaeus's autograph on them (Linda Andersson Burnett, 'Northern noble savages? Edward Daniel Clarke and British primitivist narratives on Scotland and Scandinavia *c.*1760-1822', doctoral dissertation, University of Edinburgh, 2012, p.173).

108. Dance, *A History*, p.43, 53.

"species" were in circulation before and after the publication of the *Systema*.'[109] Making new species by manipulating specimens was an activity not too different from the existing trade of the Amsterdam 'shell doctors' who filed, scraped, painted and varnished shells to make them look more attractive.[110]

How newly discovered shells and specimens could bring about a revival in the market becomes visible in London in the 1770s, as Captain Cook's three expeditions (1768-1771, 1772-1775, 1776-1779) resulted in the introduction of many new shells to England, as well as the rest of Europe.[111] The impact of Captain Cook's expeditions can be traced with the help of auction catalogues listing shells as well as other artefacts collected on the journeys. As Beth Fawkes Tobin has recently discussed, the growing cult of Captain Cook, following his death on the third expedition, helped to raise prices on any item that could be linked to him and the explorations he had been in charge of. Private museums and collections, which charged entrance fees, attracted more visitors if they had Cook-related material. Meanwhile, however, the prices of natural history specimens from the South Pacific actually declined, after the second and third Cook expeditions, as well as others setting out from France and Russia, all returned to Europe with ever more material, creating a surplus.[112]

We can trace Cook's celebrity in the sales of Christie and Manson's auction house in London from the mid-1770s and onwards as they provide one of the few remaining complete series of auction catalogues from eighteenth-century London. In November 1775, 'a valuable and curios collection of scare and beautiful shells, from the new discovered islands in the South Sea, consisting of a remarkable fine Trochus, and other different Species, entirely new' were put up for sale.[113] More material, including clothes and weapons from present-day New Zealand and Tahiti, was auctioned off in the 1780s.[114]

The reference to a scientific name, however, as in the use of the genus *Trochus* mentioned above, is not representative of the descriptions of the natural history material put up for sale at Christie's and Manson's.[115]

109. Dance, *A History*, p.54.
110. Dance, *A History*, p.55.
111. Dance, *A History*, p.67-68, 71.
112. Beth Fowkes Tobin, *The Duchess's shells: natural history collecting in the age of Cook's voyages* (New Haven, CT, 2014), p.168-73, 175-78.
113. Sales catalogue, 24 November 1775, Messrs Christie and Manson, London. This and all other references to sales catalogues published by Messrs Christie and Manson listed below are held at the Christie's Archive Department, King Street, St James's London.
114. Sales catalogue, 10 May 1782; Sales catalogue, 21 December 1785; Sales catalogue, 30 March 1787, Messrs Christie and Manson, London (Christie's Archive, London).
115. I have extracted auctions involving natural history material put up for sale at Christie

Most natural history specimens sold at the auction house were seemingly sold as part of an interior design, they included 'curious subjects in natural history' such as the 'Skin of a beautiful Zebra' or 'Two large boxes with a quantity of Minerals and Fossils for grotto work'.[116] The first explicit reference to a species authored by Linnaeus can be found under the heading 'Curious Birds and other Subjects of Natural History' in a catalogue from 1788 which lists the possessions of Benjamin Bond, Clapham, Surrey. Here we find a 'fine and extremely scarce' bird, a 'purple-breasted blue manakin' with the alternative name 'Ampelis Cotinga Linn'.[117] With a few other exceptions the sales catalogues produced by Messrs Christie and Manson between 1771 and 1810 made little use of the new binary names invented by Linnaeus.[118]

The reluctance of collectors and auctioneers to use the new scientific nomenclature was not something specific to Christie and Manson's auction houses. Dance writes about the slow acceptance of the new binominal names among shell dealers, the latter of which he described thus: 'the tardiest converts to it were those who collected shells for pleasure, who traded in them for profit, or who wrote books on them for popular guidance. To these classes the revolution brought about in zoological science by the Swedish savant made no difference whatever. To them precision in nomenclature was new-fangled and unnecessary.'[119]

One important exception to this history of resistance to the use of scientific nomenclature among those involved in the English shell trade dates back to 1785 when the duchess of Portland's collection was put up for sale at an auction that lasted thirty-eight days. In contrast to Hans Sloane's collections which, once donated to the British state, became the cornerstones of the British Museum, the duchess's collection was broken up and sold after her death to cover her debts. The most famous item of her collection was the Portland Vase, which is today displayed at the

---

and Manson's auction house before 1810 from the lists compiled by J. M. Chalmers-Hunt covering the whole of Britain; see *Natural history auctions 1700-1972: a register of sales in the British Isles*, compiled by J. M. Chalmers-Hunt (London, 1976) p.57-178.

116. Sales catalogue, 3 June 1771; Sales catalogue, 2 June, 1807, Messrs Christie and Manson, London (Christie's Archive, London).

117. Sales catalogue, 26 April 1788, Messrs Christie and Manson, London (Christie's Archive, London). The bird specimen was bought by the collector Humphreys for £2 and 14 shillings. Note that other specimens put up for sale at the same auction were also given a scientific name with Linnaeus marked as the author.

118. For other examples than the above referred to Sales catalogue, 21 December 1785; Sales catalogue, 23 May 1788; Sales catalogue, 29 April 1796. In one case Christie and Manson sold 'new' and hence (presumably) unnamed species, see Sales catalogue, 24 November 1775, Messrs Christie and Manson, London (Christie's Archive, London).

119. Dance, *A History*, p.47.

British Museum. Among the many items sold in 1785 were also the shells which Solander had been working on to provide specimen names, and grouping them into genera, thereby fixing them in the taxonomic world created by Linnaeus. The preface to the sales catalogue, probably written by John Lightfoot (1735-1788) another of the duchess's in-house naturalists as well as her chaplain, claims the shell collection to be the greatest in Europe and full of new species; as many as three quarters of the shells put up for sale had, for example, been unknown to Linnaeus.[120]

The problem, according to the author of the preface, was how to divide the shells into auction lots. A clash between the two groups existed. In the first camp 'every true lover of Science' wanted the shells to be 'allotted in Order and Method, according to *Genus* and *Species*'. The 'Majority of the World' were however 'not *Methodist*'. Rather than buying 'Twenty different Species of *Cones* or *Turbos* in One Lot', the latter group wanted 'variety' or what lovers of science called 'Promiscuous Assemblage of various Subjects'. Selling the specimens singularly or in pairs was impossible as it would prolong the sale 'to as many *Weeks*, as it consists at present of *Days*'. The compromise was to offer lots with variety while keeping the large classes, even the genera, together 'at least as far as Respect to *Size, Rarity,* and *Beauty*' would allow.

To please the '*Man* of *Science*', the duchess was obviously thought of as an exception to her gender, the catalogue carefully listed 'the proper classical, or generally received name' to many of the shells though.[121] Many of these names were provided by Solander, which was clearly indicated in the auction catalogue by placing an 'S' beside the name. The catalogue thus makes it possible to trace the work Solander had been engaged in every Tuesday for two and a half years; a lot of his taxonomic work is otherwise lost to due to the fact Solander failed to publish the results of his endeavours.

What was described as a compromise for the lovers of science in 1785, selling specimens in lots disrespecting divisions according to genera but using binominal names, seems to be have become every-day routine in the nineteenth century. Sometimes, of course, the integrity of a collection was respected when it was put up for sale, often reflecting the intention of the collector. Collectors also regularly donated the totality of their work to museums. Using genera divisions as guides, the division of collections into auction lots is, by all accounts, rare although a more systematic study of auction catalogues is needed to securely establish this.

120. 'Preface', in *A Catalogue of the Portland Museum, lately the property of the Duchess Dowager of Portland* (London, 1786), p.iii. For a discussion of the sale of the collection see also Tobin, *The Duchess's shells*, p.217-45.

121. 'Preface', in *A Catalogue of the Portland Museum, lately the property of the Duchess Dowager of Portland* (London, 1786), p.iv, (original emphasis).

However, as taxonomy evolved it is not surprising that singular species or groups of species, identified by the stable binary names, came to form the most important component in auction lots.[122] The cheapness of auction time today, when the internet allows us to bid on individual specimens 'twenty-four seven', has probably also helped to further promote the selling of singular, scientifically named specimens.

The commercial value of scientific names had already become clear by the nineteenth century to those engaged in the business of building up collections. As the British anatomist and zoologist Robert Grant (1793-1874) expressed it in 1835, at the hearing of the Select Committee on the Affairs of the British Museum:

> Now many objects of Zoology are scarcely of any appreciable value until they are identified and names are assigned to them by competent authorities. Sometimes an object, say a very rare shell, may not be of the value of one farthing until it is identified by a competent naturalist; and then being properly named, its value may be raised to 30, 40, 50 guineas. Yet it still appears no larger, not more beautiful, not heavier than a specimen which may not have the value of one farthing.[123]

Good names were costly, though, in the sense that resources were needed to name and update the names of the species represented in the collections of the British Museum. Particularly if one, like the 'ultra-radical' Grant, was keen to update names of the specimens to reflect the taxonomic changes taking place since the time of Linnaeus, thereby successfully distancing the collection from the conservative elite associated with it.[124]

Employing scholars for nomenclature revisions was one way to add value to a collection. Scholars could be motivated to (re)name species for politically infused scholarly reasons, but also because other gains could be had. As Dance has pointed out, by singling out new species, rather than adding sub-species and variations to already extant authored ones, value was added to the collections of shell dealers. Not only could new species enhance a collection, but authorship of new species offered the naturalists scholarly capital, although too many new species gave rise to accusation of 'species mongering'.[125] This of course brings us back to the importance of scholarship and naturalists keeping track of the develop-

122. The use of binominal names in the trade with specimens is only one of many examples of how the Linnaean nomenclature came to impact the handling of objects and the organisation of information at the end of the eighteenth century and beyond. See Staffan Müller-Wille, 'Names and numbers: "data" in classical natural history, 1758-1859', *Osiris* 32 (2017), p.109-28.
123. Quoted in McQuat, 'Cataloguing', p.18.
124. McQuat, 'Cataloguing', p.16-19 (16).
125. Dance, *A History*, p.149-51.

ment of species, a history in which Linnaeus's collection came to play an increasingly crucial role, as the system for holotypes evolved.

In summary, we have seen how the rise of interest in the history of natural history and collections with a Linnaean provenance corresponded with the rise of London as a staple market for natural history. If we zoom out, and take into account the long eighteenth century, we can trace several corresponding developments, changing fashions, the economic decline of the Dutch Republic, the effects of the French Revolution, and the ensuing wars, all of which explain the movement of natural history specimens to London. If we focus in on auction catalogues we can see the contours of a development; most notable is the shift from organising lots reflecting the content of, for example, different cabinets, to lot descriptions where the Linnaean binary names became the preferred choice. Name writing was not a business free of the influence of political or monetary interests. The activity of species-mongers and shell-doctors reflected the close connection between scholarly developments and the market for natural history specimens in the long eighteenth century.

Conclusion

Mapping the estimated and real (auction) values of the Linnaean collection reveal an economic history of natural history. Linnaeus's high estimations of the value of his own collection reflects the booming market for natural history specimens in mid-eighteenth-century Sweden. This is also notable in the high prices paid for natural history specimens outside Sweden, particularly in Amsterdam, in the middle third of the eighteenth century. While we can detect downward price trends and a much less buoyant Swedish market after the 1750s, these changes were maybe not considered by Linnaeus as he priced his specimens, dividing his material scholarly heritage between his children. More than anything, the price paid for Linnaeus's collection once it was sold in 1784 reflects the rise of London as a centre of commerce for natural history collections in late eighteenth-century Europe.

A focus on the material dimension reveals that different materials were more or less delicate and expensive, corresponding to their fragility and perceived beauty. If we include preservation and presentation, the material dimension provides us with an understanding of how Linnaeus's taxonomy was compatible with mid-eighteenth-century aesthetic ideals but also how Linnaean order to a certain extent trumped the contemporary fashion ideal, in which specimens were presented together with sculptures, paintings, framed by exotic and expensive wood and textile. While the Linnaean order was replicated and diffused across royal and elite collections, as well as in more humble environments, it, too, became outdated by the end of the eighteenth century. Meanwhile a growing awareness of the historical dimension to natural history evolved. London, the new home of the Linnaean collection and other collections with a Linnaean provenance, offered a place where Linnaeus's collections could change, from being modern to becoming historically important, within a relatively short time span, thereby saving it from the ravages of time, in the form of moths and dust and other destructive forces of nature. Revolutions and wars on the Continent and a growing Empire further enhanced London's position as the staple market for natural history as the nineteenth century evolved.

Late eighteenth-century commercial practices surrounding natural history, particularly the use of auctions, reveal how Linnaeus's new nomenclature became a tool for selling and/or buying natural history. The use of binary names started off as a compromise, mediating between buyers who loved order, and those who loved variety. But valid scientific names were not cheap as naturalists needed to be employed to assign them, conferring with already dedicated names and established genera. New names were also not always valid, as naturalists could be prompted by collectors to distinguish and name new species, rather than new varieties, so as to add value to a collection. This of course brings us back to the centrality of collections stretching far back in time, determining the identity of species and genera.

All in all, the development outlined above suggests that the relocation and conservation of the Linnaean collection, as well as the broadening use of Linnaean nomenclature, is embedded in a history of trade, materiality and consumption.

4. In the name of Linnaeus: Paris as a disputed capital of natural knowledge (1730-1789)

STÉPHANE VAN DAMME

According to historiographical tradition, although almost all of Europe was quickly convinced by Linnaean science, Paris remained a centre point of resistance.[1] Buffon used various well-known arguments to challenge reception of Linnaean sciences. He denounced the systematic nature of the classification, its arbitrariness (the fact that it is not a *natural* classification) and its language. Because nature is indivisible, classifications are useless. Buffon, on the contrary, prioritised the detail of nature, dense descriptions, and saw in Linnaeus a 'quasi-metaphysical error which consists in trying to judge a whole, and the combination of several wholes, by a single part, and by comparing the differences of this single part'.[2] Moreover, Pascal Duris has shown that in France the adoption of Linnaean sciences had spread across the provinces, especially Montpellier and Lyon.[3] This conflict in the process of scientific recognition persisted for several decades in Paris, and engaged not only Buffon, but also the naturalists from the Jardin du Roi such as Michel Adanson, leading to his *Famille des plantes* of 1763-1764 which contested the non-natural classification.[4] By focusing on certain botanists, intellectual history has taken for granted that Linnaeus was poorly received in Paris. French provincial naturalists were keener to adopt Linnaean sciences, as Pascal Duris has demonstrated.[5] The geographical spread of Linnaean knowledge in France included not only the two medical capital cities, Montpellier and Lyon, but also Bordeaux, Agen, Toulouse, Angers,

I would like to thank Paul Cohen, Oliver Buxton-Dunn, Nicholas Mithen, Alexia Grosjean and Trista Selous for their help in revising my English.

1. See for instance the classic study by Frans A. Stafleu, *Linnaeus and the Linnaeans: the spreading of their ideas in systematic botany, 1735-1789* (Utrecht, 1971); and more recently, Mario Beretta and Alessandro Tosi (ed.), *Linnaeus in Italy: the spread of a revolution in science* (Sagamore Beach, MA, 2007).
2. Pascal Duris, *Linné et la France (1780-1850)* (Geneva, 1993), p.33 (translation by the author): 'cette erreur quasi métaphysique qui consiste à vouloir juger d'un tout, & de la combinaison de plusieurs touts, par une seule partie, & par la comparaison des différences de cette seule partie'.
3. Duris, *Linné et la France*, especially chapter 2 on the dispute.
4. Xavier Carteret, *Michel Adanson (1727-1806) et la méthode naturelle de classification botanique* (Paris, 2014).
5. See especially Duris, *Linné et la France*, ch.3.

Strasbourg and Nancy, and emphasised the connections between provincial naturalists such as Boissier de Sauvages, Antoine Gouan, Claret de la Tourette, Jean-Emmanuel Gilibert and Carl Linnaeus: all are 'more "admirers" rather than "followers"'.[6] However, is this picture really true? When Linnaeus was in Paris between May and August 1738, he made a lot of contacts and botanised with Antoine and Bernard de Jussieu in Fontainebleau, and maintained a correspondence with Jussieu between 1736 and 1763, which increased, even flourished, after their meeting in Paris in 1738. Linnaeus visited the capital in 1738 and attended the meeting of the Academy of Sciences. Jussieu's nephew Antoine-Laurent used his nomenclature in the Jardin du Roi from 1774, and would replant the botanical school of the Jardin du Roi according to the Linnaean method. A few years earlier, Etienne-Louis Geoffroy began to publish a work on natural history devoted to Paris and the surrounding area entitled *Histoire abrégée des insectes* [An Abridged history of insects].[7] The *Abridged history of insects* is preceded by a Preliminary discourse of thirty-one pages that sets forth a statement about the historiography of natural history of insects and the methods followed in his treatise.[8] Geoffroy pleaded for the use of systematic classification in the wake of botanists like Linnaeus, stating that without any method, the zoologist is condemned to inaccuracy and risks not identifying the insect.[9] This polemical tone transformed the Preliminary discourse into a defence of systematic classification,[10] and it pitted Linnaeus's method against that of Réaumur[11] and Buffon.[12] More broadly, close attention

6. Duris, *Linné et la France*, p.53 (translation by the author): 'ils se montrent davantage les "admirateurs" que les "continuateurs"'.
7. Etienne-Louis Geoffroy, *Histoire abrégée des insectes qui se trouvent aux environs de Paris*, vol.1 (Paris, Durand, 1762), p.xix-xxj.
8. 'ces commencemens de méthode sont trop superficiels et trop peu systématiques pour être mis en usage, et on a beaucoup de peine à distinguer dans ce grand ouvrage de M. de Reaumur, l'animal dont il traite, faute de caractères suffisans et d'une bonne description'; Geoffroy, *Histoire abrégée*, p.vij.
9. 'Je sais que quelque savans de nos jours ne conviendront pas de ce que j'avance ici. Ennemis des systèmes et des ordres méthodiques, ils semblent vouloir faire retomber les sciences dans cette espèce de confusion dont elles ont eu tant de peine à sortir, et ce qui paroît encore plus étonnant, c'est que dans un siècle aussi éclairé, de pareils paradoxes trouvent des sectateurs'; Geoffroy, *Histoire abrégée*, p.x.
10. Thierry Hoquet (ed.), *Les Fondements de la botanique: Linné et la classification des plantes* (Paris, 2005); Daniel Mornet, *Les Sciences de la nature en France, au XVIIIe siècle: un chapitre de l'histoire des idées* (1911; Geneva, 2001).
11. [René-Antoine Ferchault de] Réaumur, *Mémoires pour servir à l'histoire des insectes, par M. de Réaumur*, 6 vols (Paris, Imprimerie royale, 1734-1742). See Mary Terrall, *The Man who flattened the earth: Maupertuis and the sciences in the Enlightenment* (Chicago, IL, 2002), p.315-17; Mary Terrall, *Catching nature in the act: Réaumur and the practice of natural history in the eighteenth century* (Chicago, IL, 2014).
12. 'Nous sommes donc infiniment redevables à M. Linnaeus d'avoir cherché le premier à

to the broader scene reveals to us that reference to Linnaeus circulated regularly among botanists and gardeners, and was not seen as an exclusive choice but rather took place within a large repertoire of naturalist practices.

I will argue that this long controversy was not limited to problems of classification, naming, or nomenclature; it also revealed a deeper sociological and geographical tension among Parisian naturalists between national and cosmopolitan regimes of natural knowledge, just as it did in Sweden.[13] This polarisation between pro- or anti-Linnaean philosophies also drew a line between different tensions within the Parisian naturalist world: tension between the science of the cabinet and travelling science; tension between botanists and *botanophiles*; and tension between inventorying, describing and classifying. In this respect, Paris was not a neutral place in Europe for Linnaeus; it was rather a *passage obligé* in the recognition of his system. The polarising reaction to botanical philosophy in Paris reflected the deep divisions in the Parisian naturalist world, not only in terms of intellectual framework but also in terms of practices. By comparing the different orders of knowledge, and juxtaposing 'book knowledge' with 'in-the-field science' and 'practical knowledge' with 'active knowledge', the fluidity of these concepts and the diversity of their uses by Parisian scientists will be examined. In the case of Paris – the scientific hub during the seventeenth and eighteenth centuries – I would like to investigate the process of practising local natural history in a global city, to understand the multi-faceted strategies that supported the recognition of a local regime of knowledge.[14] This controversy reflected the competition between elites of dominant European capital cities.[15] It is not by chance that the Linnean sciences returned to Paris in the late 1780s through the mediation of British naturalists – including James Edward Smith – who were in contact with French naturalists in Versailles.[16] Putting to one side the national aspects of Paris, new researches seek to understand the Parisian Enlightenment in terms of locality. This approach is not a return to local history, but rather to

ranger méthodiquement les insectes, et à trouver des caractères génériques qui les fissent plus aisément connoître'; Geoffroy, *Histoire abrégée*, p.xiij.

13. Lisbet Koerner, *Linnaeus: nature and nation* (Cambridge, MA, 1999).

14. See my attempt, Stéphane Van Damme, *Paris, capitale philosophique de la Fronde à la Révolution* (Paris, 2005); Stéphane Van Damme, *A toutes voiles vers la vérité: une autre histoire de la philosophie au temps des Lumières* (Paris, 2014). Christophe Charle and Daniel Roche (ed.), *Capitales culturelles, capitales symboliques: Paris et les expériences européennes, XVIIIe-XXe siècles* (Paris, 2002); Christophe Charle (ed.), *Le Temps des capitales culturelles, XVIIIe-XXe siècles* (Seyssel, 2009).

15. We can compare Paris with London (see chapter 3 by Hodacs in this volume), or his reception in Italy; see Beretta and Tosi (ed.), *Linnaeus in Italy*.

16. Duris, *Linné et la France*, p.60-62.

understand both the interplay of scales of sciences and the relationship with metropolitan identity. As David Garrioch stated:

> By the 1750s and 1760s, the customary, corporate, and hierarchical social organisation of Paris was being seriously challenged – though not eliminated – by changes in the city's economy and demography, by new ideologies and new social practices. [...] From the mid-century on, educated Parisians were increasingly influenced by enlightened ideas and by economic and material changes that were widening the gulf between rich and poor. [...] Both the 'middling sort' and the noble and wealthy elite of Paris began to aspire to a broader 'metropolitan' culture, a shared culture yet one that each group lived and interpreted in different ways.[17]

In this chapter, I will therefore not retrace the intellectual reception of Linnaean sciences in Paris before the creation of the first Linnaean society in Paris in December 1787; instead I would like to understand both why Linnaeus mattered to Parisian natural philosophers and why Paris is crucial in making Linnaean sciences global.[18] I will argue that natural knowledge featured in the strategies employed to re-scale the city as a powerful and attractive metropolis, an emergence that benefitted from new ways of commodifying, consuming and interpreting nature. This also requires consideration of the capacity of an urban milieu to exploit and diversify its forms of knowledge and to relocating global natural resources. During the eighteenth century, communities of scholars in Paris formed new alliances. The localisation of scientific practices has some particular effects on a new definition of the modern city. Conversely, the onset of the internationalisation of natural knowledge involves differentiating its effects on the urban hierarchy, and it reconfigures the distinction of urban greatness. Thus, this chapter must be oriented towards a more precise examination of the 'tools of scale', understood as the instruments of the conversion of a city into a universal scientific capital.[19] This involves the recognition of expertise, the allure and centralisation of information, the multiplication of locations for analysing information, the establishment of frameworks for publicity, and so on. This idea must also take seriously the 'trials of scale' that actors and institutions elaborate on from city to city. I will use the name of Linnaeus to trace and follow the localisation and globalisation of Linnaean sciences that took place in Paris between 1730 and 1780.[20]

17. David Garrioch, *The Making of Revolutionary Paris* (Berkeley, CA, 2002).
18. Duris, *Linné et la France*, ch.5. Thierry Hoquet, *Buffon-Linné: éternels rivaux de la biologie?* (Paris, 2007).
19. Van Damme, *Paris, capitale philosophique de la Fronde*.
20. Bruno Latour, *Reassembling the social: an introduction to actor-network-theory* (Oxford, 2005).

The rise of a Metropolitan community of naturalists

This division between two different strategies for the globalisation of naturalist knowledge when describing the world of naturalists also exposed social tensions between professional royal botanists and amateurs. In his *Philosophia botanica,* Carl Linnaeus published a classification of naturalist authors in which he placed botanists (professional scholars) in opposition to 'botanophiles' (amateurs), and thus divided the epistemic community.[21] Several groups are presented by Linnaeus. First are the 'phytologists' – the true botanists composed of 'collectors' and 'methodisers': 'The collectors have been concerned primarily with the number of species of vegetables. They included the commentators, illustrators, describers, monographers, the meticulous, Adonises, compilers of Floras, and travellers.'[22] Beyond this first circle, the 'amateurs of Botany are those who have produced various [works] of botany, though they do not properly pertain to botanical science, such as anatomists, gardeners, physicians, and miscellaneous'.[23] At the opposite end, this problem of demarcation within the naturalist's world is refuted by Buffon in his *Encyclopédie* article, 'Natural history', which asserted that 'the culture of flowers and of trees belongs to Botany',[24] and therefore flower merchants and botanists shared a connoisseurship. Historians of science argued conversely that the new role played by amateur culture in metropolitan contexts empowered the epistemic community of naturalists. Consequently, collecting, reading, buying or displaying natural history is not only dedicated to natural philosophers, it is a common practice shared by different epistemic communities in Paris – pharmacists, botanists and gardeners, horticulturalists, but also physicians, geologists and engineers. It opened up a vast range of practices, which reshaped different communities: a community of readers, a community of consumers, and a tasting community. Following the route of Linnaeus's name as it was actively circulated in these communities both positively and negatively, could help us to remap the circulation and appropriation of Linnaean science.

As Londa Schiebinger argued, 'Large private collections symbolized the huge fortune amassed during this period in European capitals'.[25] The making of Paris as the centre of naturalism is not only the reflection of the mobilisation of a group of collectors or scientists. The cumulative

21. Carl Linnaeus, *Linnaeus's 'Philosophia botanica'*, translated by Stephen Freer (Oxford, 2003).
22. Linnaeus, *Linnaeus's 'Philosophia botanica'* (2003), p.18.
23. Linnaeus, *Linnaeus's 'Philosophia botanica'* (2003), p.26.
24. Jean-Marc Drouin, *L'Herbier des philosophes* (Paris, 2008), p.38.
25. Londa Schiebinger, 'Nature's unruly body: the limits of scientific description', in *Regimes of description: in the archive of the eighteenth century*, ed. John Bender and Michael Marrinan (Stanford, CA, 2005), p.25-43 (26).

aspect of the naturalist inquiry based on the paradigms of accumulation, of collection, served a political economy which took Paris as a centre of a global accumulation.[26] Natural knowledge allowed not only social circulation within the city; it strongly articulated commodities and collections. In Paris, both aesthetic and cognitive goals developed the world of the amateur naturalist, which tended to become centralised.[27] Admiring, being attentive, and being surprised are states that to a certain extent (that is, without excess), 'dispose us to the acquisition of the sciences', as Descartes wrote, and transform the scientific pursuit into a moral achievement.[28] The field of science, the field of aesthetics, and social and moral norms intersected here so as to justify and stimulate the investigation of nature. They not only accumulated and collected vast amounts of objects, plants, fossils, shells, curiosities, alongside paintings and antiquities, they proposed an order of nature through the displays in their cabinets of curiosities.[29] When their collections were sold, domestic catalogues or post-mortem inventories were printed by merchants and became thereafter an epistemic genre.[30]

Parisian short networks are observable within socio-professional milieus close to urban institutions, for example apothecaries, who belonged to the corporation of grocers, and who provided seven Paris aldermen during the eighteenth century. Specific groups are particularly interesting in that they contributed both to promoting forms of knowledge, including natural history and pharmacopeia, and to the articulation of urban authorities and scientific institutions. The desire to enforce alliances beyond personal ties with other learned people often led to a certain patrimonialisation of knowledge, a valorisation of lineage and family support within the Parisian metropolis. Venerable dynasties of learned aldermen like the Geoffroy and Boulduc families contributed to the flowering of the sciences and the management of the city's

26. Lissa Roberts (ed.), *Centres of accumulation and management*, special issue, *History of science* 52:3 (2014); Lissa Roberts (ed.), *Centres and cycles of accumulation in and around the Netherlands during the early modern period* (Berlin, 2011).

27. Charlotte Guichard, 'Taste communities: the rise of the amateur in eighteenth-century Paris', *Eighteenth-century studies* 45:4 (2012), p.519-47; Charlotte Guichard, *Les Amateurs d'art à Paris au XVIII*^e^ *siècle* (Seyssel, 2008).

28. René Descartes, *The Passions of the soul*, translated by Stephen H. Voss (Indianapolis, IN, 1989), p.60 (art.76*).

29. Emma Spary, 'Codes of passion: natural history specimens as a polite language in late eighteenth-century France', in *Wissenschaft als kulturelle Praxis, 1750-1900*, ed. Hans Erich Bödeker, Peter Hanns Reill and Jürgen Schlumbohm (Göttingen, 1999), p.105-35.

30. Bettina Dietz and Thomas Nutz, '*Collections curieuses*: the aesthetics of curiosity and elite lifestyle in eighteenth-century Paris', *Eighteenth-century life* 29:3 (2005), p.44-75. On Parisian collections see also Bertrand Daugeron, *Collections naturalistes entre science et empires (1763-1804)* (Paris, 2009), and Pierre-Yves Lacour, *La République naturaliste: collections d'histoire naturelle et Revolution française, 1789-1804* (Paris, 2014).

commonweal. Simon Boulduc, followed by his son Gilles-François, occupied key positions, first in the community of apothecaries, then within urban power structures, and finally within the Royal Academy of Sciences. Beyond these two figures, family social capital pushed alliances with the Faculty of Medicine, the grocers' corporation, and judicial officialdom. Gilles-François Boulduc, who was once an associated chemist of the Royal Academy of Sciences, and exercised the office of court apothecary, created a company with another merchant apothecary, Claude Pia, to engage in the business of pharmaceutics and chemical business at their premises in the rue de la Boucherie in December 1732.

The probate inventory of this apothecary gives an overview of business practices and their extent.[31] His premises included a shop, a back room, and a laboratory. The business ran for nine years, and the company allowed Boulduc to store his goods and his laboratory tools at the Jardin du Roi, and provide medicines to the princesse de Conti, as well as all religious communities that depended on the Jardin du Roi. Several memoirs cited in the inventory also evoke drug purchases for aristocratic customers, including the Count Sabran, marshal of Berwick, the duke of Fils-James, the cardinal d'Auvergne, the duke and duchess of Bouillon, the duke and duchess of Ruffec, the duke of Gesvres and the duke of Saint-Simon.[32] Gilles-François Boulduc also served as an apothecary to the king, and became a member of the medical commission for secret remedies, which was created in 1731. He articulated as much as he could his academic investigations and his commercial benefits. He was interested in the waters of Passy, of Bourbon-l'Archambault, of Forges, as well as Epsom salt, which he sold in significant quantities.[33] The Académie published his works on plants in 1734.[34]

The Geoffroy family offers another example of this accumulative process of capital, knowledge and position. In 1753, Etienne-Louis Geoffroy was forced to sell in Paris the cabinet which belonged to his uncle Claude-Joseph, former member of the Academy of Sciences. With the help of two librarians, H. L. Guerin and L. Fr. Delatour, he published the printed catalogue,[35] but, as Etienne-Louis noted, his uncle viewed his

31. Archives nationales (Paris), MC, CXXII, 640, 21 February 1739, quoted by Christian Warolin, 'La dynastie des Boulduc apothicaires à Paris aux XVIIe et XVIIIe siècles (suite), Gilles-François Boulduc, 20 février 1675-17 janvier 1742', *Revue d'histoire de la pharmacie* 335 (2002), p.439-50 (444).
32. Archives nationales (Paris), MC, II, 484, 16 June 1741, 'constitution de rente pour le duc de Gesvres'; Archives nationales, MC, XLIX, 598, 27 March 1738, 'constitution de rentes par le duc de Saint-Simon'; both quoted by Warolin, 'La dynastie des Boulduc', p.444, 445.
33. Warolin, 'La dynastie des Boulduc', p.447.
34. [Gilles-François] Boulduc, 'Essai d'analise des plantes', *Mémoires de l'Académie royale des sciences* (Paris, Imprimerie royale, 1734), p.101-106.
35. Etienne-Louis Geoffroy, *Catalogue raisonné des minéraux, coquilles, et autres curiosités dans le*

cabinet as a book. The metaphor is common in the language of connoisseurship. This catalogue had a specific audience: Parisian amateurs of natural history who were perfectly familiar with this cabinet thanks to the ritual of visits.[36] The sale catalogue is therefore part of a genre which lies between commercial purposes (all the items are itemised)[37] and the annotated, learned catalogue. It contained descriptions and rationale of the collection. The editor warned the amateur of natural history several times to justify his choices, contrasting the practices of the *curieux* who knew the methodical classification with those of ordinary buyers:

> Perhaps it will be found that in this Catalogue we have not always followed with perfect accuracy the order which we have prescribed. A few articles contain different pieces that have little connection. But these articles make a small number, and we have been forced to unite them so as not to be too far removed from the order of the Inventory which preceded this Catalogue. Besides, the Curiosi who buy, are little troubled by this order; they arrange, according to their ideas, the pieces they have acquired, and we do not pretend to give here a methodical order, a system of natural history. Finally, we think we ought to warn that we have been extremely reserved for notes and reasonings. We explained that what we thought absolutely necessary; we have advanced nothing that is not indisputable, wishing to avoid the defect of some works of this kind in which several fables have been debated, which may well impose upon some persons, but which do not render the exhibits we want to prize more recommendable in the eyes of the true connoisseurs. For all the rest we have contented ourselves with a simple annona, leaving the Curiosi to make their own reflections, and the praise of the articles which cannot merit it.[38]

Cabinet de feu M. [Claude-Joseph] Geoffroy de l'Académie Royale des Sciences (Paris, H. L. Guerin and L. Fr. Delatour, 1753).

36. Geoffroy, *Catalogue raisonné*, p.aij.

37. On the commercial issue, we find this indication: 'Le Règne végétal n'est pas fort considérable, à l'exception des Plantes marines qui sont belles et nombreuses. De toutes les parties, c'est celle qui fournit le moins pour un Cabinet: d'ailleurs M. Geoffroy avoit réuni presque toutes les Productions végétales à son Droguier qui étoit très considérable, qu'on a été obligé de vendre avec le fond de son Commerce, dont il a été réputé faire partie'; Geoffroy, *Catalogue raisonné*, p.v.

38. Geoffroy, *Catalogue raisonné*, p.vj-viij (translation by the author): 'Peut-être trouvera-t-on que dans ce Catalogue nous n'avons pas toujours suivi avec une parfaite exactitude l'ordre que nous nous sommes prescrit. Quelques articles contiennent différentes pièces qui n'ont guères de rapport. Mais ces articles font un petit nombre, nous avons été forcés de les réunir pour ne pas trop écarter de l'ordre de l'Inventaire qui a précédé ce Catalogue. D'ailleurs, les Curieux qui achètent, s'embarrassent peu de cet ordre, ils rangent suivant leurs idées les pièces qu'ils ont acquises, nous ne prétendons point donner ici un ordre méthodique, un système d'Histoire naturelle. Enfin nous croyons devoir avertir que nous avons été extrêmement réservés pour les notes les raisonnemens. Nous avons expliqué que ce que nous avons cru absolument nécessaire; nous n'avons rien avancé que de sûr, voulant éviter le défaut de quelques ouvrages du genre de celui-ci dans lesquels on a

The figure of the amateur scientist thus seems linked to the emergence of practical bibliophile collectors in Paris. Between the beginning and the end of the century there was, in this small world of book collectors, an unprecedented transformation in the possession of science and the arts literature. While in 1730 only 9.8% of books in the inventories of private libraries were science books (against 21.7% theology books), by 1780 this figure exceeded 20% (against 11.8 % theology books). These collectors of science books belonged largely to the booming medical professions such as doctors and apothecaries, who made up 67% of this category.[39] Geoffroy and his acquaintances had a passion for natural history. In his post-mortem inventory his friend Jussieu had valued his 'droguier', his cabinet of natural history, and 'coquillé', at 7600 *livres*. These were items that Geoffroy claimed, in his will, to have 'gathered gradually with my savings'.[40] This collection is interesting, and is published in a catalogue.[41] Thirty years later, it was presented as an example to all Parisian naturalist amateurs:

> The collection of the late M. Geoffroy, of the Royal Academy of Sciences and of the Royal Society of London, extended throughout natural history. The cabinet on the ground floor contained about eighteen hundred crystal jars filled with what was most curious in the three reigns. The suite of the sigillata was the most complete, as well as that of the bezoars. The cabinet, displayed on shelves, occupied two sides of the cabinet, with serpents preserved in glass tubes, and placed in the mounts of the wooden panels. The third side contained a library of natural history and medicine. The top has lined with several crocodiles, lizards, serpents and other reptiles. The bottom of the shelves contained four rows of drawers filled with petrifications, minerals, fossils, and figurative figures. In another cabinet on the first floor, composed of several rooms, we saw a collection of selected shells tidily arranged.[42]

débité plusieurs fables, qui peuvent bien en imposer à quelques personnes, mais qui ne rendent pas les pièces que l'on veut priser plus récommandables aux yeux des véritables connoisseurs. Pour tout le reste nous nous sommes contentés d'une simple annone, laissant les Curieux faire eux-mêmes leurs réflexions, l'éloge des articles qui ne peuvent en mériter.'

39. Michel Marion, *Collections et collectionneurs de livres au XVIIIe siècle* (Paris, 1999), p.159-63.
40. 'rassemblée peu à peu et sur mes épargnes'; AN, Min. central, CVIII, 498, 9 March 1752, 'dépôt du testament de Claude-Joseph Geoffroy' (translation by the author).
41. Etienne-Louis Geoffroy, *Catalogue des livres et estampes de défunts Mrs Geoffroy* (Claude-Joseph and his son Claude-François) (Paris, Gabriel Martin, 1754).
42. Antoine Joseph Dezallier d'Argenville, *La Conchyliologie, ou Histoire naturelle des coquilles* (Paris, Guillaume De Bure, 1780), vol.1, p.234 (translation by the author): 'La collection de feu M. Geoffroy, de l'Académie Royale des Sciences et de la Société Royale de Londres, s'étendoit sur toute l'histoire naturelle. Le cabinet du rez-de-chaussée contenoit environ dix-huit cent bocaux de cristal remplis de ce qu'il y avoit de plus curieux dans les trois règnes. La suite des terres sigillées étoit des plus complètes ainsi que celle des bezoards. Le droguier, rangé sur des tablettes, occupoit deux faces du cabinet, avec des serpens conservés dans les tubes de verre et placés dans les montans de la menuserie. La troisième

The inventory also lists the inclusion in the cabinet of a priapus of a whale, heads of wild animals (tiger, rhinoceros, hippopotamus), and several arrows, bows and quivers, collected from 'the wild', and stored at Geoffroy's home in Paris and at his country residence.

Exploring the natural history of Paris: a Linnaean political economy of nature?

In this section, I would like to examine to what extent the rise of a metropolitan community of naturalists was closed to Linnaean botanical philosophy regarding the conception of local knowledge and local geography of plants. If Linnaeus himself was sceptical of the idea of using geography to study plants, he summarised his ideas in his *Philosophia botanica* in 1751:[43]

> The native location does not indicate separate species.
>
> There are many reasons to persuade us that the location ought not to determine the specific name.
>
> 1. No one could easily go to Japan, the Cape of Good Hope, or Peru to recognise a plant.
> 2. The location quite often changes; all alpine plants, especially those in the mountains, turn out to be marsh plants outside the Alps.
> 3. The location of plants of the same species is not unique: Lapland, Siberia, Canada, Asia, and America quite often produce the same types.
> 4. A Paradise garden often includes plants from all over the world.
> 5. Who would not struggle to identify a plant that was proffered to him without knowledge of its location?
> 6. Botanists are glad to recognise species in the herbarium, as are physicians in the dispensary.
>
> It is impossible to imagine a plant without its location, but it is adventitious, even though it was neighbouring; it is very variable, therefore it cannot come into the specific name.
>
> We maintain that all specific names taken from locations are erroneous, whether from the soil, the region, or the frequency.[44]

Biogeography nevertheless gained recognition, and encouraged the development of a kind of 'field science'.[45] Recently, historians including

---

face offroit une bibliothèque concernant l'histoire naturelle et la médecine. Le plafond se trouvoit garni de plusieurs crocodiles, de lézards écailleux, de serpens et autres reptils. Le bas des tablettes portoit quatre rangs de tiroirs remplis de pétrifications, de minéraux, de fossiles et de pierres figurées. Dans un autre cabinet au premier étage, composé de plusieurs pièces, on voyait une collection de coquilles choisies et très bien rangées.'

43. *Philosophia botanica*, book X, paragraph 334, p.263, quoted by Drouin, *L'Herbier des philosophes*, p.170.
44. Linnaeus, *Linnaeus's 'Philosophia botanica'*, p.226.
45. Alix Cooper, *Inventing the indigenous: local knowledge and natural history in early modern Europe* (Cambridge, 2007).

Lisbet Koerner and Alix Cooper have paid attention to locality and to the political production of local knowledge, generating discussion about the Linnaean political economy of nature. To what extent and in which forms were these new views on fieldwork or terrain present in Paris? Did the development of a natural history of Paris lead to a definition of a *terrain*? By the eighteenth century, Paris emerges as a favoured site for the observation of certain kinds of local knowledge. The natural history of the Parisian locale, for example, was understood by the observation and inventorying of plant and animal life within precise geographical limits. An understanding of nature in the metropolis could not be gained at a desk; it required individuals to move, to walk within and survey areas around the city. Far from being limited to a claim of autochthony, Paris was a space for scholars to master and survey. Urban space became the terrain for philosophical and scientific investigation. This terrain was both physical, since it became an object of knowledge, and a kind of epistemological model for the new field sciences.[46] At the crossroads of practices and field work founded upon observation, description of the natural and social world of Paris, and of the corporeal skills that enabled the scholar to embark on his discoveries, the construction of Paris as a 'terrain' for knowledge opened up an ethical and political conception of the capital.

The first element to emerge was the growing practice of botanising in Paris. In the late seventeenth century, Joseph Pitton de Tournefort gave precise botanical observations of different 'flora' in the places he visited.[47] During the whole eighteenth century we can follow the unfolding process of surveying and plotting of areas in and around Paris. Linnaeus himself went to botanise with Jussieu at Fontainebleau in the late 1730s. During one of his numerous stays in Paris, Jean-Jacques Rousseau did not hesitate to botanise in the villages surrounding Paris, in Charonne and Ménilmontant:

> On Thursday the 24th of October 1776, after dinner I wandered the boulevards up to the Chemin Vert, by which I reached the heights of Menilmontant, and from there, taking the footpath over the vineyards and fields, I crossed the pleasing landscape separating the two villages up to Charonne; then I made a detour to return by the same fields, taking a different path. I enjoyed passing through them with the pleasure and interest that agreeable places have always evoked in me, stopping a few times to take special note of individual plants within the greenery. I found two there that I have rarely seen around Paris but that were abundant in this district. One was *Picris/hierciodes*, the other *Bulpherum falcatum*, an ombelliferous plant. I delighted in this discovery, and it kept me happy for a long time, until finally

46. Henrika Kuklick and Robert E. Kohler (ed.), *Science in the field*, *Osiris* 11 (Chicago, IL, 1996).
47. Bibliothèque centrale du Muséum national d'histoire naturelle (hereafter, BCMNHN), MS 77, 'Herborisations aux environs de Paris', by Joseph Pitton de Tournefort (1656-1708).

> I discovered a plant still rarer, at least on cultivated land, known as *Cerastium aquaticum* [...] which I found in a book that I had taken along, and placed it in my herbarium.[48]

There is a clearly a bucolic tone to this description of Paris. The passage also stresses how entrenched the cantonal view of urban territory was. The task of maintaining a herbarium, like that of composing a mineralogical catalogue, suggests the panoptic character of compiling natural history collections. Rousseau is again an exemplary witness. Such attempts to delimit the terrain under investigation remained feeble and impressionistic, resembling quotidian or leisurely activities more than scientifically inspired exercises.[49]

A second element to be noticed is the influence of quantitative methods based on masses and groups of plants that emerged in the second part of the eighteenth century.[50] This was linked to the idea of a political economy of natural resources.[51] The work of counting required increasingly precise definitions of the area to be surveyed. In his project that aimed to complete Tournefort's flora, Sébastien Vaillant estimated numbers for his inventory:

> This book mentions [blank] genera of plants and 1886 species, including varieties of which [blank] grow naturally in the area around Paris, and are

48. Jean-Jacques Rousseau, *Les Rêveries d'un promeneur solitaire* (Paris, Marcel Didier, 1948), p.19-20 (translation by the author): 'Le jeudi 24 octobre 1776, je suivis après dîner les boulevards jusqu'à la rue du chemin-verd par laquelle je gagnai les hauteurs de Ménil-Montant, et de la, prenant les sentiers à travers les vignes et les prairies, je traversai jusqu'à Charonne le riant paysage qui sépare ces deux villages, puis je fis un détour pour revenir par les mêmes prairies en prenant un autre chemin. Je m'amusois à les parcourir avec ce plaisir et cet intérest que m'ont toujours donne les sites agréables, et m'arrêtant quelque fois à fixer des plantes dans la verdure. J'en apercus deux que je voyais assez rarement autour de paris, et que je trouvai très abondantes dans ce canton-là. L'une est le *Picris/ hiercioides* de la famille des composées, et l'autre le *Bupleurum falcatum*, de celle des ombellifères. Cette découverte me réjouit et m'amusa très longtemps et finit par celle d'une plante encore plus rare surtout dans un pays élevé, savoir le *Cerastium* aquaticum que, malgré l'accident qui m'arriva le même jour, j'ai retrouvé dans un livre que j'avois sur moi et place dans mon herbier'.
49. On Rousseau as botanist see Alexandra Cook, *Jean-Jacques Rousseau and botany: the salutary science* (Oxford, 2012); Jean-Marc Drouin, 'Rousseau', in *Dictionnaire culturel des sciences*, ed. Nicolas Witkowski (Paris, 2001), p.368; Jean-Marc Drouin, 'Rousseau, Bernardin de Saint-Pierre et l'histoire naturelle', *Dix-huitième siècle* 33:1 (2001), p.507-16; and Jean-Marc Drouin, 'Les herborisations savantes du promeneur solitaire', in *L'Accident de Ménilmontant*, ed. Anouchka Vasak (Paris, 2015), p.41-55.
50. New studies on eighteenth-century description include: Joanna Stanalker, *The Unfinished Enlightenment: description in the age of the 'Encyclopaedia'* (Ithaca, NY, 2010); John Bender and Michael Marrinan (ed.), *Regimes of description: in the archive of the eighteenth century* (Stanford, CA, 2005); Brian W. Ogilvie, *The Science of describing: natural history in Renaissance Europe* (Chicago, IL, 2006).
51. See on this issue Grégory Quenet, *Versailles: une histoire naturelle* (Paris, 2015).

> cultivated there annually or planted over a long period, as well as many others that I did not think worthy of mention. The history of the area around Paris identifies only [blank] plants. This is because we have continued to make observations since the printing of that History until today: 31 October 1704.[52]

These words placed at the top of the manuscript indicate the difficulty of counting, and thus figures are left blank as a task to be carried out once the new inventory was complete. The title clarifies the geographical boundaries of the census: 'Counting of plants that grow in the area around Paris, to the distance of one day or fifteen to sixteen leagues around the city: extract from the history of these plants given by M. Tournefort in 1698 with the addition of those that escaped his notice and have been observed since the printing of his book.'[53]

Antoine Nicolas Duchesne (1747-1827) was also aware of this when he wrote his *Statistique végétale des plantes de Paris* in 1771:

> It would be desirable for us to have in France a precise description of the plants that grow in the countryside within an area of 18-20 leagues around each city. The surrounding area of Paris from a distance of 20 leagues to the north to 20 leagues to the south forms a straight line of 40 leagues. Similarly, if you go from 20 leagues to the east to 20 to the west you have another straight line of 40 leagues: use these two lines to make a square and you will have a map that will contain an area of sixteen hundred square leagues. [...] As with the inhabitants of the city of Paris, so with the plants that grow in and around it; not all were born there, but by dint of long stay have each gained, as it were, burghers' rights.[54]

52. 'Ce livre fait mention de [blanc] genre de plantes et de 1886 espèces y comprenant les variétés dont [blanc] croissent naturellement dans les environs de Paris s'y cultivent annuellement ou y sont plantées de longue main ainsy que beaucoup d'autres dont je n'ay pas cru devoir faire mention. L'histoire des environs de Paris ne fait estat que de [blanc] plantes. C'est donc qu'on a observées depuis l'impression de cette Histoire jusqu'aujourd'huy 31e octobre 1704'; BCMNHN, MS 1093, 'Dénombrement des plantes qui naissent aux environs de Paris, extrait du livre de Tournefort avec des additions par Sébastien Vaillant' (1704), no folio numbers.
53. 'Dénombrement des plantes qui naissent aux environs de Paris, dans l'étendue d'une journée ou de quinze à seize lieues autour de la ville : extrait de l'histoire que Mr. Tournefort a donnée de ces plantes en 1698 avec les additions de celles qui luy ont échappées et qu'on a observées depuis l'impression de son livre'; BCMNHN, MS 1093, f.1.
54. 'Il seroit à désirer que nous eussions en France la description exacte des plantes qui croissent à la campagne dans les environs de chaque grande ville à 18 à 20 lieues à la ronde. Les environs de Paris à 20 lieues de distance au nord de 20 lieues au midi forment une ligne droite de 40 lieues. Si vous vous éloignez de même de 20 lieues au levant et 20 au couchant vous aurez une autre ligne droite de 40 lieues : formez un quarré de ces deux lignes, vous aurez une carte qui contiendra seize cent lieues quarrées de superficie. [...] Il en est des plantes des environs de Paris comme des habitans de cette grande ville, tous n'ont pas pris naissance dans le pays, mais par un long séjour les uns et les autres ont pris pour ainsi dire droit de bourgeoisie' ; BCMNHN, MS 1297, 'Statistique végétale des environs de Paris, par Duchesne' (1771), quotation f.2.

The parallel with the evolving studies of urban population here is symptomatic of the new methods of political arithmetic. The ideas of natural territory and of natural citizenship were coming to the fore, but they were not unquestioningly accepted.

Duchesne was not a scientific botanist, rather he was the quintessential gardener botanist. Provost of the king's buildings, architect and painter, he was fascinated by the natural sciences, and he often visited the royal gardens of the Trianon in Versailles. There he became a student of Bernard de Jussieu, botanist to the king who catalogued the plants in that garden.[55] As early as 1764, Duchesne published a *Manuel de botanique*,[56] which gives the common names of plants cultivated in Paris and its surrounding area. He later became known for his *Histoire naturelle des fraisiers* of 1766, and his *Histoire naturelle des courges* of 1772. This biography reveals the attention paid to the inclusion of plants grown in gardens, or under glass in cataloguing the botany of Paris.[57] The inventory of nature becomes political through the comparison of the 'citizenship' of plants with the right to membership of the Parisian bourgeoisie. The manuscript's presentation of plants in an alphabetical table follows two principles: arrangement by plant names and sizes (from over 3 inches to 4 feet), followed by plant habitats ('rivers and pools', 'streams and springs', 'banks', 'stagnant waters', 'marshes', 'meadows', 'fields and fallow land', 'gardens and vineyards', 'wasteland and thoroughfares', 'hedges and shrubbery, forest'), and flowering seasons. In addition to habitat, information is given about the place of collection (Bois de Boulogne, Meudon). However, these 'statistics' do not involve any form of quantification, as Duchesne says in his note to the reader:

> In considering these disadvantages [of other methods], we have come to see that the Botany of Paris should not be treated as a universal Botany: it was apparent that, in guiding botanical excursions within a confined area, the most obvious qualities of the plants should always be given pride of place, even though they may not be the most fundamental, and that noting the place where the plant grows, the time of flowering, usual height and various other aspects of external form and colour, would better facilitate finding the name of the plant in question than by using more methodically ranked characteristics, to find the name of the plant in question and thence the

55. Here we follow Harry S. Paris, 'La biographie de Antoine Nicolas Duchesne', in H. S. Paris, *The Drawings of Antoine Nicolas Duchesne for his 'Natural history of the gourds'* (Paris, 2007), p.21-22. On the Trianon garden, see BCMNHN, MS 1169, Bernard de Jussieu, 'Catalogus generum juxta ordinem in horto Trinanonesi institutum' (1759).

56. Antoine Nicolas Duchesne, *Manuel de botanique, contenant les propriétés des plantes utiles pour la nourriture, d'usage en médecine, employées dans les arts d'ornement pour les jardins, et que l'on trouve à la campagne aux environs de Paris* (Paris, Didot le jeune, 1764).

57. BCMNHN, MS 1297, 'Statistique végétale des environs de Paris, par Duchesne' (1771), quotation f.2.

> more detailed description provided with it in order to definitively identify the genus, species, race and even variety to which it belongs. The public will judge whether the form that the author has chosen for his tables is more or less successful in achieving the goal he has set himself. A more detailed explanation of this arrangement and the way to use the book in the countryside can be found at the bottom of the pages, where the tables leave a space followed by a glossary of terms used in the Descriptions and a précis of the lives and works of the principal botanists.[58]

This extract implies that writing up the botanies of Paris appears to be associated more with the private practice of botanical observation than with the administration of nature.

Localising the global, relocating Linnaean botany in Paris

The Swedish natural philosopher did not only justify the naturalist practices of a community of scientific amateurs who both encouraged the local practising of field-science and succeeded in commodifying Parisian nature, he also defended a science 'of the cabinet' based on the accumulation of data and collections on a worldwide scale. Beyond the polemics with Buffon, we can see to what extent Linnaean sciences also shared practices with the French naturalists by considering the relevance of the global gathering of plants and natural knowledge.[59]

During his stay in Holland, Linnaeus met the apothecary Albertus Seba (1665-1736) and the botanist Herman Boerhaave (1668-1738). These men put him in touch with the influential George Clifford III (1685-1760), director of the Dutch India Company and merchant banker. Clifford was part of a large network of Dutch merchants who were also collectors of plants from around the world. From the seventeenth

58. BCMNHN, MS 1297, 'Avertissement', separate sheet, recto-verso (translation by the author): 'En réfléchissant sur ces inconvénients [des autres méthodes], on a cru voir que la Botanique parisienne ne devoit pas être traité comme la Botanique universelle : il a semblé que pour guider dans une herborisation bornée, les qualités les plus apparantes des plantes devoient toujours prévaloir quoiqu'elles ne soient pas les essentielles : et qu'en précisant le lieu de la naissance, le tems de la fleur, la hauteur la plus ordinaire, et diverses autres nottes des formes et des couleurs extérieurs, on pouvoit parvenir plus promptement que par les caractères les plus méthodiquement à mettre le doigt sur le nom de la plante en question et de là pour la description plus approfondie qui s'y trouve de se confirmer sans retour le genre, l'espèce, la race et même la variété à laquelle on doit la rapporter. Le public jugera si la forme que l'auteur a choisie pour ses tableaux remplit à peu près le but qu'il s'est proposé. L'explication plus détaillée de cette disposition ainsi que la manière de se servir du livre à la campagne se trouve au bas des pages où les tableaux laissent du vide suivie d'un vocabulaire des termes employés dans les Descriptions et d'un précis de la vie et des ouvrages des principaux botanistes.'
59. Londa Schiebinger has made a similar argument about armchair botanists, see L. Schiebinger, *Plants and empire: colonial bioprospecting in the Atlantic world* (Cambridge, MA, 2004).

century onwards, Amsterdam had been considered an entrepôt of *naturalia*[60] but came increasingly to be challenged for this position by London and Paris. In London, the roles played by Hans Sloane, the Chelsea Physick Garden, and the Royal Society were decisive – these being so closely connected to imperial and commercial networks.[61] Where Pascal Duris insisted that the relationship between Paris and the provinces impacted on how Linnaeus was received in France, several recent works on the history of science have paid more attention to the role played by Paris as a global capital city.[62]

In order to understand this cycle of accumulation historians have moved away from an institutional framework which stressed the 'academisation' of natural knowledge, or the role played by Court society and provincial institutions, toward a more global perspective.[63] The critique proposed by Lissa Roberts against the centre and periphery model, shaped in large by Latour's concept of centres of calculation, helped to deepen the understanding of many different uses and practices of accumulation at work in the capital city.[64] The image of Paris has fundamentally altered over the course of the last decade and the focus is now instead on the description of circuits and long-distance networking in the making of natural knowledge in Paris. This approach underlined the role played by new equipment and facilities within the city which give rise to a new economy of knowledge. The inventory of equipment and facilities (such as libraries and laboratories), the institutions of knowledge, sites of teaching, spaces of sociability, the structures to attract foreign scholars, the prosopography of scholars – these all demonstrate clearly the redistribution of scholarship and knowledge in Enlightenment Paris, and moreover in relation to the new 'colonial machine' and

60. Harold Cook, *Matters of exchange: commerce, medicine and science in the Dutch Golden Age* (New Haven, CT, 2007).
61. James Delbourgo, 'Sir Hans Sloane's milk chocolate and the whole history of the cacao', *Social text* 106 29:1 (2011), p.71-101.
62. Sven Dierig, Jens Lachmund and Andrew Mendelsohn (ed.), *Science and the city*, *Osiris* 18 (Chicago, IL, 2003).
63. Alice Stroup, *A Company of scientists: botany, patronage and community at the seventeenth-century Parisian Royal Academy of Sciences* (Berkeley, CA, 1990). This monographic perspective was also shared by Yves Laissus on the Jardin du Roi in Yves Laissus and Jean Torlais, *Le Jardin du Roi et le Collège royal dans l'enseignement des sciences au XVIIIe siècle* (Paris, 1986). Pierre-Yves Lacour, 'L'administration des choses naturelles: le Muséum et ses collections autour de 1800', in *Les Lieux de savoir*, vol.2, *Les Mains de l'intellect*, ed. Christian Jacob (Paris, 2011), p.246-62. See also Elise S. Lipkowitz, 'Seized natural-history collections and the redefinition of scientific cosmopolitanism in the era of the French Revolution', *British journal for the history of science* 47:1 (March 2014), p.15-41.
64. On cycles of accumulation see Lissa Roberts (ed.), *Centres and cycles of accumulation*, and Roberts, 'Accumulation and management in a global perspective: an introduction', *History of science* 52:3 (2014), p.227-46.

bureaucracy, as James McClellan and François Regourd have recently outlined.[65]

Making the new nature visible, and exploiting natural resources, was prioritised in French foreign and domestic policy. The work of leading men of science played an important role. States mobilised citizens, missionaries, military officers, doctors and loyal indigenous peoples. By linking the Royal Navy, the Academy of Sciences, the Jardin du Roi, the Society of Agriculture, or the Royal Academy of Medicine, the 'colonial machine' thereby revealed a network of institutions, and a number of problems faced by European colonialism. Joseph Pitton de Tournefort, and his voyage in the Levant,[66] Michel Adanson's natural history of Senegal,[67] the voyages of Philibert Commerson,[68] and that of Pierre Poivre or Bernardin de Saint-Pierre in the Indian Ocean and Ile de Bourbon,[69] Pierre Sonnerat in South India, in New-Guinea, Malaysia and the Philippines[70] just to quote a few, display the relationship between travel, state-building and natural history. From the very beginning, this change of scales reflected the many different strategies of the royal monarchy (including intelligence-gathering, political economy, colonial enterprise). At the Jardin du Roi and within the network of the gardener André Thouin, to be useful these natural observations had to be reduced to standardised techniques, classified and carefully stored (as Etienne-Louis Turgot suggested in his *Instructions*).[71] This was the raw material necessary for the calculated decisions that made action at a distance possible. Information from around France and the world was relayed back, classified and stored in accessible filing cabinets housed in govern-

65. James E. McClellan III and François Regourd, *The Colonial machine: French science and overseas expansion in the old regime* (Turnhout, 2011). See also Anja Bandau, Marcel Dorigny and Rebekka von Mallinckrodt (ed.), *Les Mondes coloniaux à Paris au XVIIIe siècle: circulation et enchevêtrement des savoirs* (Paris, 2010). See also the research of Dorit Brixius, 'French empire on the ground: plants, peoples, and knowledge in the service of eighteenth-century Isle de France', doctoral dissertation, European University Institute, 2017.
66. [Joseph] Pitton de Tournefort, *Relation d'un voyage du Levant fait par ordre du Roy* (Paris, Imprimerie royale, 1717).
67. Michel Adanson, *Histoire naturelle du Sénégal: coquillages* (Paris, Bauche, 1757); Xavier Carteret, 'Michel Adanson au Sénégal (1749-1754): un grand voyage naturaliste et anthropologique du siècle des Lumières', *Revue d'histoire des sciences* 65:1 (2012), p.5-25.
68. BCMNHN, MS 301, Philibert Commerson, 'Mémoires pour servir à l'histoire du Voyage fait Autour Du Monde par les Vaisseaux Du Roi La Boudeuse et l'Estoile Pendant les années 1766-1768'.
69. Pierre Poivre, *Voyages d'un philosophe, ou Observations sur les mœurs et les arts des peuples de l'Afrique, de l'Asie et de l'Amérique* (Lyon, J. de Ville and L. Rosset, 1769).
70. Pierre Sonnerat, *An Account of a voyage to the Spice-Islands and New Guinea* (London, W. Green, 1781).
71. Emma Spary, *Utopia's garden: French natural history from the old regime to Revolution* (Chicago, IL, 2000); Emma Spary, 'The nature of Enlightenment', in *The Sciences in Enlightenment Europe*, ed. William Clark, Jan Golinski and Simon Schaffer (Chicago, IL, 1999), p.272-306.

ment departmental offices.[72] Items were put on display in cabinets, and entombed in the memoirs of various scientific societies, which subsequently gave way to imperial and metropolitan storehouses. The globalisation of knowledge was not a simple process of universalisation, but was 'instrumentalised' by the royal power. They not only mobilised institutions, a web of royal botanists, and created a network of correspondents and exchange, but also provided new tools and new technologies for gathering botanical information. Mapping, drawing, classifying, listing and note-taking were enriched by colonial practices of description.[73]

However, the 'colonial machine' was neither very efficient nor mechanistic, and left space for the agency of local intermediaries or go-betweens who were barely recognised by Parisian scholars, like the naturalist Nicolas L'Empereur in Madras, whose herbarium returned to Paris, but remained unpublished.[74] The Jardin du Roi itself was not only a place for centralisation or accumulation, but also an institution for redistribution at the European level. André Thouin left several lists and letters, which testified to his exchange of grains. In 1779, sixty-four people received eighty-one grains or plants from across Europe.[75] Paris, by virtue of this circulation, became a centre for the redistribution of plants for Europe as a whole, even if the French Revolution would reduce the number and the range of distribution of these donors, offering us a picture of a more centralised and nationalised network.[76]

Moreover, bioprospecting extended French expertise beyond the French Empire proper and built a form of invisible empire. Similar to the web of scientists and objects, instruments which Robert Boyle called the 'Invisible college' in 1646 and 1647 with reference to what became the Royal Society before its actual creation, I use the same metaphor to signal the rise of a virtual globalisation beyond the walls of the Jardin du Roi in Paris.[77] This invisible empire is based on the creation of a long-

72. Marie-Noëlle Bourguet, 'La collecte du monde: voyage et histoire naturelle (fin XVIIe siècle-début XIXe siècle)', in *Le Muséum au premier siècle de son histoire*, ed. Claude Blanckaert *et al.* (Paris, 1997), p.163-96. Denis Lamy, 'Le dessin botanique dans la transmission des connaissances', in *Passions botaniques: naturalistes voyageurs au temps des grandes découvertes* (La Roche Jagu, 2008), p.139-55.

73. James Delbourgo, 'Listing people', *Isis* 103:4 (2012), p.735-42; Marie-Noëlle Bourguet, 'A portable world: the notebooks of European travellers (eighteenth to nineteenth centuries)', *Intellectual history review* 20:3 (2010), p.377-400.

74. Kapil Raj, 'Surgeons, fakirs, merchants and craftspeople: making L'Empereur's *Jardin* in early modern South Asia', in *Colonial botany: science, commerce and politics in the early modern world*, ed. Londa Schiebinger and Claudia Swan (Philadelphia, PA, 2005), p.252-69.

75. BCMNHN, MS 313: 'Etats des distributions végétales du Muséum' (1790-1823).

76. See more broadly Pierre-Yves Lacour, *La République naturaliste.*

77. See the classical study by Diana Crane, *Invisible colleges: diffusion of knowledge in scientific communities* (Chicago, IL, 1972).

distance network of agents and field-botanists able to identify natural rarities. I would like to contrast the emergence of an invisible empire built by the circulation and gathering of exotic nature transformed into natural resources on a global scale by people such as scholars, adventurers, collectors, Jesuits and botanists who were employed or used by the French India Company, using the ideal of the universalism of the colonial machine which is commonly used by eighteenth-century philosophers to designate the Enlightenment project and led to an unprecedented process of capitalising botany.[78] This invisible empire also became a virtual and potential empire, not only comprising the usual lists of books, reports, instructions and provisional catalogues, but also techniques of storage which represented both the archive of this long-distance action and the instruments of the utopia of a universal or infinite natural world. From the French Revolution onwards, royal institutions were primarily critically viewed as all-encompassing institutions which set up a very complex machinery of regulations and practices to control the readers and information. But the utopian idea of creating an infinite King's garden was also from the very beginning counter-balanced by scepticism. Confronted by the proliferation of books and *naturalia*, and the circulation of information, Enlightenment utopia underlined the predicament of a new order of knowledge characterised by the paradigm of accumulation. These criticisms distanced themselves from a new economy of knowledge based on predatory practices and called into question the epistemological foundation of the Enlightenment, making it important methodologically to distinguish projects from efficient achievements, to include limits and failures.

In that context, Michel Adanson, considered to be one of Linnaeus's main opponents who stayed in Senegal between 1749 and 1754 with the French Indian Company, was not persuaded of the nonsensical artificiality of the system used to describe tropical flora. In a letter to Bernard de Jussieu, he expressed his admiration for Linnaeus and sent information about the Baobab: 'I leave you free to communicate the character of the Baobab to Mr Linnaeus. I think this will not be disadvantageous or rebound badly on me. I further ask you to kindly ensure that this learned man learns of the infinite esteem I have for him and his works.'[79]

78. Simone Balayé, 'La bibliothèque du Roi, première bibliothèque du monde (1664-1789)', in *Histoire des bibliothèques françaises*, vol.2, *Les Bibliothèques sous l'ancien régime, 1530-1789*, ed. Claude Jolly (Paris, 1988), p.209-34.

79. Letter, 20 February 1752, quoted in Xavier Carteret, 'Michel Adanson au Sénégal', p.15 (translation by the author): 'Je vous laisse la liberté de communiquer à M. Linnaeus le caractère du Baobab; je pense que cela ne peut m'être désavantageux ni tirer à aucune

Later, Adanson's classification could be seen as response to the anxiety raised by the influx of information and the proliferation of new classifications as well as the tensions between different regimes of collecting and describing nature. As for Linnaeus's endeavours, it is important to acknowledge the role of such far-flung imperial laboratories in the production and legitimation of new scientific practices in strengthening the authority of French classifications as well.[80] Thanks to his travels and the collections he gathered, Adanson was able to produce his *Famille des plantes* in 1763. In this context of growing connections between institutions, Buffon possessed important rhetorical skills, and employed the familiar quarrelsome polemics.[81] In the history of natural sciences, literary questions have long been perceived as playing a minor part, mainly considered with regard to their role in the dissemination and vernacularisation of a given body of knowledge. Over the past two decades, thanks to studies on the rhetoric of science that followed the seminal work of Jeff Hoveland and Thierry Hoquet, a different direction has been pursued: the role of writing, rhetorical figures, and the notion of genre have become central to reflections on the extent to which natural knowledge constituted a creative process. The constitution of Buffonian authority was not only a matter of arguments and epistemological claims, it was also dependent on his network of institutions and royal botanists, and of his powerful rhetorical skills, defending French eloquence against the crude Linnaean Latin.[82] Restaging the quarrel between Buffon, his disciples and Linnaeus in this context is also appropriate when considering the obstacles to recognition that the Swedish naturalist faced in Paris, despite his growing favourable reception by amateur botanists on Linnaeus's nomenclature. The strong local connection between Buffonian theories, Court society (both in Versailles and in the Parisian salons), and the state certainly played a role in the controversy by marginalising Linnaeus's vision.[83] As Pascal Duris has

mauvaise conséquence: je vous prie même de vouloir bien assurer ce savant de l'estime infinie que j'ai pour sa personne et ses ouvrages.'

80. On the relationship between natural history and visual culture in Spanish America see Daniela Bleichmar, *Visible empire: botanical expeditions and visual culture in the Hispanic Enlightenment* (Chicago, IL, 2012). On material culture see Daniela Bleichmar and Peter C. Mancall (ed.), *Collecting across cultures: material exchanges in the early modern Atlantic world* (Philadelphia, PA, 2011).

81. Jeff Loveland, *Rhetoric and natural history: Buffon in polemical and literary context*, *SVEC* 2001:03 (Oxford, 2001); Thierry Hoquet, *Buffon: histoire naturelle et philosophie* (Paris, 2005).

82. Duris, *Linné et la France*, p.34-37.

83. On the recognition of Buffon in the salons see Antoine Lilti, *The World of the salons: sociability and worldliness in eighteenth-century Paris*, translated by Lydia G. Cochrane (New York, 2015); and Antoine Lilti, '"Une autre paire de manches": Buffon et la sociabilité des Lumières', *Lumières* 8 (2006), p.235-39.

noticed, the diffusion of Linnaean ideas in France was largely implemented by foreign naturalists like James Edward Smith who visited Paris in 1786 and met Louis-Guillaume Le Monnier, Louis XV's doctor, at Versailles. He also met André Thouin and the traveller Joseph Dombey on his return from South America. Le Monnier was in contact with Linnaeus about duke of Noailles's garden at Saint-Germain-en-Laye.[84] The same network also included Claude Richard, another correspondent of Linnaeus, head of the Trianon Gardens. Smith developed his Linnaean conversations with Thouin and even with Jussieu. All these anecdotes underlined the role played by a network of courtly naturalists which developed in parallel with the Jardin du Roi, relying on aristocratic or royal gardens. It draws attention to a more cosmopolitan culture, which was influenced by the British naturalists and the Grand Tour. Intellectual history has only indirectly acknowledged the importance of such networks linked to court society in the circulation of Linnaean practices in Paris.[85] The distinct bodies of natural knowledge produced at the Jardin du Roi and at Versailles lead us to question whether there were two methods of acclimating far-flung knowledge in Paris. Moreover, behind the cliché that celebrates the conflict between Parisian naturalists and Linnaeus, there were not only rivalries for the monopoly of the globalisation of knowledge but also difficulties when seeking to collectively impose a new image of nature, and these episodes revealed also a strong competition between the various empires of sciences.

Branding Linnaean classification or practices?

The debate between classification and nomenclature as highlighted by Geoffroy in his *Abridged history*, revealed a tension between empirical works and theoretical genres. Within this framework we should also be particularly attentive to scientific controversies, to the knotty polemics that arose from knowledge, as well as the ways in which knowledge and the links between the world of learning and urban authorities were evaluated. The Buffon dispute probably corresponded to this higher standard. By taking into account both the interplay of scales and metropolitan context I wanted to re-stage Linnaeus in Paris before the creation of the Linnaean society in 1787 and to locate where Linnaean sciences mattered in two ways.

First, by moving from an epistemological history to a history of scientific naturalist practices, I suggested that Linnaean sciences had

84. Duris, *Linné et la France*, p.61. See also Tom Kennett, *The Lord Treasurer of botany: Sir James Edward Smith and the Linnaean collections* (London, 2016).

85. Duris, *Linné et la France*, p.65.

made a multi-faceted impact on Parisian naturalists. The conflictual reception of Linnaeus mirrored the singularity of the recognition of Paris as a centre of natural knowledge in the mid-eighteenth century, and the circulation of his name helped to map different networks operating within the city. The antagonistic representation was the result of a long, intense and uncertain process, sometimes highly polemical, sometimes contested and fragile. A new set of research reflects upon Linnaean sciences not only in terms of concepts but in terms of knowledge practices which tended to encompass not only professionals in the natural sciences, but also amateurs such as gardeners, horticulturalists, pharmacists, engineers and physicians. In the field of the history of science, this question of Linnaean sciences should be treated as part of a broader interrogation into the nature of the city.

Second, I argued that the definition of Paris as a problematic Linnaean capital city should not be considered to be obvious – as an automatic reflection of the metropolitanisation or globalisation of the Parisian elites. If metropolises and capitals in general are condemned to be global cities with universal knowledge – forgetting their local identity and blurring their local attachment – then using the Linnaean sciences can be seen as a means of relocating Parisian nature by revealing in the natural history of Paris a way of commodifying and consuming natural knowledge. Linnaean science, through its attention to political economy and globalisation, could become a new repertoire for Parisian naturalists. The Parisian case illustrated the possibility of being universal by defending localism, and to assess what the anthropologist Arjun Appadurai called the 'global production of locality'.[86] More recently, historians of science have focused on the emergence of Linnaean sciences as such, and their relationship to various settings (mainly empires and cities). My starting point was to break both with the teleological approach (that is, the advent of Linnaean sciences in Paris) and with the intellectual genealogy by re-contextualising Linnaean authority in a different perspective.

Nonetheless, far from opposing local and global recognition, the aim is to rearticulate the social and cultural effects of a widespread social acceptance of Linnaean natural philosophy with the metropolitan definition and representation of naturalist knowledge, even if this was largely based on a conflicted relationship. By officially criticising Linnaean classifications at the close of the *ancien régime*, it seems clear that Paris had affirmed itself as the tribunal of the natural sciences by its capacity to declare whether or not something was scientific. The Buffonian camp embodied a local way of producing classification

86. Arjun Appadurai, *Modernity at large: cultural dimensions of globalisation* (Delhi, 1997).

following Tournefort, which amounted to a national claim in the European competition between capital cities. For a long period of time the use of 'Linnaean' as a qualifier indicated highly controversial knowledge. A wide range of discourses and polemics situated Paris as a capital of natural knowledge by offering representations of Parisian natural sciences structured around sites of knowledge, including great libraries, cabinets of curiosities, as well as around aristocratic sociability. After the death of Buffon and Linnaeus, the conversion of the Linnaean sciences into symbolic and metropolitan representations useful for Parisian scholars did not result in a simple accumulation of references, exchanges, sociability, local interaction or facilities. We must therefore reinterpret the notion of Linnaean *hauts lieux*, embodied in the late 1780s by the multiplication of Linnaean societies in France (notably the Société linnéenne de Paris in 1787, created by André Thouin),[87] through their dynamic and conflictual dimensions, and account for the specific actions of certain networks in Paris. These both participated in the territorialisation of naturalist activities in Paris and proclaimed their membership within a universal 'Republic of Linnaean Sciences' established through the correspondences between Linnaean societies and the global topics discussed above.[88]

87. After Pascal Duris's thesis, the most important work is Jean-Luc Chappey, *Des naturalistes en révolution: les procès-verbaux de la Société d'histoire naturelle de Paris (1790-1798)* (Paris, 2009).
88. In Paris, the *Mémoires* left traces of a great range of interests in colonial or far-flung territories and most of the *mémoires* start with Linneaus: *Mémoires de la Société d'histoire naturelle de Paris* 1 (1799), p.1, 3, 4, 12, 14, 29, 34, 60, 64, 65, 111, 120, 128, 142. About the conchyliology for instance, they wrote: 'Cet illustre naturaliste, qui, dans tous ses ouvrages, a laissé l'empreinte d'un génie supérieur, et qui a institute l'art utile d'exposer avec concision et avec une précision scrupuleuse les caractères distinctifs des objets en histoire naturelle, n'a pas été moins utile à la conchyliologie dans le peu qu'il a fait pour elle, qu'il l'a été dans les autres parties de la zoologie et dans la botanique. On peut dire qu'il a établi les vrais principes qu'on doit suivre dans l'étude et la détermination des coquilles, et qu'il a posé les bases de cette intéressante partie de nos connoissances' (64).

5. On the use and abuse of natural history: Linnaean science in Kant's Königsberg

JONAS GERLINGS

The science of man and the man of science

In January 1738, Carl von Linnaeus wrote to his 'distinguished friend' Albrecht von Haller: 'Bartsch has been about a month at Surinam. He is indebted to me for his appointment, as well as for what he knows of botany; and he is one of the most diligent of men. I expect much from him. He is by birth a German.'[1] Like several others, Johann Bartsch (1709-1738) had come from Königsberg to study with Herman Boerhaave in Leiden. Here he had befriended Linnaeus who had been offered a position as physician for the Dutch West India Company in Surinam. Linnaeus had declined but recommended Bartsch in his place. In return, Linnaeus expected knowledge, accurate descriptions and samples of the local flora in Surinam. Sadly, Bartsch died soon after arriving in Surinam and with him so did the Königsberg contribution to the Linnaean exploration of the foreign parts of the world.

Despite his dire fate, Bartsch had already sparked an interest in Linnaeus for Bartsch's homeland. In 1737, Linnaeus had taken the liberty of writing to Franciscus Ernst Brückmann in Wolfenbüttel enquiring about scholars in Prussia engaged with natural history. Brückmann responded, listing Jacob Theodor Klein, a diplomat from Königsberg residing in Danzig, and Königsberg local Karl Heinrich Rappolt among the top.[2] Klein and Rappolt were part of a local tradition of natural historians in and around Königsberg. Their predecessor Johann Gottsched (1668-1704) had already completed a catalogue of *Prussian plants*,[3] and had intended to expand it with sections on the animal kingdom and mineralogy to make a complete natural history of Prussia.

Unless otherwise indicated, the use of bold and italics in quotations reflects the original text. Unattributed translations are the work of the author.

1. Carl Linnaeus, *A Selection of the correspondence of Linnaeus and other naturalists*, ed. James Edward Smith, vol.2 (London, 1821), p.315.
2. Franciscus Ernst Brückmann to Carl Linnaeus, 18 September 1737, *The Linnaean correspondence*, http://Linnaeus.c18.net, letter L0199 (consulted 28 April 2016).
3. Johann Gottsched, *Flora Prussica, sive plantae in regno Prussiae sponte nascentes* (Königsberg, sumptibus Typographiae Georgianae, 1703). Gottsched's work was largely based on Johann Lösel's *Plantae in Borussia sponte nascentes* (Königsberg, typis Paschalii Mensenii, 1654).

Klein, who excelled in the study of the animal kingdom, and Rappolt, who excelled in mineralogy, found themselves in the middle of a period of transformation which, over the long eighteenth century, would literally translate the nomenclature of this local tradition into Linnaeus's global nomenclature. The task was finally completed in 1818 by Karl Gottfried Hagen with his publication of *Prussian plants*.[4]

This long history of the institutionalisation of Linnaean science in Königsberg has, however, been overshadowed by one name: Immanuel Kant. Ernst Cassirer characterised Kant as 'the *logician* of Linnaeus' descriptive science', stating that Kant in the second part of his *Critique of the power of judgment* (1790) had 'tracked down the hidden problem contained in Linnaeus' work.'[5] In contrast, Michel Foucault understood Kant's critical philosophy as something that would 'blur the old space of natural history'.[6] These studies, focusing on the discourses of natural history, have created a certain chronology, temporally separating Linnaeus and Kant, as captured within the title of James Larson's book, namely *from Linnaeus to Kant*.[7] In later studies, however, the chronology has been confronted with a geographical dimension: from a post-colonial perspective this scientific ordering of things has been analysed as a world order with a European centre and an imperialist agenda.[8] As one critic has noted, 'Linnaeus's table of the races' framed Kant's 'transcendental philosophy of race'.[9] The Enlightenment debates on the nature of man was one of the most significant and complex

4. Karl Gottfried Hagen, *Preußens Pflanzen* (Königsberg, 1818). Latin version *Chloris Borussica* (Königsberg, 1819).
5. Ernst Cassirer, *The Problem of knowledge: philosophy, science and history since Hegel* (New Haven, CT, 1978), p.127.
6. Michel Foucault, *The Order of things: an archaeology of the human sciences* (New York, 2005), p.176.
7. James L. Larson, *Interpreting nature: the science of living form from Linnaeus to Kant* (Baltimore, MD, 1994). See also John H. Zammito, *The Genesis of Kant's critique of judgment* (Chicago, IL, 1992), and Vesa Oitinen, 'Linné zwischen Wolff und Kant: zu einigen Kantischen Motiven in Linnés biologischer Klassifikation', in *Kant's 'Philosophie der Natur': ihre Entwicklung im Opus Postunum un Ihre Wirkung*, ed. Ernst-Otto Jan Onnasch (Berlin, 2009), p.51-78.
8. See for example Mary Louise Pratt, *Imperial eyes: travel writing and transculturation* (New York, 1992), and Londa Schiebinger, *Plants and empire: colonial bioprospecting in the Atlantic world* (Cambridge, MA, 2004).
9. Emanuel C. Eze, *Achieving our humanity: the idea of the post-racial future* (New York, 2013), p.103. For other critics see Robert Bernasconi, 'Who invented the concept of race? Kant's role in the Enlightenment construction of race', in *Race*, ed. Robert Bernasconi (Malden, 2001), p.11-36; Ian Hunter, 'Kant's regional cosmopolitanism', *Journal of the history of international law/Revue d'histoire du droit international* 12:2 (2001), p.165-88; and James Tully, 'The Kantian idea of Europe: cosmopolitan and critical perspectives', in *Public philosophy in a new key*, vol.2, ed. James Tully (Cambridge, 2008), p.15-42. Defenders of Kant are represented by Sankar Muthu, *Enlightenment against empire* (Princeton, NJ, 2003); and Pauline Kleingeld, *Kant and cosmopolitanism: the philosophical ideal of world citizenship* (Cambridge, 2012). See also Katrin Flikschuh and Lea Ypi (ed.), *Kant and colonialism* (Oxford, 2014).

subjects in eighteenth-century intellectual history. This is primarily due to two opposing trends: firstly, the apparent interdisciplinary character of the focus, which made the science of man a unifying endeavour among a variety of disciplines; and secondly, the vast empirical data collected in these studies, which demonstrated a broad variety of perspectives, challenging the very notion of a unified human race.

It is an inherent part of this post-colonial criticism that the dissemination of European values and ideas goes hand in hand with the European expansion and exploration of the world. However, this Eurocentric critique has also slowly become more nuanced. The Baltic Sea region contained local centres that in themselves were part of the European periphery and, as Sverker Sörlin has pointed out, the nations in this region 'were at the same time colonial, with their own empires, and colonised in the sense that within their own boundaries, they had large hinterlands with ethnic groups and minority populations'.[10] What needs to be taken into account is the local particularity of Königsberg. Such an analysis cries out for the present intellectual map of the Baltic Sea region to be redrawn. This intellectual map outlines a divide between Linnaeus's Uppsala and Kant's Königsberg, each on opposite shores of the Baltic Sea. For, while the Baltic Sea Region in general has been characterised as an Enlightenment culture concerned with the utility of science,[11] Königsberg has been singled out as a rival 'metaphysical enlightenment'.[12] This characterisation, however, largely hinges on the historical reconstruction of Kant's intellectual environment within the confined disciplinary space of philosophy, retrospectively presupposing that the primary context for understanding Kant's work is that of metaphysics. Kant's introduction of natural history into the *Critique of the power of judgment* is described by John Zammito as an *intrusion* to Kant's actual interest: *aesthetics*.[13] New research analysing the local intellectual conditions at Königsberg, however, seems not only to support that Königsberg had its own tradition of natural history, which ultimately became Linnaean, but also that Kant was part of this tradition.[14] The

10. Sverker Sörlin, 'Ordering the world for Europe: science as intelligence and information as seen from the northern periphery', in *Nature and empire: science and the colonial enterprise*, *Osiris* 15 (2000) p.59-69 (53).
11. See Lisbet Koerner, 'Daedalus hyperboreus: Baltic natural history and mineralogy in the Enlightenment', in *The Sciences in Enlightened Europe*, ed. William Clark, Jan Golinski and Simon Schaffer (Chicago, IL, 1999) p.389-422 (389).
12. See Ian Hunter, *Rival Enlightenments: civil and metaphysical philosophy in early modern Germany* (Cambridge, 2001).
13. John H. Zammito, *The Genesis of Kant's critique of judgment* (Chicago, IL, 1992), p.2.
14. See Reinhard Brandt, Werner Euler and Werner Stark, *Studien zur Entwicklung preussischer Universitäten* (Wiesbaden, 1999); Michael Oberhausen and Riccardo Pozzo, *Vorlesungsverzeichnisse der Universität Königsberg (1720-1804)* (Stuttgart, 1999); Michael Oberhausen

untimely coexistence of Linnaean science in Kant's Königsberg suggests that Kant's relation to Linnaeus should not be perceived as a tale of two cities on opposite shores of the Baltic coast but rather as an engaged reflection of the globalisation of Königsberg. The aim of this chapter is to contrast the intellectual history of contestation with the long history of institutionalisation of Linnaean science in Königsberg. This will be done by juxtaposing the intellectual debates on the science of man with the two distinct epistemologies personified by Linnaeus and Kant. By following this simultaneous development, the chapter will demonstrate how these two notable 'men of science' set out with similar research interests but each came to personify different disciplinary personas and research practices that could be disciplined and professionalised as the naturalist and the philosopher.

The cultivation of natural history in eastern Prussia

In January 1725 Professor Christian Gabriel Fischer (1686-1751) presented a project to the king of Prussia, Friedrich Wilhelm I, on how the 'poor conditions of the Academy, the medical- and school-system could be emphatically improved'.[15] The project envisioned a possible future for the Albertina, by transforming it from a backward institution into a progressive research university.[16] Previously, Friedrich Wilhelm I had issued an inspection order and requested a lecture catalogue[17] from the president of the Albertina, the pietist Heinrich Lysius. Instead of submitting course descriptions, Fischer responded with an open criticism of the current teaching at the Albertina. In return, Friedrich Wilhelm I put Fischer in charge of a small committee tasked with improving the teaching syllabus at the Albertina.

Fischer's critique highlighted two general problems to be improved: firstly, the material and institutional basis of the university; and secondly, the lack of an overall structure of the sciences. Fischer was professor of physics at the Albertina. He taught natural history and natural philosophy, subjects that required a practical approach through observations

---

and Riccardo Pozzo, 'The place of science in Kant's university', *History of science* 40:3 (2002), 353-68; Werner Stark, 'Naturforschung in Königsberg: ein kritischer Rückblick aus den Präliminarien einer Untersuchung über die Entstehungsbedingungen von Kant's *Vorlesung über Physische Geographie*', *Estudos Kantianos* 2:2 (2014), p.29-60.

15. Albert Predeek, 'Ein verschollener Reorganisationsplan für die Universität Königsberg aus dem Jahre 1725', *Altpreußische Forschungen* 4:2 (1927), p.65-107 (70): 'Verworrene Zustand der Akademie, des Medizinal- und Schulwesens nachdrücklich zu bessern sei'.
16. On the origins of the research university see William Clark, *Academic charisma and the origins of the research university* (Chicago, IL, 2006).
17. The entire lecture catalogue for Königsberg University (1720-1800) has recently been published. See Oberhausen and Pozzo, *Vorlesungsverzeichnisse*.

and experiments. As the first professor of physics in Königsberg he had introduced excursions as part of his lectures on natural history, but excursions were by no means sufficient. Fischer complained that the Albertina had no anatomical theatre, botanical garden, observatory, chemical laboratory, apothecary, cabinet of curiosities or natural history collection. Hence, Fischer asserted, his teaching had become restricted to theory and dogmatic disputes. Drawing on the philosophy of Christian Wolff, Fischer believed that metaphysical disputes should be based on concepts abstracted from observations and experiments. Fischer therefore introduced an economic plan to fund the scientific institutions lacking at Königsberg. With regard to the second problem Fischer believed that Wolffian metaphysics had the advantage of providing a hierarchical order to the sciences. In Fischer's eyes the syllabus at the Albertina did not offer a strict distinction between different disciplines and least of all a progression within the courses, making it difficult to teach at an advanced level. The existing disciplinary distinctions rested on the crumbling order of Aristotelian metaphysics. However, the advantage of the dire state in which Fischer found the university to be was that it was ripe for reforms. Fischer formulated a plan for reforming the university, which he presented to Friedrich Wilhelm I. Its overall aim was to restructure the university to maximise its utility to the state, giving the king increasing power at the Albertina – evidently it appealed to Friedrich Wilhelm I. Fischer not only presented a plan which provided a progressive order and a useful purpose to the teaching, but also a plan to fund new buildings and equipment, which would ultimately save money.

Fischer's plan had the consequence that it would marginalise the increasingly influential pietist group at the Albertina. Fischer had openly criticised and even ridiculed the pietists, but they responded by using their connections in Halle to eventually get him banished from Prussia. Frederick Wilhelm I had already banished Christian Wolff from Prussia in 1723 and, when notified of Fischer's Wolffian sympathies, he gave Fischer twenty-four hours to leave Prussia, who subsequently fled to Danzig (present day Gdansk), a free city under protection of the Polish-Lithuanian Commonwealth. This meant that Fischer's talent for natural history was re-located to the neighbouring city Danzig, where he assisted the famous Jacob Theodor Klein (1685-1759) in building one of the most extensive private natural history collections of the eighteenth century, a collection which Fischer believed surpassed the Tournefortian Museum in Paris.[18] In the meantime, the Albertina scholars working on natural history who were associated with Fischer became marginalised. Fischer's

18. Christian Gabriel Fischer, 'Ob sich nach herrn Linnäus System ein Naturalienkabinet einrichten lasse?', *Neue Gesellschaftliche Erzählungen für die Liebhaber der Naturlehre, der Haushaltungswissenschaft der Arzneykunst und der Sitten* (Leipzig, Holle, 1758), p.163-66 (166).

perhaps most important student, Karl Heinrich Rappolt (1702-1753), was rejected for the chair of natural philosophy in favour of the pietist candidate Johann Gottfried Teske. Disappointed, Rappolt journeyed through Danzig to London, his friend Johann Christoph Bohl (1703-1785) went to Leiden, and others followed suit. Through the late 1720s to the early 1730s the constellation of scholars linked with Fischer spread, moving all over Europe from London to St Petersburg.[19] By the mid-1730s the political climate had cleared slightly. Fischer was permitted to return to Königsberg although he was no longer allowed to teach at the Albertina. Rappolt was appointed to take up Fischer's former chair, followed later by Bohl, who was appointed professor at the medical faculty in 1741. They had achieved international fame and could no longer be overlooked. Their teaching on specific branches of natural history, physical geography and anatomy, in conjunction with lectures on metaphysics and psychology, contributed to a philosophy of man.[20] Their work was conducted partly in collaboration with Klein and others in Danzig who, in 1743, established the Society for Natural Research in Danzig.[21]

With the publication of the *Systema naturae* in 1735 Linnaeus was becoming increasingly known in eastern Prussia. Upon his return to Königsberg Fischer dedicated part of his time to investigating 'Whether a natural history collection can be organised according to Linnaeus's method?',[22] while, at the Albertina, Rappolt later became the first to teach using Linnaean textbooks.[23] The introduction of Linnaean science was, however, not without controversy: Fischer answered the above question critically, and Klein positioned his system explicitly in opposition to Linnaeus. Fischer's critique was one of epistemology; Klein's critique was one of anthropology.

By contrasting Linnaeus's project to create a universal catalogue of nature with the local catalogue of the museum, Fischer was able to distinguish between two different research practices, each with a distinct objective. While the museum catalogue dealt with objects immediately assessable for observation, Linnaeus's universal catalogue sought to logically organise and categorise vast quantities of continually accumulating information, which in scale far exceeded the holdings of any local

19. Regrettably, a comprehensive study of this network is still a task for the future; however, the letter correspondence of Johann Christoph Gottsched does provide the contours of this constellation. See the ongoing Johann Christoph Gottsched, *Briefwechsel, Historisch-kritische Ausgabe*, ed. Detlef Döring and Manfred Rudersdorf, vol.1 (Berlin, 2007).
20. See the 1730s to 1750s in Oberhausen and Pozzo, *Vorlesungsverzeichnisse*, in particular p.74-214.
21. See Andrzej Januszajtis, 'Societas physicae experimentalis, the first physics society in Poland: the origins of the society', *Task quarterly* 6:2 (2002), p.323-30.
22. Fischer, 'Ob sich nach herrn Linnäus System'.
23. Oberhausen and Pozzo, *Vorlesungsverzeichnisse*, p.188.

museum.[24] The realisation of such a catalogue rested on what Bettina Dietz has identified as a 'collaborative culture',[25] the mobilisation and co-ordination of a global network of reliable collectors and observers, whose observations and descriptions could be trusted. This, however, implied a practice of categorisation detached from observation. 'Natural history collections are for the world and the senses. The systems are for the learned and for the understanding',[26] Fischer wrote. This identification of two incompatible research practices drove a wedge in between the faculties of the senses and the faculty of understanding, compromising Fischer's own scientific aim of abstracting general categories and laws from singular observations.

Klein's criticism was of a different kind.[27] In the first edition of *Systema naturae* Linnaeus had classified humans along with apes among the anthropomorphic, manlike, creatures, hence classifying man along with animals, minerals and plants. In addition to the categorical placement of man within the system of nature, he had followed his subdivision of classes, orders, genera, species and variety and sub-divided man into four different varieties; spatially ordered[28] according to their geographical location: *Europaeus*, *Americanus*, *Asiaticus* and *Africanus*.[29] A criticism of Linnaeus's categorisation of man among apes rippled along the shores of the Baltic Sea from Johann Gottschalk Wallerius in Uppsala to Johann Georg Gmelin in St Petersburg. Even so, the anatomical comparison between man and ape was not entirely rejected. Fischer noted in 1743 that 'Our body is by design quite similar to the body of animals'.[30] But the same year Klein explicitly formulated his critique in a

24. See Staffan Müller-Wille and Isabelle Charmantier, 'Natural history and information overload: the case of Linnaeus', *Studies in history and philosophy of science*, part C, *Studies in history and philosophy of biological and biomedical sciences* 43:1 (2012), p.4-15.
25. Bettina Dietz, 'Contribution and co-production: the collaborative culture of Linnaean botany', *Annals of science* 69:4 (2012), p.551-69.
26. Fischer, 'Ob sich nach herrn Linnäus System', p.166.
27. For the controversy between Klein and Linnaeus see Theodore W. Pietsch and Hans Aili, 'Jacob Theodor Klein's critique of Peter Artedi's *Ichthyologia* (1738)', *Svenska Linnesällskapets årsskrift* (2014), p.39-84.
28. Staffan Müller-Wille has pointed to Linnaeus's use of a spatial order in *Botanik und Weltweiter Handel: zur Begründung eines Natürlichen Systems der Pflanzen durch Carl von Linné (1707-1778)* (Berlin, 1999), p.90.
29. See Gunnar Broberg, *Homo sapiens L.: Studier i Carl von Linnés naturuppfattning och människolära* (Stockholm, 1975), p.222.
30. [Christian Gabriel Fischer], *Vernünftige Gedanken von der Natur. Was sie sey? Daß sie ohne Gott und seine allweise Beschränkung unmächtig sey? Und wie die einige, untheilbare göttliche Kraft, in und durch die Mittelursachen, nach dem Maß ihrer verliehen Würkbarkeit oder Tüchtigkeit, hie in der Welt alles alleine thätig würke? Durch fleißiges Nachsinnen, überlegen und schließen gefasset Und zur Verherrlichung göttlicher Majestät, auch Förderung wichtiger Wahrheiten herausgegeben von einem Christlichen Gottes-Freunde* (Königsberg, Christlichen Gottes-Freunde, 1743), p.699: 'Unser Leib ist der Thiere Leibern, der Bauart nach, ganz ähnlich.'

Treatise on the doubts about the classes of quadrupeds and amphibians in Carl Linnaeus's 'Systema naturae'.[31] Opposite Man ('Homo'), Linnaeus had written the phrase '*nosce te ipsum*', or 'know thy self'.[32] For Klein this formulation appeared to be there only to make the less honourable reader wonder whether he was man or man-like: 'Whether man, who arranges animals according to their places, should be placed among Anthropomorphs, is a matter of dispute. The warning of a good author suits those of bad habits: know thyself whether you are a man or an anthropomorph',[33] Klein wrote. Klein had questioned Linnaeus on two levels. Firstly he questioned whether the anatomical similarities between man and ape justified categorising them together. Secondly, however, Klein questioned the principal problem of operating with man on two levels: both as the object and subject of study, placing man within an order he himself had created. Although he refrained from making a public response, Linnaeus was privately troubled by Klein's critique, noting, 'If Klein knew himself he would not have included himself among the apes, if he knew his own strength he would have been satisfied with noticing his own gullibility'.[34] In this critical defence Linnaeus did not attack Klein's system but his self-conception as a scholar. A personal rivalry had erupted between the two men positioning them as competing authorities on opposite sides of the Baltic Sea; as Johann Samuel Haller, a former student from Königsberg, later noted in Berlin, 'The Swedish Knight Linnaeus and the famous scholar Mr Klein from Danzig have each, in accordance with their type, manufactured a [system] from the reservoir of their predecessors'.[35] Despite the personal rivalry Linnaeus continued to develop his system. He maintained that 'it had been impossible to discern a single mark by which one could distinguish

31. Jacob Theodor Klein, *Summa dubiorum circa classes quadrupedum et amphibiorum in Caroli Linnaei 'Systemate naturae'* (Leipzig, J. F. Gleditschium, 1743).

32. Caroli Linnaei, *Systema naturae, sive regna tria naturae systematice proposita per classes, ordines, genera, & species* (Lugduni Batavorum [Leiden], Theodorum Haak, 1735).

33. Klein, *Summa dubiorum*, p.10: 'An homo de animalibus disponens inter *Antropomorpha* Iocandus, sub judice lis est; ad male moratos homines quadrat monitum *optimi autoris:* Nosce te ipsum, an sis homo vel antropomorphus.' I am thankful to Kaarlo Havu for his help with the translation.

34. Pehr Löfling, 'Contra Kleinii animadversions', *c.*1743, Library of the Linnean Society of London, LM/MA/LOE/1: 'Hade Klein Kändt sig, hade han ikke bort räknas inter Simias, hade han kandt sin egen styrka, hade han visserligen varit nögd, att betrakta sin egen enfaldighet, den han visat i alla sina misun'. I am thankful to Gunnar Broberg and Isabelle Charmantier for their help in finding these notes. See also Broberg, *Homo sapiens L.*, p.171-72.

35. Johann-Samuel Haller, *Naturgeschichte der Thiere in Sistematischer Ordnung. Die Vierfüssigen Thiere, welche lebendige Jungen zur Welt bringen; nebst der Geschichte des Menschen* (Berlin, Christian Friedrich Boß, 1757), preface: 'Der Schwedische Ritter Linnäus, und der berühmte Danzigergelehrte Herr Klein, beide haben aus dem Vorrate ihre Vorgänger, ein jeder nach seiner Art eins verfertigt.'

between man and the apes'[36] but, in a revised edition of the *Systema naturae* from 1758, Linnaeus changed two classifications, substituting the quadruped with Mamalia and the Anthropomorpha with Primate.[37] Only one year earlier Johann Samuel Haller (1727-1810), attempting to tread carefully between the two authors, had sided with Klein in his *Natural history of the animals in systematic order. The quadrupeds who give birth to their young, along with the history of Man.*[38] Haller's work formulated the first comprehensive treatment of human nature based on Klein's methodology and, although he confirmed the anatomical similarities between man and ape, he expanded the essence of Klein's critique, pointing to the limits of an anthropology based on anatomy. 'It is understanding alone and not the body that separates Man from the vile community with animals',[39] he wrote. The comparison between man and ape was therefore restricted to the purely anatomical comparison.

Despite the limited influence of Haller's book, it was adopted by one of his contemporaries in Königsberg, a young scholar by the name of Immanuel Kant (1724-1804).[40] Kant had dedicated his first work to Rappolt's friend Bohl and, after the untimely death of Rappolt in 1753, Kant continued Rappolt's lectures on physical geography in 1756. Kant initially taught metaphysics, logic and mathematics, but his newly adopted lectures on physical geography would soon become his most popular.[41] Kant's general teaching practice followed Fischer's programme, approaching abstract metaphysical questions on cosmology and psychology through examples, experiments and observations. Kant developed the lectures on physical geography as a consequence of this teaching practice, as he explained that his students 'learned the art of subtle argumentation but they lacked any adequate knowledge of historical matters which could make good their lack of *experience*'.[42] Unlike

36. Carl Linnaeus, *Fauna Suecica Sistens Animalia Sveciae Regni: quadrupedia, aves, amphibian, pisces, insecta, vermes, distribute per classes & ordines, genera & species, cum differentiis specierum, synonymis auctorum, nominibus incolarum, locis natalium, descriptionibus insectorum* (Stockholm, L. Salvius, 1746), preface: 'nullum characterem hactenus eruere potui, unde Homo a Simia internoscatur.' Translation from John Zammito, *Kant, Herder and the birth of anthropology* (Chicago, IL, 2002), p.446.
37. See Londa Schiebinger, 'Why mammals are called mammals: gender politics in eighteenth-century natural history', *The American historical review* 98:2 (1993), p.382-411.
38. Haller, *Naturgeschichte der Thiere*, preface.
39. Haller, *Naturgeschichte der Thiere*, p.26: 'Der Verstand allein, und nicht der Körper entfernt den Menschen, von der niederträchtigen Gesellschaft der Thiere.'
40. Werner Stark, 'Einleitung', in *Vorlesungen über Physische Geographie: Kant's gesammelte Schriften*, ed. Werner Stark, vol.26 (Berlin, 2009), p.v-lxxx (xii-xiii).
41. Charles W. J. Withers, *Placing the Enlightenment: thinking geographically about the age of reason* (Chicago, IL, 2007), p.223-24.
42. Immanuel Kant, 'M. Immanuel Kant's announcement of the programme of his lectures for the winter semester 1765-1766', in *Theoretical philosophy, 1755-1770* (Cambridge, 1992)

Haller, Kant did not position himself between opposing systems but, following Fischer's critique, positioned the system of nature against geographical description: 'Division of knowledge according to concepts is **logical**; according to time and space it is **physical**. By means of the former, we obtain a **system of nature** (*Systema naturae*), as for example that of **Linnaeus**. With the latter, we obtain a **geographical description of nature**.'[43]

In a sense he was mirroring Fischer's critique by using the stark contrast between a logical and a sensible order; however, Kant was no longer referring to the local space of the museum but the global space of the traveller. Kant admitted that geography lacked the 'philosophical precision' of physics and natural history, but maintained that it should be approached 'with the reasoned curiosity of a traveller who everywhere seeks rather the noteworthy, the strange, and the beautiful, compares the observations he has collected, and revises his ideas accordingly'.[44]

Capitalising nature in Königsberg

In the year of Linnaeus's death, 1778, the Swiss mathematician Johann Bernoulli (1744-1807) travelled from St Petersburg through Europe. He stopped over in Königsberg where he attended a dinner party at the home of Count von Keyserling. Here he met with the now famous Immanuel Kant, who told Bernoulli about his current work on metaphysics which he hoped to finish soon. After the dinner, Kant showed Bernoulli and some of the other guests the castle library. In his travel book Bernoulli wrote: 'From the castle library I went with the gentlemen

p.287-300 (298): 'daß sie frühe vernünfteln lernt, ohne gnugsame historische Kenntnisse, welche die Stelle der Erfahrenheit vertreten können'; Immanuel Kant, 'Nachricht von der Einrichtung seiner Vorlesungen in dem Winterhalbjahre von 1765-1766', in *Kant's gesammelte Schriften*, vol.2 (Berlin, 1912), p.303-14 (312).

43. Immanuel Kant, 'Physical geography', in *Natural science* (Cambridge, 2012), p.434-679 (448): 'Die Eintheilung der Erkenntnisse nach Begriffen ist die logische, die nach Zeit und Raum aber die physische Eintheilung. Durch die erstere erhalten wire in Natursystem (*Systema naturae*), wie z.B. das des Linne, durch die letztere hingegen eine geographische Naturbeschreibung'; Immanuel Kant, *Physische Geographie*, in *Kant's gesammelte Schriften*, vol.9 (Berlin, 1923), p.151-438 (159-60).

44. Immanuel Kant, 'Plan and announcement of a series of lectures on physical geography with an appendix containing a brief consideration of the question: whether the West winds in our regions are moist because they travel over a great sea', in *Natural science* (Cambridge, 2012), p.386-95 (388): 'mit der vernünftigen Neubegierde eines Reisenden, der allenthalben das Merkwürdige, das Sonderbare und Schöne aufsucht, seine gesammelte Beobachtungen vergleicht und seinen Plan überdenkt', 'Entwurf und Ankündigung eines Collegii der physischen Geographie nebst dem Anhange einer kurzen Betrachtung über die Frage Ob die Westwinde in unsern Gegenden darum feucht seien, weil sie über ein großes Meer streichen', in *Kant's gesammelte Schriften*, vol.2 (Berlin, 1912), p.1-12 (3).

Bode [Bock], Kant, and others to the Sartoriusschen garden where the aforementioned big museum of natural history was stored.'[45] The Saturgus family had built a rococo-style palace with a garden containing a sun clock, a water garden and a maze, and inside the palace they had their famous collection of curiosities on display. In 1766, Kant had become library-assistant to Friedrich Samuel Bock (1716-1785), who had been the caretaker of the Saturgus collection since 1764. The Saturgus family, one of the wealthiest merchant families in Königsberg, had bought several private collections – including the private collections of Fischer and Rappolt.[46] Bock integrated the collection into his teaching at the Albertina; it was a spectacle of aesthetic display. Here the 'rare specimens of natural things [...] collected from the three kingdoms of nature' were used 'not only to delight the eyes' but also to demonstrate their 'purpose and advantage' and in finally to 'stir up the veneration and worship of the supreme deity'.[47] In this place, secluded from matters of daily life, aesthetic appreciation merged with scholarly observation. The collection had been founded on the prosperity of the Saturgus family's local industry and interwoven with its commercial networks. At the time of Bernoulli's visit the Saturgus family's wealth was in decline and the collection no longer belonged to them. As Bernoulli noted: 'Mr D. Bode is no longer in charge of the collection, and with the change in propietorship even the Cabinet has fallen into other hands.'[48]

Bock's central position as caretaker of the Saturgus collection had been pivotal to the transformation of Königsberg into a Linnaean capital. Bock belonged to the circle of scholars associated with Fischer, Rappolt and Klein, but Klein's system was slowly disappearing, literally being translated into the Linnaean nomenclature,[49] while Bock taught according to Linnaeus. The transformation in Königsberg was largely engineered by Bock and completed by Karl Gottfried Hagen (1749-1829), a local pharmacist and later professor at the Albertina. Both epitomised the entanglement between science and industry but, while Bock was a professor of Greek philology, Hagen personified the archetype of a naturalist, making the Linnaean system the basis of the *Materia medica.*

45. Johann Bernoulli, *Reisen durch Brandenburg, Pommern, Preußen, Curland, Rußland und Polen in den Jahren 1777 und 1778*, vol.3 (Lipzig, Fritsch, 1779), p.66: 'Von der Schloßbibliothek begab ich mich mit dem Herrn Bode, Kant und andern, nach dem sartoriußchen Garten, wo das schon erwähnte große Naturalienkabinet aufbewahret wird.'
46. Stark, 'Naturforschung in Königsberg', p.38.
47. Oberhausen and Pozzo, *Vorlesungsverzeichnisse*, p.257.
48. Bernoulli, *Reisen durch Brandenburg*, p.66: 'Herr D. Bode hat es nicht mehr unter seiner Aufsicht, weil sich die Umstände des Eigenthümers verändert hatten, wodurch auch das Kabinet selbst in andere Hände gerathen dürfte.'
49. Johann Beckmann, professor in Göttingen, published: 'Linneische Synonymie zu J.T. Kleins verbesserter Historie der Vögel', *Der Naturforscher* 1 (1774), p.65-78.

During the prosperous 1760s close relations arose between the nobility, the civic administration, merchants and the intelligentsia of Königsberg. In this constellation Bock, as the administrator of the museum, served as a go-between. In 1773, encouraged from the 'highest office' by Johann Friedrich von Domhardt, Bock established a weekly journal entitled *The Prussian collector*. It was practically oriented and dealt with cameral matters.[50] At the same time, Bock started working on his *Industrial natural history of the kingdom of East- and West Prussia* in which he noted: 'The entire industry is in its extensive scale nothing other than applied natural history, for the promotion of anything useful to the economy, or the avoidance of that which could be disadvantageous.'[51] Bock used what he considered to be the success of the Swedish state to endorse the Linnaean method: 'The Swedish example in modern times shows how natural goods from other areas [of the world] can be put to use. [...] Should it not be possible to accomplish in other countries what the Swedes have achieved in Sweden?'[52] Through Linnaeus's efforts Sweden had saved a lot of money as he had discovered medicinal plants in Sweden that used to be imported at great cost.[53] The same could be done in Prussia, Bock asserted. Ultimately, however, Bock found that the industrial use of natural history could only satisfy our brutish needs and desires, while the aesthetic and scientific appreciation spoke to our human minds.

Despite Bock's enthusiasm it was Karl Gottfried Hagen who would systematically introduce new scholars to the Linnaean system. His obituary stated that: 'His lectures, in which he introduced the Linnaean system to Prussia [...] were so well attended that they were overcrowded.'[54] Hagen lectured on all parts of the natural kingdom and,

50. Theodor Gottlieb von Hippel, *Sämtliche Werke*, vol.13 (Berlin, 1838), p.171-72. On the relation between natural history and cameral sciences see Lisbet Koerner, *Linnaeus: nature and nation* (Cambridge, MA, 1999); and Maragret Schabas and Neil De Marchi (ed.), *Oeconomies in the age of Newton* (Durham, NC, 2003).
51. Friedrich Samuel Bock, *Versuch einer wirtschaftlichen Naturgeschichte von dem Königreich Ost- und Westpreussen*, vol.1 (Dessau, Buchhandlung der Gelehrten, 1783), p.xxix: 'Die ganze Wirtschaft ist in ihrem weitläufigen Umfange nichts anders, als eine angewendete Naturgeschichte, zur Beförderung des der Oekonomie irgend nützlichen, oder Abwendung dessen, was derselben nachtheilig seyn könnte.'
52. Friedrich Samuel Bock, *Betrachtung über das Nutzbare und Anmuthige in der Naturgeschichte* (Königsberg, Zeis, 1767), p.11-12: 'Schwedens Beyspiel in den neuern Zeiten erweiset, wie wir uns anderer Gegenden Naturgüter zu Nutze machen könnten. [...] Sollte das, was in Schweden zu bewerkstelligen möglich gewesen, nicht auch in andern Ländern möglich seyn, die sich noch zum Theil unter einem günstigern Himmelstrich befinden?'
53. Friedrich Samuel Bock, *Betrachtung*, p.44.
54. *Neues Journal der Pharmacie für Ärzte, Apotheker und Chemiker* 18 (1829), p.273: 'Seine Vorlesungen, in denen er das Linné'sche System in Preußen einführte [...] wurden gleich so stark besucht, daß es an Raum fehlte.'

although he distinguished himself from Bock by being an educated naturalist, his academic practice was intertwined with commerce. He was the son of the pharmacist Heinrich Hagen who, in 1772, had unfortunately died when Karl Gottfried was still young, leaving him the business. Having proved himself to be an accomplished natural historian K. G. Hagen, a student of Kant and Bohl, among others, was left in charge of Rappolt's unpublished manuscripts. The overlap between commercial and scientific networks provided by the Saturgus collection enabled Hagen to excel by publishing a study of the Hottentot flower in 1777.[55] Baron von Großmann had collected the flower on his travels in Africa and upon his stay in Warsaw he had donated it to the Saturgus collection.[56] Bock had given Hagen permission to describe the flower and make it known. Hagen, who attempted to categorise it according to Linnaean taxonomy, declared that someone 'who a short while ago had spoken to the knighted Linnaeus and showed him a drawing of the flower was assured that the great naturalist was not aware of the flower and keenly desired a description of it.'[57] Under Bock's wings Hagen saw himself as a contributor to the Linnaean collection community. Just as Hagen straddled international commercial and scientific networks, his teaching practice merged scientific scholarship with local industry as he moved the teaching of pharmacy and experimental chemistry from the lecture halls of the Albertina to his private laboratory in his apothecary. His *Textbook on the pharmaceutical art*[58] would appear in nine continually revised editions, demonstrating its provisional character and its close connection to his research.

The merger between science and industry, the objectivity and utility in the study of nature became central in the investigation of the scientific foundation of natural history. In a prize-winning essay the Swiss naturalist Jean Senebier (1742-1809) had presented a study on the *Art of observation*, a study which *observed the observer* so that the art and practices of the scientist could be understood, reproduced and cultivated. In the essay Senebier noted how Klein had observed that shellfish emerge out of their eggs with a shell entirely similar to their grown shell. 'How did one, however, in the midst of this multitude of unknown or badly explained

55. Karl Gottfried Hagen, 'Beschreibung der Hottentottenblumen, oder einer neuen Art wieder auflebender Pflanzen', *Berlinische Sammlungen* 9 (1777), p.113-27 (116).
56. Hagen, 'Beschreibung', p.125.
57. Hagen, 'Beschreibung', p.123: 'jemanden, der vor kurzem den Ritter von Linné gesprochen, und ihm eine Abbildung dieser Pflanze vorgezeiget, versichert worden, daß dieser große Naturkundige von derselben gar noch nichts gewußt und sehnlichst eine Beschreibung von ihr gewünscht habe.'
58. Karl Gottfried Hagen, *Lehrbuch der Apothekerkunst* (Königsberg and Leipzig, Hartung, 1778).

occurrences decide upon the object of one's investigation?', Senebier asked rhetorically, only to answer that:

> One would without a doubt have to be able to observe everything, but according to the philosophical notes on the pattern of all observers Art takes a long time while life is very short. So what will someone do who thinks he can become an interpreter of nature? He will follow the instinct of his genius, but he will for the most part observe those occurrences which may be useful to man, and even then only observe single particular occurrences.[59]

As the naturalist cannot begin with a complete knowledge of the totality of nature, the observer must begin his observation process with what he finds useful. As Senebier concluded: 'Science, which is not useful, must indeed be laborious.'[60] This, however, came to be at the core of what may be regarded as the most severe criticism of natural history and in particular that of Linnaeus, presented by Comte de Buffon (1707-1788): if the observation of nature is founded on what is most useful to us, how can we know that it says something essential about nature? Bohl's good friend Albrecht von Haller (1708-1777), who had written the preface to the German translation of Buffon's *Natural history*,[61] agreed with Buffon that Linnaeus had based his system on an *arbitrary* principle.[62] Buffon's criticism held that the naturalist investigated nature in relation to himself, prioritising what was most familiar and useful to him.[63] Von Haller had not entirely discarded Linnaeus, but instead advocated the necessary use of a total system for any science to advance: had nature not been ordered in a *totality* a science of things in nature would not be possible – hence even arbitrary systems were considered quite useful, as they not only brought order to an increasing amount of information, but also standardised practices of observation for further studies.

Hagen's treatment of the animal and plant kingdoms was based on Linnaeus's system.[64] Kant called it a 'logical masterpiece', upholding the

59. Jean Senebier, *Die Kunst zu beobachten* 1 (Leipzig, Weygand, 1776), p.67-68: 'Wie hat man sich aber mitten in dieser Menge unbekannter, oder übel erklärter Begebenheiten zu betragen, um den Gegenstand seiner Untersuchung zu wählen? Man müßte ohne Zweifel alles beobachten können, aber nach der philosophischen Anmerkung des Musters aller Beobachter ist die Kunst sehr lang und das Leben sehr Kurz. Was wird also der thun, der glaubt er könne in die Natur bringen? Er wird dem Triebe seines Genies folgen, aber er wird überhaupt die Begebenheiten betrachten, die den Menschen nützlich seyn können, und dann erst insbesondre einzelne Begebenheiten.'
60. Senebier, *Die Kunst*, p.69: 'Die Wissenschaft, die nicht nützlich ist, muß mühsam werden.'
61. Georges Louis LeClerc de Buffon, *Allgemeine Historie der Natur* (Hamburg, Grund and Holle, 1750).
62. Buffon, *Allgemeine Historie*, p.xvi.
63. See Phillip R. Sloan, 'The Buffon-Linnaeus controversy', *Isis* 67:3 (1976), p.356-75 (359).
64. Hagen, *Lehrbuch*, p.17-18.

contrast between the system of nature and geography as two different modes of practising natural history. In 1775, however, Kant would revisit this epistemological critique. He had published an essay, 'On the different races of human beings', which introduced a new perception of nature. In this attempt to analyse man he used neither the principles of logic nor space, but time. In the essay he revised the earlier spatial division between four human *varieties* by introducing the concept of *race*. Although the concept of variety did refer to a spatial and hence a real and not purely logical division, it was limited to describing something relative to its environment. By following the spatial organisation one ran the risk of dealing with an infinite number of varieties, while following a categorical division, 'according to *resemblances*'[65] one ran the risk of creating a system founded on arbitrary marks, as Comte de Buffon had interjected against Linnaeus. In order to navigate between the two poles Kant proposed a temporal perspective, which could direct the newly introduced concept of race. Doing this Kant juxtaposed natural history with natural description. Natural history was the study of how the nature of things is reproduced in time, meaning the study of 'hereditary marks',[66] which Kant, in the study of races among humans, suggested was skin colour. Observing hereditary marks had the advantage of neither being arbitrary nor individual characteristics, and they could therefore become the focus of scientific studies.

The study of the multiple races of humans did, however, not lead to a universal understanding of mankind; if anything it contributed to its relativisation. In his private reflections Kant made a note of the difference between man and ape: 'Linnaeus cannot find any in the body. If one took all reason from him: what sort of animal would he be. Beautiful or ugly [...] quadruped or biped.'[67] For Kant the distinction between man and ape was not to be found in any observable characteristic. At the dinner table Kant had told Bernoulli that his lectures on physical geography had been complemented with a lecture course on pragmatic anthropology, a lecture course which took man's self-consciousness as its point of departure – 'know thy self'.

65. Immanuel Kant, 'Of the different races of human beings', *Anthropology, history and education* (Cambridge, 2007), p.82-97 (84): 'nach Ähnlichkeiten'. Immanuel Kant, 'Von den verschiedenen Racen der Menschen', in *Kant's gesammelte Schriften*, vol.2 (Berlin, 1912) p.427-44 (429).

66. Immanuel Kant, 'Of the different races', p.85: 'erblichen Merkmale'; Immanuel Kant, 'Von den verschiedenen Racen der Menschen', in *Kant's gesammelte Schriften*, vol.2 (Berlin, 1912) p.427-44 (430).

67. Immanuel Kant, *Kant's handschriftlicher Nachlaß: Anthropologie*, in *Kant's gesammelte Schriften*, vol.15 (Berlin, 1923), p.778: '*Linnäus* kann keinen im Korper finden. Wenn man ihm alle Vernunft nähme: was würde er vor ein Thier seyn. Schön oder heslich (Bart. Kahl *etc.*) Vierfüßig oder zweyfüßig (wild gefundene Menschen, Bergbewohner von Madegascar).'

Having left Königsberg Bernoulli continued to Danzig, where he was appointed an honorary member of the scientific academy along with a local scholar, the Linnaean naturalist Johann Reinhold Forster (1729-1798). Johann Reinhold and his son Georg had both become famous for accompanying Captain Cook on his second voyage.

In contrast to the Forsters, Kant had not travelled far beyond Königsberg and the region was no longer a flourishing centre for natural history as it had been in his youth. Kant did not possess a network of trusted collectors, unlike Linnaeus; his lectures on geography and anthropology primarily relied on his extensive reading of a heteronomous magnitude of literature. As a reader he was faced with the challenge of organising this material, that is, to be able to critically distinguish between contradictory information as well as to compare opposing systems. The work on metaphysics that Kant had informed Bernoulli of was such a book. This work, which he titled *Critique of pure reason*, did not presume to deal with divine concepts, but instead turned to investigate the limits of human reason as the metaphysical basis for any metaphysical system. In this endeavour he followed one of his favourite poets, Alexander Pope, in his interpretation of the dictum *know thy self!*: 'Know then thy-self, presume not God to scan; / The only Science of Mankind is Man.'[68] His investigation was not an *observation of the observer*, as had been the case with Senebier, but rather what Kant himself described as a '*metaphysics of metaphysics*'.[69] In it he introduced a new tone which would distance him from his former work: 'Examples and illustration always appeared necessary to me, and hence actually appeared in their proper place in my first draft.' Kant noted. However, due to the sheer size of the project Kant removed the examples as he 'became aware that this alone, treated in a dry, merely **scholastic** manner, would suffice to fill an extensive work'.[70] With this new tone in his writing Kant introduced a style and persona that was in stark contrast to how he presented himself in his lectures.

Characterisation of nature: between caricature and character

In 1787 the Göttingen naturalist Johann Friedrich Blumenbach (1752-1840) received a letter from his friend in Königsberg, Professor of

68. Alexander Pope, *An Essay on man: in epistles to a friend, Epistle II* (London, J. Wilford, 1733), p.23.
69. Immanuel Kant, *Kant's Briefwechsel*, in *Kant's gesammelte Schriften*, vol.10 (Berlin, 1922), p.269: 'Metaphysik von der Metaphysik'.
70. Immanuel Kant, *Critique of pure reason* (Cambridge, 1998) p.103-104; Immanuel Kant, *Kritik der reinen Vernunft*, in *Kant's gesammelte Schriften*, vol.4 (Berlin, 1911), p.1-252 (12).

Medicine Johann Daniel Metzger (1739-1805). Blumenbach had written to enquire about Kant, who had made himself noted by entering into disputes with both his former student Johann Gottfried von Herder (1744-1803) and the now famous Georg Forster (1754-1794).[71] Metzger responded that Kant 'unquestionably deserves praise as an independent thinker and genius'. However, according to Metzger, Kant had become 'reluctant to tolerate contradiction. As even now he is disgruntled with me because of my article on the various races man'.[72] Metzger, whom Herder had praised as an early developer of a physical anthropology,[73] had entered into the debate in 1786 by writing a harsh criticism of Kant's essay 'Determination of the concept of a human race' from 1785.

By the mid 1780s the conditions for the debate had changed. The Albertina was becoming a Linnaean institution due to the efforts of Hagen, who had been working to strengthen the material conditions of the university. The Saturgus family zoological collection had been appropriated by the university[74] and Hagen had also initiated plans to establish a botanical garden for the Albertina.[75] Despite his praise of Klein's study of the animal kingdom, Metzger maintained that 'Klein's methodology was too complicated to be a success. It has been repressed by the Linnaean.'[76] But while Linnaean science was being institutionalised locally, its public status was subject to controversy. Around Kant a new constellation of academics took shape, which would sideline him as a natural historian. Apart from the already mentioned Herder, Georg Forster and Metzger, the constellation also included the anatomist Samuel Thomas Sömmerring (1755-1830). Central to the debate was the question of how to distinguish man from ape as raised by Linnaeus; however, it was not so much a question of observation as a question of

71. See Manfred Riedel, 'Manfred Riede und Kritizismus: Kants Streit mit G. Forster und J. G. Herder', in *Deutschlands kulturelle Entfaltung: die Neubestimmung des Menschen*, ed. Bernhard Fabian, Wilhelm Schmidt-Biggemann and Rudolf Vierhaus (Munich, 1980), p.31-48; Wolfgang Proß, '"Ein Reich unsichtbarer Kräfte": was kritisiert Kant an Herder?', *Scientia poetica* 1 (1997), p.62-119; Tanja van Hoorn, *Dem Leibe abgelesen: Georg Forster im Kontext der physischen Anthropologie des 18. Jahrhunderts* (Tübingen, 2004); Hans Erich Bödeker, 'Georg Forsters Entwurf einer "Wissenschaft vom Menschen"', *NTM International journal of history and ethics of natural sciences, technology and medicine* 18 (2010), p.137-67; Han F. Vermeulen, *Before boas: the genesis of ethnography and ethnology in the German Enlightenment* (Lincoln, 2015).
72. Johann Friedrich Blumenbach, *The Correspondence of Johann Friedrich Blumenbach*, vol.3 (Göttingen, 2010), p.106.
73. Johann Gottfried Herder, *Ideen zur Philosophie der Geschichte der Menschheit* (Leipzig, 1841), p.233.
74. Fritz Gause, *Die Geschichte der Stadt Königsberg in Preussen*, vol.2 (Graz, 1968), p.185.
75. Karl Gottfried Hagen, *Programma primum de plantis in Prussia cultis* (Königsberg, Hartung, 1791).
76. Johann Daniel Metzger, *Skizze einer pragmatischen Literärgeschichte der Medicin*, vol.1 (Königsberg, Nicolovius, 1792), p.367-68: 'Kleins Methodologie war zu verwickelt, um ihr Glück machen zu können. Sie wurde durch die Linneische verdrängt.'

categorisation, as Blumenbach had noted: 'a student of nature will in practice hopefully not be in the embarrassing situation of confusing men and apes.'[77]

Kant's definition of the concept of race had been an attempt to clarify it as a concept but in such a way that it could be useful for observation. Kant suggested that to test whether skin colour could serve as an hereditary marker one should observe Africans in Europe, that is, outside their natural environment. Forster responded that: 'a negro is actually only a true negro when in his own country. Each essence of nature is only what it should be within the place from which it originated, a truth that we see confirmed in menageries and botanical gardens daily. The Negro, born in Europe, is like a greenhouse plant, a modified creature.'[78] Forster did not simply reject Kant's suggestion of skin colour as a signifying marker of race, rather he rejected his priority of a temporal mode of observation, favouring the spatial concept of environment. Instead of four set categories of race Forster argued for an infinite number of varieties, which could be studied empirically. For Kant, however, clear concepts were necessary to guide the investigations, earning him the nicknames 'the arch-sophist and arch-scholastic of our time'.[79] Nevertheless, Kant's critics did not avoid the concept of race; indeed, they suggested an anatomical distinction between two races – the Negro and the European – with an infinite possibility of environmentally dependent varieties in between.

After returning to Europe from his journey with Captain Cook in 1775, Forster visited the Collegium Carolinum in Cassel from 1778-1784 where he met the anatomist Samuel Thomas Sömmerring. In Cassel Duke Frederick II had established a colony of Africans[80] and Sömmerring had been able to perform autopsies on the deceased. Based on his work in Cassel Sömmerring had published *On the Moors bodily difference from the European* in 1784, revised and re-titled *On the Negros bodily difference from the European* the following year.[81] In his descriptions

77. Johann Friedrich Blumenbach, *Handbuch der Naturgeschichte*, 5th edn (Göttingen, Dietrich, 1797), p.7: 'Aber auch ohne dieselben wird doch hoffentlich nie ein Naturforscher in praxi in Verlegenheit gekommen seyn, Menschen und Affen etwa zu verwechseln.'
78. Georg Forster, 'Noch etwas über die Menschenraßen', *Teutsche Merkur* (1786), p.57-86 (72): 'ein Neger sey eigentlich nur in seinem Vaterlande ein rechter Nager. Ein jedes Wesen der Natur ist, was es seyn soll, nur an dem Orte, für den sie es entstehen ließ; eine Wahrheit, die man in Menagerien und botanischen Gärten täglich bestätigt sieht. Der Neger, in Europa geboren, ist wie eine Treibhauspflanze, ein modificirtes Geschöpf.'
79. Georg Forster, *Georg Forster's Sämmtliche Schriften*, vol.7 (Leipzig, 1843), p.364.
80. See Londa Schiebinger 'The anatomy of difference', *Eighteenth-century studies* 23:4 (1990), p.387-405.
81. Samuel Thomas von Sömmerring, *Über die körperliche Verschiedenheit des Mohren vom Europäer* (Mainz, 1784), and Samuel Thomas von Sömmerring, *Über die körperliche Verschiedenheit des Negers vom Europäer* (Frankfurt and Mainz, Barrentrapp Sohn and Wenner, 1785).

Sömmerring had followed Petrus Camper (1722-1789), his 'great teacher and kindest friend',[82] who had made an aesthetic comparison between the head shape of the ape, 'the negro' and the European, hierarchically ordering them in a line of perfection (see Fig.7, p.156). The European was represented by the Greek ideal of beauty, while 'the negro' was aesthetically positioned between the Greek and the ape. It was particularly the flat nose of 'the negro', allegedly allowing less space for the brain, which placed him closer to the ape than the European. Although Sömmerring noted that the body of 'the negro' made 'him the most perfect creature for his climate'[83] and that there were many individual negroes that were more beautiful or smarter than the *average* European, he concluded that the *average*[84] African negro was closer to the ape than the European was, but that he was by all means human. Sömmerring stressed that he, to the best of his efforts, had approached the matter as a question of scientific investigation; however, he had explicitly framed it as a political argument against slavery. The conflation of natural history and economy had, perhaps unintentionally, made man a subject to commerce. Sömmerring noted with regret that 'In America one pays 10 pounds sterling for every imported negro, similar to a commodity or an animal'. At the same time, Sömmerring was embarrassingly aware of the inhumane conditions which had enabled his dissections: 'Only the European can treat his slaves so badly, he buys them to collect them.'[85]

The ongoing debate led Kant to revisit natural history as part of a broader investigation, firstly in a text on *The Use of teleological principals in philosophy* and secondly in his third work of critical philosophy, *Critique of the power of judgment* from 1790. Instead of returning directly to the debate Kant turned to the presuppositions of natural history in order to investigate how characterisation and classification at all was possible. This investigation both pointed to the legitimacy of natural history as well as an inherent criticism of its objectification of man. Although Bock is not mentioned anywhere, his combination of economy, aesthetics and natural history seems to function as an interpretive key to the work; the paradigm for understanding Kant's investigation of aesthetics is just as much the chemist's laboratory or the natural history museum as the art gallery or the theatre. However, unlike Bock, Kant restrains himself from

82. Sömmerring, *Über die körperliche Verschiedenheit des Mohren*, p.7.
83. Sömmerring, *Über die körperliche Verschiedenheit des Negers*, p.79.
84. Sömmerring, *Über die körperliche Verschiedenheit des Mohren*, p.32.
85. Sömmerring, *Über die körperliche Verschiedenheit des Negers*, p.vii-viii: 'Man zahlte in Amerika zehen Pfund Sterling für jeden eingebrachten Neger [reference to Doctor Franklin], gleich einer Waare oder Thier', 'Nur der Europäer könne seine Sklaven so übel behandeln, er kauft sie, um durch sie zu sammeln'.

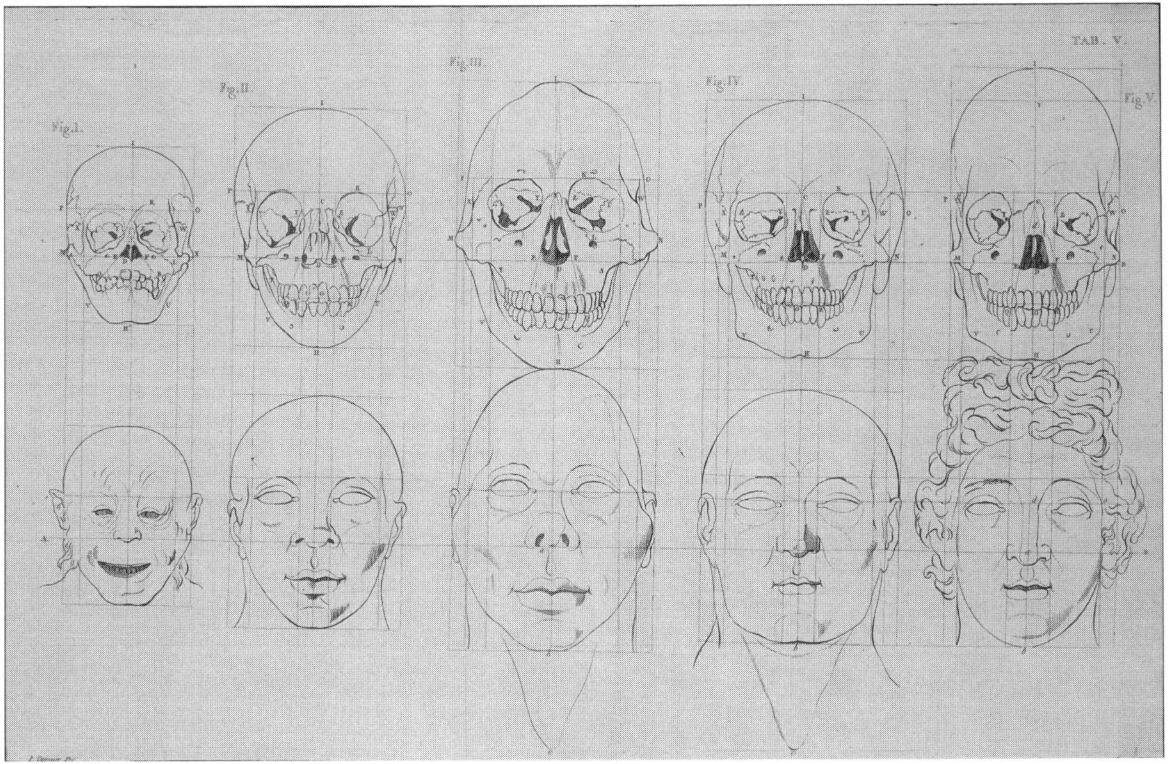

Figure 7: Drawings of 'the development of facial and cranial expressions' from Petrus Camper, *Dissertation sur les variétés naturelles qui caractérisent la physionomie des hommes des divers climats et des différens ages* (Paris, H. J. Jansen, 1791), Tab.V (unpaginated). Courtesy of the Getty Research Institute.

making metaphysical assumptions by viewing nature as purposeful due to the design of God. Instead he maintains that man views nature as purposeful due to his own interest. It is only by analogy that we may understand nature as purposeful in itself, by understanding natural objects in analogy to tools or techniques. However, despite our primary interest in the nature of a thing probably originating from its use we also have the capacity for aesthetic appreciation, or what Kant calls a 'disinterested interest'. While we may find the first interest within the walls of the factory or the workshop, the disinterested interest resides within the halls of the museum, the collections of art or natural history. The notions of beauty and aesthetic appreciation are not opposed to natural science; on the contrary it is constitutive for any kind of disinterested observation. Kant distinguished between two kinds of beauty. The first is one used in natural history, for example by Sömmerring. As Sömmerring had argued, his primary interest was the general characteristics of the *average* Negro or European. What Kant made clear was how the gaze of the natural historian rested on a statistical gaze.[86] Camper and Sömmerring had applied what Kant calls a 'normal idea' of beauty. This, Kant holds, is empirically derived by induction within a clearly defined totality, such as a country, so that one may talk of 'the average man, the average head, the average nose, etc.'. Kant points out that 'under these empirical conditions a Negro must necessarily have a different normal idea of the beauty of a figure than a white person, a Chinese person a different idea from a European.'[87] The concept of beauty inherent in the 'normal idea' is that which does not stand out or offend as in Camper's drawings:

> One will find that a perfectly regular face, which a painter might ask to sit for him as a model, usually says nothing: because it contains nothing characteristic, and thus expresses more the idea of the species than anything specific to a person. What is characteristic in this way, when it is exaggerated,

86. On the observation in general see Lorraine Daston, 'Observation and enlightenment' in *Scholars in action: the practice of knowledge and the figure of the savant in the eighteenth century*, ed. André Holenstein, Hubert Steinke and Martin Stuber (Leiden, 2013), p.657-78. On the gaze of the natural historian see Phillip R. Sloan, 'The gaze of natural history', in *Inventing human science: eighteenth-century domains*, ed. Christopher Fox, Roy Porter and Robert Wokler (Berkeley, CA, 1995), p.112-51. And on the statistical gaze see Hans Erich Bödeker, 'On the origins of the "statistical gaze": modes of perception, forms of knowledge and ways of writing in the early social sciences', in *Little tools of knowledge: historical essays on academic and bureaucratic practices*, ed. Peter Becker and William Clark (Ann Arbor, MI, 2001), p.169-96.
87. Immanuel Kant, *Critique of the power of judgment* (Cambridge, 2000) p.119: 'ein Neger nothwendig unter diesen empirischen Bedingungen eine andere Normalidee der Schönheit der Gestalt haben muß, als ein Weißer, der Chinese eine andere, als der Europäer.'; Immanuel Kant, *Kritik der Urteilskraft*, in *Kant's gesammelte Schriften*, vol.5 (Berlin, 1913), p.165-486 (234).

i.e., when it itself breaks with the normal idea (of the purposiveness of the species), is called **caricature**.[88]

The normal idea of beauty has a scientific purpose for determining something as characteristic to a species; however, 'there is still a distinction between the **normal idea** of the beautiful and its **ideal**'. Kant notes and adds: 'the ideal consists in the expression of the **moral**'.[89] This moral sense of beauty 'is the form of the **purposiveness** of an object, insofar as it is perceived in it **without representation of an end**'.[90] From the perspective of natural history we may regard humans as a species, which can be described, compared and classified according to an aesthetic idea of the normal. However, being characterised as a species reduces humanity to a caricature. The ideal of human beauty lies in a human's moral character, not in a determining nature but in a human's capacity for self-determination. This ideal beauty, however, reveals itself through communication with other humans and not through observation of them:

> The propaedeutic for all beautiful art, so far as it is aimed at the highest degree of its perfection, seems to lie not in precepts, but in the culture of the mental powers through those prior forms of knowledge that are called *humaniora*, presumably because **humanity** means on the one hand the universal **feeling of participation** and on the other hand the capacity for being able to **communicate** one's inmost self universally, which properties taken together constitute the sociability that is appropriate to humankind, by means of which it distinguishes itself from the limitation of animals.[91]

88. Immanuel Kant, *Critique of the power of judgment*, p.119: 'Man wird finden, daß ein vollkommen regelmäßiges Gesicht, welches der Maler ihm zum Modell zu sitzen bitten möchte, gemeiniglich nichts sagt: weil es nichts Charakteristisches enthält, also mehr die Idee der Gattung, als das Specifische einer Person ausdrückt. Das Charakteristische von dieser Art, was übertrieben ist, d. i. welches der Normalidee (der Zweckmäßigkeit der Gattung) selbst Abbruch thut, heißt Caricatur'; Immanuel Kant, *Kritik der Urteilskraft*, in *Kant's gesammelte Schriften*, vol.5 (Berlin, 1913), p.165-486 (235).
89. Immanuel Kant, *Critique of the power of judgment*, p.120: 'Der sichtbare Ausdruck sittlicher Ideen, die den Menschen innerlich beherrschen'; Immanuel Kant, *Kritik der Urteilskraft*, in *Kant's gesammelte Schriften*, vol.5 (Berlin, 1913), p.165-486 (235).
90. Immanuel Kant, *Critique of the power of judgment*, p.120: 'Schönheit ist Form der Zweckmäßigkeit eines Gegenstandes, sofern sie ohne Vorstellung eines Zwecks an ihm wahrgenommen wird'; Immanuel Kant, *Kritik der Urteilskraft*, in *Kant's gesammelte Schriften*, vol.5 (Berlin, 1913), p.165-486 (236).
91. Immanuel Kant, *Critique of the power of judgment*, p.229: 'Die Propädeutik zu aller schönen Kunst, sofern es auf den höchsten Grad ihrer Vollkommenheit angelegt ist, scheint nicht in Vorschriften, sondern in der Cultur der Gemüthskräfte durch diejenigen Vorkenntnisse zu liegen, welche man *humaniora* nennt: vermuthlich weil Humanität einerseits das allgemeine Theilnehmungsgefühl, anderseits das Vermögen sich innigst und allgemein mittheilen zu können bedeutet; welche Eigenschaften, zusammen verbunden, die der Menschheit angemessene Geselligkeit ausmachen, wodurch sie sichvon der thierischen Eingeschränktheit unterscheidet'; Immanuel Kant, *Kritik der Urteilskraft*, in *Kant's gesammelte Schriften*, vol.5 (Berlin, 1913), p.165-486 (355).

Kant's purpose was not to separate human nature from culture, as Hans Peter Reill has noted: 'Enlightenment thinkers never conceived of separating the humanities from the study of nature'.[92] Rather than separating, his purpose was to distinguish between different disciplinary perspectives. Having his authority as a physical geographer challenged and to some degree undermined, this rhetorical move allowed Kant to shift the perspective from practical questions regarding the characterisation of man to a more principal question regarding what makes man human, a shift of focus which allowed him to reposition himself as an authority on anthropology.

The end of nature

In 1796 the Göttingen natural historian Christoph Girtanner (1760-1800) wrote *On the Kantian principle for natural history*, in which he noted that Kant, the great Königsberg philosopher, had expressed 'highly perceptive thoughts' in three essays on the human race, 'which, if they were subjected to careful examination, would necessarily give natural history a whole new direction. Even so [...]', he added, 'I do not find that more recent naturalists have taken these thoughts into consideration'.[93] Kant's parting from natural history, in the broad sense of the word, did not simply develop throughout his work – on the contrary natural history remained a key interest during his life – rather it was through his engaged discourse on the science of man that he came to be positioned opposite natural history as the 'arch-scholastic'. In this discourse Kant's trust and authority within the so-called moral economy of knowledge was put into question. Even so, the criticism did not question his *personal* integrity and authority as such, but simply his capabilities as a *professional* natural historian. Unlike 'the queen of the universities', as Metzger called the Georg-August University in Göttingen, with its wonderful library, observatory and museum, Kant would put Königsberg on the map thanks to his role as a teacher and a philosopher there. As Metzger noted, with a hint of envy over the conditions in Göttingen, 'I have reasons to believe, that this external splendour, however useful it essentially also is, has an even greater influence on the opinion abroad than the aforementioned merits of the teachers'.[94]

Despite the contested discourse on Linnaean science, Königsberg did,

92. Peter H. Reill, *Vitalising nature in the Enlightenment* (Berkeley, CA, 2005), p.2.

93. Christoph Girtanner, *Über das Kantische Prinzip für die Naturgeschichte: ein Versuch diese Wissenschaft philosophisch zu behandeln* (Göttingen, Vandenhoek and Ruprecht, 1796), p.i.

94. Johann Daniel Metzger, *Über die Universität zu Königsberg: ein Nachtrag zu Arnold und Goldbeck* (Königsberg, 1804), p.19: 'Ich habe Ursache zu glauben, daß dieser äußere Prunk, so nützlich er auch im wesentlichen ist, einen noch größern Einfluß auf das Urtheil des Auslandes hat, als die vorerwähnten Verdienste der Lehrer.'

in practice, become Linnaean. Along with Hagen, Kant worked to reform the Albertina into a modern research institution providing it with the necessary institutions to conduct natural science. These reforms sought, in many ways, to realise Fischer's failed visions for Königsberg, even integrating former private initiatives. Fischer and Rappolt's private collections became part of the university collection and in 1787 Hagen introduced the idea of establishing a proper botanical garden where Linnaean taxonomy would be applied. It was realised in 1811 and followed by his *Prussian flowers*.

In accordance with the idea of a university dedicated to the utility of the state, Kant in his vision for the university compared it to 'a factory'[95] and described the scholars at the three higher faculties – theology, law and medicine – as '*businesspeople* or technicians of learning'. Their primary concern being the wellbeing of the state, they served as 'tools of the government'.[96] The single-minded orientation through the lenses of utility had its limits, as Sömmerring had noted the science of man had capitalised on man, not only by enslaving people but by making them part of a global economy of trade. Kant had presented a moral criticism of this view, articulated as a *conflict of faculties*, in general between the higher and the low faculty and more specifically between the faculty of medicine and the faculty of philosophy: 'a university must have a faculty of philosophy. Its function in relation to the three higher faculties is to control them and, in this way, be useful to them, since *truth* [...] is the main thing, whereas the *utility* the higher faculties promise the government is of secondary importance.'[97] While Linnaeus's natural history had become central to the *science of man*, challenging the very notion of man, Kant had taken the dictum 'know thy self!' even further and pointed it back to the *man of science* calling for a reflection of science, its limits and possibilities.

95. Immanuel Kant, 'The conflict of the faculties', in *Religion and rational theology* (Cambridge, 2001), p.233-328 (247): 'fabrikenmäßig'; Immanuel Kant, *Der Streit der Facultäten*, in *Kant's gesammelte Schriften*, vol.7 (Berlin, 1917), p.1-116 (17).

96. Immanuel Kant, 'The conflict of the faculties', p.248: 'die man also Geschäftsleute oder Werkkundige der Gelerhsamkeit nennen kann', 'Werkzeuge der Regierung'; Immanuel Kant, *Der Streit der Facultäten*, in *Kant's gesammelte Schriften*, vol.7 (Berlin, 1917) p.1-116 (18).

97. Immanuel Kant, 'The conflict of the faculties', p.255: 'Auf einer Universität muß aber auch solches Department gestiftet, d. i. es muß eine philosophische Facultät sein. In Ansehung der drei obern dient sie dazu, sie zu controlliren und ihnen eben dadurch nützlich zu warden, weil auf Wahrheit [...] alles ankommt; die Nützlichkeit aber, welche die oberen Facultäten zum Behuf der Regierung versprechen, nu rein Moment vom zweiten Range ist'; in *Kant's gesammelte Schriften*, vol.7 (Berlin, 1917) p.1-116 (28).

6. The Edinburgh connection: Linnaean natural history, Scottish moral philosophy and the colonial implications of Enlightenment thought

LINDA ANDERSSON BURNETT *and* BRUCE BUCHAN

Linnaean natural history, through its classification of humans and ethnographic observations collected by Linnaeus and his students, had a formidable influence on how human difference was discussed and classified in the eighteenth century. Linnaeus taught his students the importance of collecting ethnographic data during their domestic and overseas expeditions and Linnaeus himself, particularly during his journey to Sápmi (referred to then as Lapland) in 1732, described in detail the customs and physiology of the peoples he encountered.

In Swedish historiography there has been a tendency to play down the colonial implications of both Linnaeus's travels to Sápmi and that of his students' – often referred to as his 'apostles' – overseas voyages despite the latter travelling on transnational imperial expeditions to different countries in Africa, Asia and America. The Linnaean travellers have instead been celebrated and romanticised as intrepid explorers who put Swedish science on the world map. Swedish research has also been 'person centred' in that it has focused on the relationship between Linnaeus and individual apostles.[1]

This is in sharp contrast to international scholarship that emphasises how natural historians exemplified a process of colonial appropriation of seemingly new lands and peoples. Mary Louise Pratt, Londa Schiebinger and others have interpreted the overseas expeditions of Linnaean students as examples of how botanical studies facilitated colonialism. In the Linnaean naturalists' mercantile hunt for new botanical specimens, which they classified, universalised and exploited, they

We would like to express our gratitude to the Swedish Foundation for the Humanities and Social Sciences, the Linnaeus University Centre for Concurrences in Colonial and Post Colonial Studies and the School of Humanities, Griffith University, for providing funding to support our research. We are especially grateful to the editors and other contributors to this volume, and the participants at the 'Nature's Empire' workshop on the 14th of November 2014, European University Institute for their feedback and suggestions.

1. For a discussion of these themes see Hanna Hodacs and Kenneth Nyberg, *Naturalhistoria på resande fot: om att forska, undervisa och göra karriär i 1700-talets Sverige* (Lund, 2007), p.18-35.

often overwrote indigenous knowledge and paid little attention to the impact their expeditions had on local landscapes and populations. As Pratt puts it: 'the planet's life forms were [...] drawn out of the tangled threads of their life surroundings and rewoven into European-based patterns of global unity and order.'[2] In this way, she contends, eighteenth-century scientific systems of knowing, which Linnaeus pioneered, not only silenced indigenous narratives outside Europe – by re-labelling plants with Latin names, for example – but also had the same effect on local and rural narratives and repositories of knowledge in Europe.[3] Pratt, in other words, presents the Linnaean natural historian as a key agent for European colonialism through his mapping, classification and narration.

In this chapter we reflect further on this process of intellectual colonisation, but in doing so we seek to modify Pratt's account. Beinart argues that Pratt's notion of a 'Linnaean watershed' in the 1760s after which 'travellers took far less notice of people' and 'naturalized the travel zone' rendering it an uninhabited landscape 'unpossessed, unhistoricized, unoccupied' and ready for European intellectual authority to claim it, is exaggerated.[4] The development of colonial knowledge needs to be understood in context, and this requires being attuned to revealing moments of encounter. As Bronwen Douglas has recently argued, notions of human diversity and racial classification were applied in local settings as European travellers interacted with indigenous peoples whose agency has routinely been overlooked. Douglas seeks to correct this tendency by identifying 'indigenous countersigns', or the 'oblique traces of the imprint of local agency on foreign perceptions and representations of encounters'.[5] Thus, the colonial uses of terminology – of race, nation or savage – need to be understood in relation to episodes of encounter in which 'the concept of countersign supplies the building blocks for more informed, more subtle ethnohistories which decentre Europeans and personalize encounters as multifaceted, open-ended, embodied episodes, rather than essentialize them as generalized clashes of two opposed, homogenous cultures.'[6]

We look beyond the Uppsala-trained Linnaean naturalists and their personal relationships with Linnaeus. Instead, we analyse the role of

2. Mary Louise Pratt, *Imperial eyes: travel writing and transculturation*, 2nd edn (Abingdon, 2008), p.31.
3. Pratt, *Imperial eyes* (2008), p.12.
4. William Beinart, 'Men, science, travel and nature in the eighteenth- and nineteenth-century Cape', *Journal of southern African studies* 24:4 (1998), p.775-99 (777).
5. Bronwen Douglas, 'Philosophers, naturalists and antipodean encounters, 1748-1803', *Intellectual history review* 23:3 (2013), p.387-409 (392).
6. Douglas, 'Philosophers, naturalists and antipodean encounters', p.408.

Scottish-trained Linnaean natural historians in facilitating British colonialism and in contributing to the creation of racial discourses. In doing so, we argue that historians need to explore how the entanglement of multiple intellectual traditions in colonial contexts shaped ethnographic texts. This entanglement of Linnaean natural history, Scottish Enlightenment thought and colonial expeditions was powerfully suggested in the anonymous English preface to the Linnaean apostle Anders Sparrman's *A Voyage to the Cape of Good Hope* (1785):

> every authentic and well-written book of voyages and travels is, in fact, a treatise of experimental philosophy. From these sources natural history derives its most copious streams [...] With respect to moral philosophy and metaphysics, it may be sufficient to mention the works of PAUW, RAYNAL, KAIMES, and ROBERTSON, to evince the utility of journals and descriptions of voyages in these and similar researches.[7]

Although frequently quoted in academic works on eighteenth-century travel writing and natural history, the significance of this quote rests on the connection between Sparrman's Linnaean natural history and his endorsement of Scottish moral philosophy.[8] His references to the prominent Scottish Enlightenment thinkers Lord Kames and William Robertson underscores the 'striking extent' to which Scottish Enlightenment stadial assumptions about human progress and sociological explanations of human diversity were shared among natural historians in the late eighteenth century.[9] In addressing this connection, we will show the significance of the encounter between Linnaean natural history and the Scottish Enlightenment both in Scottish university curricula and in colonial settings. The latter will be accomplished through an analysis of the ethnographic reflections of two 'travellers' trained in both 'traditions' at the University of Edinburgh: Mungo Park (1771-1806) and William Somerville (1771-1860). Their work highlights the complicated circulation of knowledge, and the conceptual fluidity of ethnographic knowledge in particular, in the context of colonial encounter and exchange.[10]

7. Anders Sparrman, *A Voyage to the Cape of Good Hope*, 2nd edn (London, G. G. J and J. Robinson, 1786), vol.1, p.iii-iv.
8. This Preface appears only in the English edition. Authorship of the English Preface is obscure, although some scholars have identified Georg Forster (1754-1794) as the most likely candidate. See John Gascoigne, 'Crossing the Pillars of Hercules: Francis Bacon, the Scientific Revolution and the New World', in *Science in the age of Baroque*, ed. Ofer Gal and Raz Chen-Morris (Dordrecht, 2013), p.233.
9. László Kontler, 'Mankind and its histories: William Robertson, Georg Forster and a late eighteenth-century German debate', *Intellectual history review* 23:3 (2013), p.411-29 (416).
10. This circulation of knowledge has recently been insightfully explored in: László Kontler, Antonella Romano, Silvia Sebastiani and Borbála Zsuzsanna Török (ed.), *Negotiating knowledge in early modern empires: a decentred view* (Basingstoke, 2015).

Linnaean classification and ethnography

In the first edition of *Systema naturae* (1735) Linneaus included humans in his grand taxonomy project of classifying the world's plants, minerals and animals. Humans were now for the first time systematically organised and ranked. Linnaeus sub-divided people into four varieties in accordance with their colour and geography. Referring to his four groups as *varieties* rather than races, the latter term not becoming dominant until the end of the eighteenth century, Linnaeus classified humans according to the four continents: *Europaeus*, *Americanus*, *Asiaticus* and *Africanus*.[11]

For his canonical tenth edition of 1758/59, he added more information on the purported differences between the varieties including hair and eye colour, temperament, body type, traits, clothing and customs. The finest attributes were allocated to the 'European' who was described as 'inventive' and '*[g]overned* by laws', whereas the 'American' was '*obstinate*' and '*[r]egulated* by customs'. The 'Asiatic' was 'haughty' and '*[g]overned* by opinions' while the 'African' was classified in the worst terms as 'indolent' and '*[g]overned* by caprice'. In this edition Linnaeus also added a new variety of *Homo sapiens*: *Ferus* (four-footed, mute and hairy). He also included a new sub-human species *Troglodytes* (a nocturnal species) designed to bridge the gap between man and ape.[12]

In his classification of *Homo sapiens* Linnaeus was informed by classical literature, myths, folklore and travel accounts. Linnaeus and his students also produced their own travel reports from their expeditions and the act of travelling was considered an essential part of being a Linnaean naturalist. 'Experience', as Linnaeus told his students, should come 'first' and then 'reasoning should follow'.[13] During his own expedition to Sápmi, Linnaeus both described the region's plants and observed the northern inhabitants. He was particularly interested in the indigenous Sami people, whose customs, beliefs, health and physiognomy he described in detail. His students started off with smaller field trips

11. This chapter's discussion of Linnaeus is partly drawn from Linda Andersson Burnett's 'Northern noble savages? Edward Daniel Clarke and British primitivist narratives on Scotland and Scandinavia *c.*1760-1822', unpublished doctoral dissertation, University of Edinburgh, 2012. For a discussion of Linnaeus's taxonomy see also Gunnar Broberg, *Homo sapiens L.: Studier i Carl von Linnés naturuppfattning och människolära* (Stockholm, 1975), p.157, 222. For discussion of race in the eighteenth century see Roxann Wheeler, *The Complexion of race: categories of difference in eighteenth-century Britain* (Philadelphia, PA, 2000).
12. Carl Linnaeus, *A General system of nature through the three grand kingdoms of animals, vegetables and minerals*, translated by W. Turton (London, 1802), vol.1, p.9-10.
13. Carl Linnaeus, 'An oration concerning the necessity of travelling in one's own countrey, made by Dr Linnaeus at Upsal, Oct. 17, anno 1741, when he was admitted to the royal and ordinary profession of physic', in B. Stillingfleet, *Miscellaneous tracts relating to natural history, husbandry and physick* (London, 1759), p.4.

around Uppsala before setting off on larger domestic expeditions, and for around twenty students imperial arenas became their destinations.[14] To guide his students on their travels, and to mould their interpretations of what they encountered, Linnaeus gave a published lecture on domestic travel and wrote a dissertation, *Instructions for travelling natural historians* (1759), on overseas expeditions. In the latter, Linnaeus urged his disciples to daily put down their many observations about the landscapes, cultures and peoples they encountered and studied. They should, for example, study what diseases native populations suffered from, their diet, their consumption of luxury articles, their work, customs, their reproductive patterns and their characters. Linnaeus also instructed his students to visit not only natural historians but also other educated men, and to learn about the history of the places they visited.[15] Although not the pioneer, Linnaeus was very much at the forefront of establishing a model in which the complete naturalist included the study of human societies in his natural history project.[16]

Linnaeus returned to the topic of his own expedition during his inaugural lecture at Uppsala University in 1741 in which he preached about the importance of natural historians carrying out domestic expeditions before setting out on overseas trips. In his oration, which was translated into English and published in Stillingfleet's *Miscellaneous tracts* in 1759, he informed his audience about his hardships in Sápmi and stressed the utilitarian aspects of the expedition which had 'furnished' him 'with a hundred rare plants'.[17] Linnaeus also told his listeners about the importance of being open to learning from 'barbarians' and he refuted the notion that their knowledge was 'not [...] worthy to be communicated to the rest of mankind'.[18] To back up his argument, he referred to early colonial accounts such as Dampier's *Voyage* and its description of the 'skill of savages of America in the chirurgical way'.[19] When a Swedish immigrant in Dutch Surinam, a Mr D'Alberg, informed Linnaeus in 1761 that a new root with medical properties had been discovered by a free slave, Kwasímukámba of Tjedúa, Linnaeus named

14. For Linnaeus's Sápmi expedition see for example Carl-Otto von Sydow, 'Linnaeus and the Lapps', *Taxon* 25:1 (1976), p.9-14; Roger Jacobsson (ed.), *Så varför reser Linné?: perspektiv på Iter Lapponicum 1732* (Stockholm, 2005); Andersson Burnett, 'Northern noble savages?', p.38-54. For a general discussion of Linnaean travel, see Hodacs and Nyberg, *Naturalhistoria på resande fot*.
15. Carl Linné, 'Instruktion för resande naturforskare', in Hodacs and Nyberg, *Naturalhistoria på resande fot*, p.199-210.
16. John Gascoigne, *Joseph Banks and the English Enlightenment: useful knowledge and polite culture* (Cambridge, 2003), p.137.
17. Linnaeus, 'An oration', p.21.
18. Linnaeus, 'An oration', p.27.
19. Linnaeus, 'An oration', p.27.

the tree *Quassia amara* after him.[20] Having a utilitarian agenda, Linneaus's prime concern was for his scholars to gather plants and ethnographic data, particularly about medical practices. Linnaeus wanted specimens and information that would benefit the Swedish State and its inhabitants financially and medically, and he appears to have given little, if any, thought to the impact that natural-history expeditions had on local populations and their landscapes. This applied both to Sápmi, where Linnaeus's expedition was part of the Swedish State's colonial project to research and exploit Sami land and people, and to his students' participation in overseas colonial expeditions.[21]

Linnaeus and Britain

Linnaeus quickly won admirers in Holland and Britain for his Sápmi expedition. When he briefly visited Britain in 1736 to promote his *Systema naturae* (1735) it helped to open doors to the British natural-history elite. The German botanist and professor at Oxford, Johan Jacob Dillenius, wrote a letter to the influential botanist Dr Richard Richardson introducing Linneaus as a 'new botanist [who] is arisen in the North' and who had both founded a new taxonomy and 'travelled over Lapland'.[22] During his visit Linnaeus met and formed close links with prominent natural historians and with Hans Sloan, the President of the Royal Society.[23]

20. Susan Scott Parrish, *American curiosity: cultures of natural history in the colonial British Atlantic* (Chapel Hill, NC, 2006), p.1.
21. A discussion of the colonial implications of Linnaeus's Sápmi expedition is unfortunately beyond the scope of this chapter. For discussions on the colonial relations in Sápmi see for example Andersson Burnett, 'Northern noble savages?', p.38-51; Gunlög Fur, *Colonialism in the margins: cultural encounters in New Sweden and Lapland* (Leiden, 2006); Sverker Sörlin, 'Rituals and resources of natural history: the North and the Arctic in Swedish scientific nationalism', in M. Bravo and S. Sörlin (ed.) *Narrating the Arctic: a cultural history of Nordic scientific practices* (Canton, MA, 2002) p.73-122; Magdalena Naum and Jonas M. Nordin, *Scandinavian colonialism and the rise of modernity* (New York, 2013).
22. Wilfrid Blunt, with William T. Stearn, *The Compleat naturalist: a life of Linnaeus* (London, 1971), p.119; David Elliston Allen, *The Naturalist in Britain: a social history* (London, 1976), p.31.
23. Blunt, *The Compleat naturalist*, p.109-14. Linnaeus's descriptions of his Sápmi expedition were either quoted or referred to in prominent texts such as Stillingfleet's *Miscellaneous tracts on natural history* (1759) which included his lecture on domestic travel; Richard Pulteney's *A General view of the writings of Linnaeus* (1781); Lord Kames's *Sketches of the history of man* (1774); Thomas Pennant's *Arctic zoology* (1785-1787); and George Shaw's *General zoology* (1800-1812). Linnaeus's diary of his Sápmi expedition was translated into English in 1811 as *Lachesis Lapponica, or A Tour in Lapland*; see Linda Andersson Burnett, 'Selling the Sami: Nordic stereotypes and participatory media in Georgian Britain' in *Communicating the North: media structures and images in the making of the Nordic region*, ed. Jonas Harvard and Peter Stadius (Farnham, 2013), p.171-96.

The actual uptake of Linnaeus's botanical taxonomy in Britain was nevertheless quite slow due to the strong adherence to the classification system of John Ray. Linnaeus's big breakthrough in this regard happened in the 1750s with the publication of *Philosophia botanica* (1751), which gave a good overview of Linnaean botany and discussed all his key principles, followed by his *Species plantarum* (1753), which listed all the plants he knew of and constituted the formal introduction of his binomial system.[24] Benjamin Stillingfleet's popular *Miscellaneous tracts on natural history* (1759) also contributed to Linnaeus's fame by introducing in the vernacular the Linnaean system and by including translated essays by Linnaeus and several of his students. By the 1760s, important works modelled on Linnaean principles were published in Britain, and Linnaean classification was introduced to curricula at the universities of Cambridge, Edinburgh and Glasgow.[25]

Linnaeus never visited Britain again. However, of key importance for the dissemination of his taxonomy was the arrival in England of one of his most gifted students, Daniel Solander, in 1760.[26] In England, Solander was initially employed to organise the museum of the duchess of Portland before he was appointed as assistant librarian at the British Museum – the first institution to adopt the Linnaean system.[27] The popular Solander explained Linnaeus's works clearly and convinced many who were not yet converts.[28] Solander met Joseph Banks in 1764 and quickly became the latter's mentor in Linnaean taxonomy. The two also developed a lifelong friendship, becoming one of the most influential intellectual partnerships of the eighteenth century. They placed Linnaean botany at the centre of imperial activities when they travelled together with Cook on the *Endeavour* to the Pacific in 1768-1771. Banks surrounded himself with other Linnaean students, including Herman Spöring, an illustrator aboard the *Endeavour*, and Johann Christian Fabricius, who organised Banks's insect collection upon his return. Jonas Dryander meanwhile, became Banks's resident librarian after Solander's death.[29] Linnaean expertise was also present on Cook's second ex-

24. Blunt, *Compleat naturalist*, p.215.
25. These works included: James Lee's *Introduction to botany* (1760); William Hudson's *Flora Anglica* (1762); Thomas Lightfoot's *Flora Scotica* (1777); and Philip Miller's eighth edition of his popular *Gardener's dictionary* (1768). Allen, *The Naturalist in Britain* (1976), p.42; D. J. Mabberley, 'Hope, John (1725-1786)', in *Oxford dictionary of national biography* (Oxford, 2004), http://www.oxforddnb.com/view/article/13730 (last accessed 8 October 2016).
26. Blunt, *The Compleat naturalist*, p.193; Elizabeth Baigent, 'Solander, Daniel (1733-1782)', in *Oxford dictionary of national biography* (Oxford, 2004), http://www.oxforddnb.com/view/article/25985 (last accessed 15 December 2016).
27. Patricia Fara, *Sex, botany and empire* (Cambridge, 2004), p.39.
28. Gascoigne, *Joseph Banks*, p.101.
29. Gascoigne, *Joseph Banks*, p.105.

pedition (1772-1775) in the person of Anders Sparrman, the assistant to the father and son naturalists Johann Reinhold and Georg Forster.

Although British ethnography in the early eighteenth century was very unsystematic, the Cook expeditions between 1768 and 1779 served as a vehicle for a more thorough approach. As Bronwen Douglas has commented, by 'the mid-eighteenth century, human taxonomy was still unusual, despite the pioneering efforts of Bernier and Linnaeus'.[30] She contends that it was Buffon's publication of the 'Varieties of the human species' in 1749 that gave new impetus to systematic studies of human differences and to the taxonomic urge to describe and classify the various 'races'.[31] While academic institutions in France and Germany, as Jonas Gerlings' chapter in this volume discusses, lent unparalleled support to the development of this systematic study, British ethnography was not so well organised and remained an 'unspecialised' activity.[32] Nonetheless, Britain was not without significant support for ethnographic studies, especially in the focus on natural history at the University of Edinburgh.[33] This support, we argue, exemplified a complicated intellectual inheritance, blending Linnaean and Buffonian natural histories along with moral philosophy and historical conjecture.

Edinburgh, medicine and natural history

There were 'several generations of Scottish physicians' who received training in 'practical natural history' at the universities of Edinburgh and Glasgow in the late eighteenth and early nineteenth centuries and who subsequently travelled to make ethnographic 'observations' in a wide variety of colonial contexts.[34] These physicians often found employment in an official capacity as military or civilian administrators, or contracted 'explorers' throughout the Empire. Their writings serve as a unique window into the complicated intellectual influences they imbibed, and their efforts to apply and adapt them in varied locales. Though the Scottish Enlightenment stretched well beyond Edinburgh, the intellectual context of the city and its university was vibrant and influential. Edinburgh University not only hosted the appointments of several leading philosophers and natural historians – among whom William

30. Douglas, 'Philosophers, naturalists and antipodean encounters', p.388.
31. Douglas, 'Philosophers, naturalists and antipodean encounters', p.389.
32. Gascoigne, *Joseph Banks*, p.34, 158-59.
33. It must also be said that eighteenth-century Scottish natural historians built on a solid tradition of English/British natural history writing stretching back to the mid-seventeenth century. Alix Cooper, *Inventing the indigenous: local knowledge and natural history in early modern Europe* (Cambridge, 2007), p.118-20.
34. Richard Drayton, 'Knowledge and empire', in *The Oxford history of the British Empire*, vol.2, *The Eighteenth century*, ed. Peter J. Marshall (Oxford, 1998), p.231-52 (240).

Robertson, Adam Ferguson, William Cullen, John Hope, Joseph Black and John Walker. Within the wider city these figures were brought into contact with leading intellectuals outside the university such as Adam Smith and David Hume, as well as leading figures in the legal establishment who also produced texts on the 'science of man' such as Lords Kames and Monboddo. The university thus provided a focal point for the intellectual life of an exceptionally vibrant community of Scots scholars, many of whom clustered in and around the School of Medicine.

Founded in 1726, the School of Medicine soon earned a reputation as the leading medical school in Britain offering a systematic education in all the branches of medicine, including *Materia Medica* (for which botanical knowledge was essential), Chemistry, Medical Theory and Practice, Midwifery, Anatomy and Surgery, as well as a round of clinical lectures.[35] For all that, there was no rigid order to the curriculum. Students could attend any number of classes inside or outside the school, and in any given order. Because it was not necessary to obtain a full degree in order to practice medicine, it has been estimated that as few as 20 per cent of students (between 1765 and 1825) actually graduated.[36] This meant that while the school produced a small number of highly qualified graduates, many more students passed through with one or more years of study undertaken to improve their knowledge or broaden their education.

The university already had a chair in botany prior to the creation of the Medical School, at least a part of whose functions was to offer courses in the pharmacological properties of plants.[37] Botany was thus both a practical course of study within the medical curriculum, as well as having a significant prestige value in linking students and practitioners to what was rapidly becoming an extended global network of amateur and professional botanists at the cutting edge of 'scientific' knowledge. Classification of the kind practised by botanists was also, as Pickstone has suggested, a 'key' to medicine – a method of ordering knowledge about disease and treatment.[38] Linnaeus's *Systema naturae* was among the standard texts in the teaching of botany, and the Linnaean influence at Edinburgh was further consolidated by the appointments of Dr John Hope (1725-1786) in 1769 as Professor of Botany and *Materia Medica*, and the Reverend Dr John Walker (1731-1803) in 1779 to the chair in Natural

35. Lisa Rosner, 'Thistle on the Delaware: Edinburgh medical education and Philadelphia practice, 1800-1825', *Social history of medicine* 5:1 (1992), p.19-42 (20-21).
36. Rosner, 'Thistle on the Delaware', p.22.
37. Roger L. Emerson, 'The founding of the Edinburgh Medical School', *Journal of the history of medicine and allied sciences* 59:2 (2004), p.183-218 (188-89).
38. John V. Pickstone, *Ways of knowing: a new history of science, technology and medicine* (Manchester, 2000), p.71.

History, a position which was attached to the Medical School and which he held until he died in 1803.[39]

Walker's familiarity with natural history predated his appointment to the university. Walker, through his close patron relationships with improvement-minded aristocrats such as Lord Kames but also through his connections to the Church of Scotland and the Board of Annexed Estates, had in the 1760s and 1770s carried out a number of natural history expeditions to the Scottish Lowlands and Highlands.[40] Walker modelled his expeditions in the Scottish Highlands on Linnaeus's expedition to Sápmi.[41] Constructing himself as an intrepid explorer, Walker presented the Highlands, as Linnaeus had depicted Sápmi, as a wild and unmapped region that was worthy of the attention of the natural historian. Writing to Kames in 1764, Walker described for example the Highland landscape as 'inanimate & unfrequented as any in the Terra australis'.[42]

Walker also shared Linnaeus's belief in the importance of incorporating ethnographic observations into his natural history reports. Tellingly, he introduced his eventual report, *Natural history of the inhabitants of the Highlands*, by reflecting on the importance of ethnographic observation:

> The first Object of Nature, which is Man himself, has been too much neglected. For tho' the Naturalist claims not the extensive Study of human Nature and human Manners, as a Part of his Science; yet the natural History of a Country can be reckoned but incomplete, if it does not exhibit the most remarkable Phenomena, and all the striking Pecularities, to be observed among its Inhabitants.[43]

In surveying 'a sort of new World', as he described it to Kames, Walker expressed a profound interest in the racial identity of the inhabitants.[44] The Highlanders were, he supposed, 'one of the few Nations in Europe, who are aborigines of the Country in which they live', 'unconquered and

39. Emerson, 'The founding', p.193; Matthew D. Eddy, *The Language of mineralogy: John Walker, chemistry and the Edinburgh Medical School, 1750-1800* (Aldershot, 2008), p.25; Henry J. Noltie, *John Hope (1725-1786): Alan G. Morton's memoir of a Scottish botanist* (Edinburgh, 2011).

40. Fredrik Albritton Jonsson, *Enlightenment's frontier: the Scottish Highlands and the origins of environmentalism* (New Haven, CT, 2013), p.63-64.

41. Linda Andersson Burnett, 'An eighteenth-century ecology of knowledge: patronage and natural history', in *Culture unbound* 6 (2014), p.1275-97 (1279).

42. Reverend Dr John Walker, Correspondence, Letters, 1762-1779, Edinburgh University Archive, LAIII352, Centre for Research Collections (CRC), University of Edinburgh (EUL).

43. John Walker, 'Natural history of the inhabitants of the Highlands', Walker Papers, Dc. 1.18 (CRC, EUL).

44. Walker to Kames, Moffat, 10 December1764, in Reverend Dr John Walker, Correspondence, Letters, 1762-1779, LAIII352.1 (CRC, EUL).

unmixed' with other races.[45] Only by 'slow Degrees' is any 'considerable alteration of aspect' achieved in 'any Race of Men', hence their correct observation and classification according to a systematic scheme was an important indication of the deep history of a people.[46]

Once Walker obtained the chair in Natural History, with Kames and William Cullen's help, he delivered lectures in the broad ethos of natural history by teaching his students botany, meteorology, geology, zoology, ethnography and racial classification. Walker began his lectures by emphasising the utility of the Linnaean system of classification, which he presented as: 'our guide in preference to all others [...] a system [which is] the best adapted to use [...] the most accurate in its arrangement: and a knowledge of its terms & characters is indispensibly necessary to every one that would become proficient in the study of natural history.'[47] According to Walker, the Linnaean system was clearly a superior method to that proposed by Buffon. 'Mons. Buffon', Walker propounded, 'undervalues all the advantages of a systematic arrangement and proceeds in his history without either classes or order.'[48]

Walker's employment of Linnaean taxonomy went beyond a merely scholarly interest in classification. Linnaean natural history represented a practical and adaptable methodology for natural historians, orienting them not just to the study of phenomena, but their direct observation 'in the field'. Walker's 'Queries concerning the North of Scotland', for example, begins with a Latin quotation from Linnaeus himself on the importance of domestic travel from Linnaeus's own lecture on this topic.[49] Though he did not himself travel to the colonies, Walker was exposed to various sources of ethnographic knowledge through his membership of the Royal Society of Edinburgh, and from his copious notes taken from the publications of Linnaean 'travellers' Kalm and Osbeck.[50] Walker was also a conduit through which colonial ethnographic knowledge circulated back to the heartland of the Scottish Enlightenment. Among his students was the American, Benjamin Smith Barton, who had recorded ethnographic information about Native Americans during a visit to a Native American village on the Beaver River in Ohio in 1785 and who presented his findings in an

45. John Walker, 'Natural history of the inhabitants of the Highlands', Walker Papers, Dc. 1.18 (CRC, EUL).
46. This was a matter of considerable importance to Kames and a number of Scots intellectuals who were enthused by James Macpherson's Ossianic publications of alleged ancient Highland poetry. Walker was one of these prominent Lowland supporters of Macpherson and he referred to, for example, Ossianic place names in his inventories.
47. John Walker, MS Lectures, Lecture 3, Dc. 2.17 (CRC, EUL).
48. John Walker, MS Lectures, Lecture 53, Dc. 2.17 (CRC, EUL).
49. John Walker, 'Queries concerning the North of Scotland', Dc. 2. 36 (CRC, EUL).
50. Reverend Dr John Walker, Walkers Papers, MS Dc. 1. 57 (CRC, EUL).

intemperately worded dissertation 'The natural history of the North American Indians', which he presented to the Royal Medical Society of Edinburgh in 1788.[51]

Intellectual historians have begun to emphasise the importance of, and to explore the manifold connections between Scottish moral, historical, economic and social thought, and the development of natural history.[52] It was the influential Edinburgh professor of moral philosophy, Dugald Stewart, who gave Scottish moral and historical thought a distinctive identity by describing it as 'conjectural history'. He first formulated the notion in his lectures on moral philosophy in 1789-1790 when he maintained that the 'chasms' of knowledge in ancient written records and in contemporary travellers' observations of 'man in a Savage State' required the 'Philosopher' to supply the deficiency 'by conjecture & such kind of writing may be called conjectural history. It has been successfully used by D'allembert [...], Mr Hume in his Natural History of Religion, Mr Smith, Lord Kaimes Etc.'[53] For Stewart, the Scottish contribution to knowledge lay precisely in providing a natural history of humanity that drew equally on written records, on direct observation by 'philosophical travellers', and informed conjecture.

By these means, Scottish philosophers developed an influential model of stadial historical progress, in which all human societies progress through stages of development from 'savagery' to 'civilisation'. This progressive view of history was developed by thinkers such as Adam Ferguson, William Robertson and Lord Kames. Underpinning stadial theory was the belief that societies in different stages of development could coexist at the same time and that more or less all people passed through the same stages of development which could be discerned and studied by means of natural historical observation and classification. The strength of Linnaean taxonomy as an exemplar of natural history was its simplicity and brevity, which left little space for theoretical arguments. Linnaeus's interest and forte was in classifying human varieties, not in explaining their origin or development.

Both natural history and stadial history were communicated to generations of medical and moral philosophy students at Edinburgh and other universities. Although natural history and moral philosophy were taught as separate subjects within different faculties in the Scottish universities, students could attend any lecture on any subject by paying

51. Joseph Ewan and Nesta Ewan, *Benjamin Smith Barton: naturalist and physician in Jeffersonian America* (St Louis, MO, 2007), p.97, 120.

52. Silvia Sebastiani, *The Scottish Enlightenment: race, gender and the limits of progress* (New York, 2013); Jonsson, *Enlightenment's frontier*, p.130-31.

53. 'Lectures on moral philosophy delivered by Professor Dugald Stewart, session 1789 & 1790', vol.1, Gen. 1987-89 (CRC, EUL).

the professor's fee for his lectures, and students often studied the same 'canonical' literature. Despite appearing to be two binary Enlightenment traditions of thought, natural history and stadial theory can be seen as two entangled intellectual orientations that borrowed from each other and enriched each other even though they provided, at times, conflicting interpretations of the world.[54] The teaching of moral philosophy at Edinburgh was, for example, framed as 'the science of man' with more in common with the orientation of natural history than one might suppose. Indeed, Silvia Sebastiani has defined stadial history as a 'natural history' in that it classified and described creation and highlighted *natural* rather than divine causes of the development of societies. Lord Kames, for instance, intended that his *Sketches of the history of man* should be a natural history of how 'mankind' had progressed, while Adam Ferguson adopted the method of natural history to combine 'conjecture' with factual 'observation' in his *An Essay on the history of civil society* (1767).[55] In his lectures on moral philosophy Ferguson likewise presented 'Pneumatics [...] or the physical science of mind' as the 'foundation of moral philosophy', and his lectures incorporated racial taxonomies drawn from both Linnaeus and Buffon.[56] Natural historians were correspondingly inspired by their colleagues and friends who taught stadial theory. Walker noted, for instance, the importance of studying not only the 'progress of human life from Infancy to manhood' but also of studying the development of countries.[57] Occasionally, natural historians could feed back critical ethnographic material questioning stadial assumptions, as in Barton's intemperately worded attack on William Robertson's celebrated account of Native Americans (in his 1777 *History of America*) as 'copiously' built on 'whatever' his 'imagination suggested'.[58] Among

54. Although in this article we are looking at the Edinburgh-trained natural historians, Linnaeus's student, Anders Erik Sparrman, demonstrated his own blending of Scottish stadial assumptions along with a Linnaean taxonomic approach. Further examination of Sparrman and his context lies beyond the scope of the present paper.

55. Adam Ferguson, *An Essay on the history of civil society* (1767; Edinburgh, 1967), p.2. Ferguson's distinction was similar to that drawn by Linnaeus between 'reason' and 'experience' as two pathways to knowledge that the successful natural historian must learn to integrate; see Linnaeus, 'An oration', p.5-6. We would like to thank Kenneth Nyberg for making this observation.

56. Adam Ferguson, *Institutes of moral philosophy* (1769; Basil, James Decker, 1800), p.10, 13, 16-17. See, for example, Paul B. Wood, 'The science of man', in *Cultures of natural history*, ed. Nicholas Jardine, James A. Secord and Emma C. Spary (Cambridge, 1996), p.197-210 (205).

57. 'An inquiry into appearances that generally precede the downfall of states, January 1774', Reverend Dr Walker, Walkers Papers, MS Dc. 1. 59.

58. Robertson also made a highly selective use of questionnaire responses returned to him from America, suggesting that he down-played ethnographic knowledge developed 'in the field'. See Robertson-Macdonald Papers, Materials collected by Principal Robertson, National Library of Scotland (NLS), MS 3954. Similar criticisms have been made of Adam Smith's use of ethnographic sources in C. Marouby, 'Adam Smith and the anthropology of

Barton's criticisms was that Robertson and other speculative historians were 'deficient' in the 'Science of Nature'. By contrast, Walker schooled his students in the rigorous Linnaean classification of humanity into different varieties in his lecture on mammals – a lecture he delivered over many years.[59]

This is not to say that all eighteenth-century Scottish natural historians or moral philosophers adhered to Linnaean racial taxonomy. The rigidity of Linnaeus's human varieties was criticised in particular by those who preferred the more theoretical Buffon, who argued there was no such thing as fixed identities and who instead put differences between humans down to climate, diet and habitation.[60] Other Scottish philosophers such as Ferguson believed that 'race' may be useful as a categorisation but not as an explanation for progress or lack of it. Ferguson argued in his lectures in the 1770s, for example, that racial differences may be no more significant than national differences. As he put it: 'by the same reason that the European is distinguished from the Tartar, the Italian, tho in a less degree may be distinguished from the Swede'.[61] Historical progress was instead interpreted as a question of continuous improvement driven by universal human ingenuity. Humans, he lectured, 'set out with practices suited to their circumstances & which they can change with their circumstances'.[62] Linnaeus's controversial taxonomy therefore also had an impact by encouraging the adaptation, blending and amendment of rival taxonomies.

Ferguson retired in 1785 and was succeeded by his former student and friend, Dugald Stewart. The notes from Stewart's lectures on moral philosophy attest to the continuing significance placed on race and racial taxonomies at Edinburgh University. In 1789/90, Stewart employed a similar though seemingly more elaborate model of racial description to that employed by his predecessor. Stewart disparaged the idea that climate and geography could be used to explain racial differences. Rather, Stewart 'conjectured' that physical differences between populations emerged 'spontaneously' but only gradually developed into racial differences.[63] Nonetheless, in the following years he also asserted,

---

Enlightenment: the "ethnographic" sources of economic progress', in *The Anthropology of the Enlightenment*, ed. Larry Wolff and Marco Cipollini (Stanford, CA, 2007), p.87-91.

59. John Walker, MS Lecture 53, Dc. 2.20 (CRC, EUL).
60. Ivan Hannaford, *Race: the history of an idea in the West* (Baltimore, MD, 1996), p.205.
61. Adam Ferguson, Lecture 12, 1776/1779, Adam Ferguson, 'Lectures in pneumatics', 3 vols, Dc. 1. 84-86 (CRC, EUL).
62. Adam Ferguson, Lecture 29, 1784, Adam Ferguson, 'Lectures in pneumatics', 3 vols, Dc. 1. 84-86 (CRC, EUL).
63. 'Lectures on moral philosophy delivered by Professor Dugald Stewart, session 1789 & 1790', Gen. 1987-1989 (CRC, EUL), vol.1, no pagination.

following Ferguson, that historical progress was driven more by 'moral' factors – by human ingenuity and temperament – than by physical endowments.[64] Stewart's references to race exemplify the entanglement of natural history with moral philosophy, 'conjectural' history and travel writing. The concept of race played an important part in his lectures. It was a loosely defined concept, and served as both a physical descriptor and a subject for philosophical speculation.[65] This entanglement remained a characteristic feature of the ethnographic writings of former students of the University who travelled further afield.

Natural history students in the colonies

Mungo Park and William Somerville both studied under Walker at the University of Edinburgh in the early 1790s.[66] Their education and subsequent writings were produced at a time of considerable conceptual and discursive malleability and intellectual fluidity in the understanding of human diversity. Although a biologically determinist concept of race did not predominate over universalist assumptions of stadial progress until the 1820s, there is a detectable difference between Park and Somerville in their uses of the concept. Neither subscribed to a rigid definition of race, nor did they conceptualise human difference exclusively by reference to it. Nonetheless, their intellectual backgrounds and subsequent careers may be used to chart the pre-history of what later became a dominant discourse of race.

Through connections to Sir Joseph Banks, Park obtained his first position as assistant surgeon aboard an East India Company ship to Sumatra in 1793. One outcome of the expedition was a publication by Park in the *Proceedings of the Linnaean Society* on some species of marine fish. In 1794 he was selected as the leader of an expedition on behalf of the African Association. This expedition was undertaken between 1795

64. Dugald Stewart's 'Lectures on moral philosophy', 1793-1794, Dc. 4. 97 (CRC, EUL), p.290.
65. Stewart, for example, refuted Lord Kames' theory that climate wrought differences between races, by arguing that such differences spring up spontaneously 'as a kind of *Lusus Naturae*' (or freak of nature), and that by 'intermixing' and locating themselves in different parts of the world, different races thus developed. This, he claimed, was a 'conjecture' but not an 'unphilosophical' one. To prove the point, Stewart cited Captain Cook's expeditions to the Pacific to refute Kames' polygenist argument (for multiple racial origins of humanity) in favour of a monogenist view that all the Pacific Islanders were descended from the same 'Mallais Race'; 'Lectures on moral philosophy delivered by Professor Dugald Stewart, session 1789 & 1790', Gen. 1987-1989 (CRC, EUL), vol.1, no pagination.
66. Park attended Walker's classes in 1791, while Somerville attended them in 1793; see Eddy, *The Language of mineralogy*, p.245, 247. Another of Walker's former students who went on to travel to the colonies was Robert Brown (1792); see Eddy, *The Language of mineralogy*, p.232. Brown accompanied Matthew Flinders on his circumnavigation of Australia in 1799-1802 and went on to become the doyen of British botany in the early nineteenth century.

and 1797, and the journal of his travels along the Gambia and Niger rivers, *Travels in the interior districts of Africa* which appeared in 1799, was an enormous success. The journal provided evidence of Park's familiarity with Linnaean taxonomy and European botanical scholarship. At one point, for instance, he described the Africans harvesting berries from the plant named by Linnaeus as *Rhamnus lotus* for making into bread, and also mentioned the illustrations of related plants by Desfontaines in the *Mémoires de l'Académie royal des sciences*.[67] When it came to describing the human populations in Africa, Park did not employ a rigid taxonomy of any kind. Although he employed the terminology of race, it encompassed both physical and social characteristics. In this vein, he wrote that the 'Jaloffs (or Yallofs)' were 'an active, powerful, and warlike race' differing from the 'Mandingoes' in 'language, but likewise in complexion and features'.[68] It is significant that Park here linked physical characteristics such as the profile of the face and the complexion of the skin alongside language as features of 'race'. Much later in his narrative, he spoke of the 'Mandingoes' as a 'gentle race', though prone to stealing his goods.[69] The thievery of the 'Mandingoes' was not a quality of their biology, but their social condition. It was explicable, he argued, by the fact that the laws offered him, a European traveller carrying valuable goods, no protection. If a foreign traveller with valuable goods arrived in England and received no security under the law, Park speculated that the same fate would befall him.

Park's references to race thus indicate that the term carried a certain ambivalence. The terminology of race did not imply an absolute biological endowment, but an interplay between physical features and social influences. Among the most significant of these social influences was the institution of slavery. In Park's account, slavery was ubiquitous throughout Africa, accounting for up to 'three-fourths' of the entire population who exist in a 'state of hopeless and hereditary slavery [...] employed in cultivating the land, in the care of cattle, and in servile offices of all kinds, much in the same manner as the slaves in the West Indies'.[70] The antiquity of African slavery made it a defining characteristic for the great majority among the native population:

> A state of subordination and certain inequalities of rank and condition are inevitable in every stage of civil society; but when this subordination is carried to so great a length, that the persons and services of one part of the

67. Mungo Park, *Travels in the interior districts of Africa performed under the direction and patronage of the African Association in the years 1795, 1796 and 1797*, introduction and notes by Bernard Waites (1799; Ware, 2002), p.90-91.
68. Park, *Travels*, p.14.
69. Park, *Travels*, p.242.
70. Park, *Travels*, p.19.

community are entirely at the disposal of another part, it may then be denominated a state of slavery; and in this condition of life a great body of the Negro inhabitants of Africa have continued from the most early period of their history, with this aggravation, that their children are born to no other inheritance.[71]

Park's attitude to slavery and its role within his text became a source of some controversy immediately after publication. Park returned from Africa with copious notes that he originally struggled to form into a compelling narrative. A decisive role was played by the MP and Secretary of the African Association, Bryan Edwards, who acted as 'ghost writer' for the first ten chapters of the journal, and editor of the remaining chapters. Edwards was a vigorous opponent of the growing Abolitionist movement in Britain, and his 1793 *History of the British West Indies* retailed a variety of justifications of African slavery, and its economic value to British and colonial commerce.[72] The controversy over Park's published journal centred on its notable rejection of the progressive narrative the Abolitionist movement sought to project that halting the African slave trade would drive up demand for labour, bolster the money economy, and lead the development in Africa of a commercial society based on demand for European imports. Park's journal however, struck a discordant note by arguing that slavery in Africa pre-dated the European slave trade, and that African slavery would persist beyond British abolition of the trade. While some attributed these sentiments to Edwards rather than Park, it appears that Park's observations of African slavery bore little resemblance to Edward's specious justifications in his earlier *History*.[73] While we cannot be sure of the exact extent to which Edwards may have shaped Park's text, there are good grounds for supposing that Park's observations of African slavery were shaped more decisively by his engagement with Scottish Enlightenment social theory and moral philosophy.[74]

71. Park, *Travels*, p.279.
72. See, for example, Bryan Edwards, *The History, civil and commercial, of the British colonies in the West Indies*, 3rd edn (1793; London, 1801), vol.2, p.39-42, 129-30.
73. Lester has argued that such arguments, premised on 'African racial inferiority', were developed by 'West Indies slave owners [...] forced to articulate a *theory*' justifying slavery because the 'anti-slavery lobby challenged their existing *practices*'. In a similar way, he argues, 'settlers on the frontiers of Britain's expanding empire during succeeding decades were compelled to construct more theoretically consistent and empirically referenced ideas about racial differences because of humanitarian criticisms of their capitalist projects'. As we argue here, however, this view pays scant regard to the complex formulation of such ideas within a diverse discursive environment. See Alan Lester, 'Colonial settlers and metropole: racial discourse in the early nineteenth-century Cape Colony, Australia and New Zealand', *Landscape research* 27:1 (2002), p.39-49 (40).
74. Edwards had also been a critic of William Robertson's 'conjectural' history of the 'undifferentiated character' of American 'savages' in his *History of America*. See Neil

It is well known that Park left for Africa with a copy of Dugald Stewart's *Elements of the philosophy of the human mind*, though no scholars have yet paid attention to its possible resonances in Park's published journal.[75] While we cannot be sure that Park attended any of Stewart's courses, Stewart's lecture notes indicated that, in common with Ferguson and other Scottish Enlightenment thinkers, the development of moral principles was considered an index in the natural history of man. It was a fundamental axiom of Stewart's political thought that the progress of humankind toward 'perfection' was a 'gradual' process requiring no sudden innovations, but the cautious and prudent shepherding by 'statesmen' toward that ultimate aim.[76] In Stewart's scheme, gradual progress charted a path from mere savage and brutish 'prejudice' to enlightened and civilised knowledge. To this end, he speculated that a 'history of human prejudices [...] may, at some future period, furnish to the philosopher, a subject of speculation, no less pleasing and instructive, than that beneficent wisdom of nature, which guides the operations of the lower animals'.[77] It is in this sense that we might understand Park's analysis of the effects of African slavery not as a justification for an odious economic institution, but as a constitutive element in the natural history of African populations. Seen in this light, the aim of Park's natural history was to explain the causes for what he considered to be the relative lack of moral and intellectual progress in Africa. His explanation was not dependent on biological essentialism, but on the social and cultural forces prevailing in Africa.

Park's most extended reflection on African slavery occurred in chapters 20-22 and prompted considerable controversy at the time. These chapters represented an interruption in the narrative flow of the text, and were represented not as a first person narrative, but as an objective social analysis. Significantly, this analysis was introduced at the end of chapter 19 by an exchange that, following Bronwen Douglas, we might call an 'indigenous countersign'. During an extended stay in Kamalia, Park recalled that he encountered an enslaved warrior begging for food. Park was shocked to discover that he had previously encountered the warrior and benefitted from his hospitality. Now Park could provide only a poor return to the warrior's hospitality by offering a pitiful handful of food. In return he was told the warrior's story of how he and his

---

Hargraves, 'Beyond the savage character: Mexicans, Peruvians and the "imperfectly civilised" in William Robertson's *History of America*', in *The Anthropology of the Enlightenment*, ed. Larry Wolff and Marco Cipolloni (Stanford, CA, 2007) p.103-18 (104-105, 108).

75. Park, *Travels*, p.ix.

76. Dugald Stewart, *Elements of the philosophy of the human mind* (London, Strahan and Cadell, 1792), p.260.

77. Stewart, *Elements*, p.271.

companions were taken as 'prisoners of war' and would now be sold as slaves.[78] The encounter is significant in two senses. First, Park was moved by the humanity of this encounter – his pity for the warrior, his shame at the inability to adequately return the warrior's original hospitality, and his powerlessness to alter the course of events. Indeed, here as elsewhere, Park's text gave ample evidence not only of African agency, but African moral sentiment frequently expressed by African women. On another occasion, for instance, after having been robbed of his goods in the kingdom of Kajaaga, Park recalled that:

> Towards evening, as I was sitting upon the Bentang [his bed] chewing straws, an old female slave passing by with a basket upon her head asked me *if I had got my dinner*. As I thought she only laughed at me, I gave no answer; but my boy, who was sitting close by, answered for me, and told her that the king's people had robbed me of all my money. On hearing this, the good old woman, with a look of unaffected benevolence, immediately took the basket from her head, and showing me that it contained ground-nuts [peanuts], asked me if I could eat them; being answered in the affirmative, she presented me with a few handfuls, and walked away before I had time to thank her for this seasonable supply. This trifling circumstance gave me peculiar satisfaction. I reflected with pleasure on the conduct of this poor untutored slave, who, without examining into my character or circumstances, listened implicitly to the dictates of her own heart. Experience had taught her that hunger was painful, and her own distresses made her commiserate those of others.[79]

For present purposes however, the second feature of Park's narrative that we wish to emphasise here is that Park's own poignant powerlessness to help the enslaved warrior established the absolute force in Park's mind of the connection between slavery and war. The tenor of Park's discussion in the following chapters was such that the twin institutions of war and slavery operated as homogenising forces across the face of the continent. Despite the very great advantages of fertile lands and abundant water, he supposed that Africa remained sparsely populated: 'I found many extensive and beautiful districts' he wrote, 'entirely destitute of inhabitants [...] thinly peopled or entirely deserted.'[80] A large part of the reason, he argued, was that Africa was riven by wars whose sole or predominant purpose was plunder and enslavement of the vanquished. Park supposed that war was the most probable origin of slavery and was

78. Park, *Travels*, p.237.
79. Park, *Travels*, p.62.
80. Park, *Travels*, p.242. In this sense, Park might be said to have exemplified the shift in European writing toward seeing equatorial lands in the 'torrid zone' as possessing 'overall abundance' and having the capacity to support civil societies. Neil Safier, 'The tenacious travels of the Torrid Zone and the global dimensions of geographical knowledge in the eighteenth century', *Journal of early modern history* 18:1-2 (2014), p.141-72 (147).

still the cause of most slavery in Africa.[81] Park was emphatic about the ubiquitous nature of war:

> In a country divided into a thousand petty states, mostly independent and jealous of each other, where every free man is accustomed to arms and fond of military achievements, where the youth, who has practised the bow and spear from his infancy, longs for nothing so much as an opportunity to display his valour, it is natural to imagine that wars frequently originate from very frivolous provocation. When one nation is more powerful than another a pretext is seldom wanting for commencing hostilities.

Park concluded that there were two kinds of war in Africa. The first kind, which he called 'killi', was most similar to European war 'because such wars are openly avowed and previously declared'. There the similarity ended, because these wars operated as an 'exterminating system', the object of which was to enslave as many of the vanquished as possible, killing all those not serviceable as slaves. The second form of war was called 'tegria' and consisted in undeclared raiding, plundering and looting of enemy territory in a perennial cycle of vengeance driven by 'hereditary feud'. The pervasive destruction of these devastating wars and the dominance of the institution of slavery throughout Africa played a key role in the text. The recurrent destruction of war prevented the intergenerational intensification of the rational exploitation of nature. Villages burned and abandoned in war were soon rebuilt, but the rebuilding and resulting devastation acted as a powerful check on progress while enslavement acted as a constant drain on population. The result, as Park so imperiously declared, was that 'all the Negro nations that fell under my observation, though divided into a number of petty independent states, subsist chiefly by the same means, live nearly in the same temperature, and possess a wonderful similarity of disposition.'[82]

Park's pessimistic account of African homogeneity and abasement by war and slavery was the reverse image of Stewart's optimistic projection of the interconnection of 'natural causes' that drove the rapid advance of Europe beyond other parts of the globe. As Stewart described it:

> The alterations which have taken place in the art of war, in consequence of the invention of firearms [...] have given to civilised nations a security against the irruptions of barbarians, which they never before possessed. The more extended, and the more constant intercourse, which the improvements in commerce and in the art of navigation have opened, among the distant quarters of the globe, cannot fail to operate in undermining local and national prejudices, and in imparting to the whole species the intellectual

81. Park, *Travels*, p.282. The following discussion is based on p.282-85.
82. Park, *Travels*, p.242.

acquisitions of each particular community. The accumulated experience of ages has already taught the rulers of mankind, the most fruitful and most permanent sources of revenue, are to be derived, not from conquered and tributary provinces, but from the internal prosperity and wealth of their own subjects – and the same experience begins to teach nations, the increase of their wealth, so far from depending on the poverty and depression of their neighbours, is intimately connected to with their industry and opulence.[83]

Above all these forces for progress, however, Stewart identified the art of printing as making by far the most significant contribution not only in stimulating literature and the arts, but in allowing the intergenerational accumulation of knowledge. So profound was its effect, Stewart argued, that the 'art of printing [...] is entitled to be considered as a step in the natural history of man'. To doubt its powerful effects is to 'reason as unphilosophically, as the member of a savage tribe, who, deriving his own acquaintance with former times from oral tradition only, should affect to call in question the efficacy of written records, in accelerating the progress of knowledge and civilisation'. Once again, Park exemplified Stewart's method by providing a reverse image: 'the Negroes have no written language of their own' he wrote, and 'the general rule of decision is an appeal to *ancient custom*'.[84] For Park, the absence of writing not only explained Africa's lack of progress, it accounted for the remarkable homogenisation among Africa's population.

The homogeneity supposed by Park to subsist among West African populations contrasts with the picture presented by Somerville in South Africa. After taking his medical degree, Somerville joined the army as a surgeon. In this capacity he participated in the 1795 British conquest of the Dutch Cape Colony under Major General James Craig, whom he also served as private secretary. He went on to serve Craig's successor, Lord Macartney, as Secretary of the Court of Appeal. Although British takeover of the Cape Colony had been acknowledged by the Dutch, unrest persisted among the various Xhosa, Khoi and San (Bushman) tribes and nations to the north of the colony. Having been appointed Assistant Resident Commissioner in 1799, Somerville undertook expeditions in 1799 and again in 1801-1802 to negotiate a peace along this fractious frontier. In his father's autobiography, the 'object' of William's mission was described in terms that it is conceivable to think the son provided for his father:

to prevent that barbarous treatment of the native savages that had never been restrained by the preceding government of the Dutch, and to obtain a

83. Stewart, *Elements*, p.264-65.

84. Park, *Travels*, p.16.

> more intimate knowledge of the more distant parts of the country and of their inhabitants, as the basis of future plans for the extension of civilization, as well as the enlargement of the commerce of the British empire.[85]

Although Somerville's journal of his mission remained unpublished in his lifetime, like Park's more celebrated publication, it provides a window into the complicated discursive patterning of 'stadial' assumptions about human progress and invariable physical classifications of human races. Over the course of his expedition, it appears that Somerville's perspective on the tribes he observed seemed to shift. In one of his earliest observations on the situation of the San and Khoi tribes, he reflected that the 'savage' cruelty and deceit ascribed to them by the Boer and Dutch farmers was in fact learned from themselves. According to Somerville, it was the settler's avidity for pastoral land, and their violent dispossession that drove the San and Khoi to cattle raiding, and a 'plundering life' that became 'habitual'.[86] Somerville here exemplifies a familiar pattern of stadial thinking. His positive evaluation of African simplicity, in contrast to Boer cruelty, confirmed rather than undercut the stadial assumption that social progress from primitive simplicity is a learned attribute (thought subject to the ever present possibility of the degeneracy or cruelty among those deemed more 'civilised').[87] On succeeding pages, however, Somerville's analysis wavers between an emphasis on specifying the physical characteristics of the 'diminutive race' of San Bushmen and their particular origins, and a looser description of the 'general character of the nation' motivated not by biology but 'necessity' and 'exertion' illustrating the entire 'history of activity and energy throughout mankind'.[88]

Somerville often reflected on the effects of the barren environment on the conduct and dispositions of the indigenous peoples. He argued that the 'roving disposition' so apparent among indigenous peoples and settlers proceeded from the 'little inducement' that the 'country yields to excite what the inhabitants of all the known parts of the world is a

85. Thomas Somerville, *My own life and times 1741-1814* (Edinburgh, 1861), p.296. Somerville's father, the minister of Jedburgh, was well connected to the Edinburgh intellectual elite through the Kirk and through his former studies at the university. Among his 'intimate friends' and associates he cites Hume, Ferguson, Smith, Robertson, Kames, Stewart and others.

86. William Somerville, *William Somerville's narrative of his journeys to the Eastern Cape Frontier and to Lattakoe, 1799-1802* (Cape Town, 1979), p.26.

87. Clifton C. Crais, 'The vacant land: the mythology of British expansion in the Eastern Cape, South Africa', *Journal of social history* 25:2 (1991), p.255-75 (258-59). For corresponding French views see David Johnson, 'Representing the Cape "Hottentots": from the French Enlightenment to post-apartheid South Africa', *Eighteenth-century studies* 40:4 (2007), p.525-52 (534).

88. Somerville, *William Somerville's narrative*, p.27-30.

strong and frequently instinctive passion [of] local attachment'.[89] The vast extent of territory and the propensity of the San to roam across it in isolated tribes and parties was adduced by Somerville to explain how physical differences among them had arisen, leading to the speculation that there was 'more than one Stem' of Khoi and San descent 'who have now branched out into those hordes which are now known', differentiated by physical and moral traits – by stature and strength as well as valour and ingenuity.[90] Somerville's text indicates that he maintained a greater interest than Park in the physical properties of race. The question of racial origins continued to interest him, and the racial diversity and distinctions between the San and Khoi was a subject of repeated reflection. His propensity for physical description and delineation was most evident in the appendix to the journal entitled 'On the structure of Hottentot women' which was presented as an exercise in comparative physiognomy.[91] Nonetheless, it is telling that he sought for clues to the natural history of Khoi and San peoples not only in their physical characteristics, but in the linguistic connections between the various tribes and nations.[92]

It is crucial to remember that when Somerville writes about indigenous nations, their social structures, beliefs, the powers of their chiefs and kings, and their hunting techniques, he relied upon unnamed indigenous informants. As the chapters in this volume by Nyberg and Giraldo, and by Jensen each suggest, indigenous presence is everywhere in the texts of European colonists. That presence is unavoidable and obvious, but has been deliberately rendered silent. Much more than Park, it is Somerville who best exemplifies Pratt's observation about the colonising attitude of Enlightenment natural history. Indigenous voices are not recorded; their conversations and questions are not heard. This silent presence has direct implications for the way in which Somerville represents himself and his position as a communicator of ethnographic knowledge about other peoples who do not speak for themselves. One of the implications is that Somerville attempted to explain indigenous difference by reference to speculations about their racial origins that he thought were detectable in their physical characteristics and linguistic traits. Somerville was never in any doubt about the accuracy of his speculations, nor of the supposed superiority of Europeans and their culture over those he found in South Africa. It was he said, a 'poetical fancy of a Rousseau' to suppose 'the state of nature' as the 'only one in which true happiness is to be attained, and every deviation from this to be a

89. Somerville, *William Somerville's narrative*, p.69, 84.
90. Somerville, *William Somerville's narrative*, p.74.
91. Somerville, *William Somerville's narrative*, p.236-41.
92. Somerville, *William Somerville's narrative*, p.93, 118.

corruption of the original purity of the human race...'[93] In the 'state of Nature' that Somerville supposed was revealed all around him in Africa, want rather than plenty was the rule. This led to what he presumed to be the 'greatest enjoyment' of which a 'savage' is capable – a full stomach: 'appeased by the stimulus of agreeable food the business of the day is over the [...] savage is at peace with the world, he then exerts little attention to external objects' until hunger awakens exertion once again. Interestingly, Somerville here cited the authority not of his own scientific observations, but of (an as yet unidentified source by) Erasmus Darwin. Somerville pessimistically concluded that in contrast to Rousseau, savage virtues are not of the high-minded 'disinterested' type, but limited to those necessary for 'mutual security', while savage vices are at least as bad as those that 'disgrace humanity in civilized society'. 'Many and certain are the wants they daily feel and the ills to which they are constantly exposed. Few I am persuaded of those whose lot in polished society is deemed the most unhappy would willingly change conditions with the most envied of the African Chiefs. An idea of pastoral life formed from the alluring descriptions of Arcadian happiness would be found realized only in the elegant imagination of Fenelon.' Somerville's speculations here echoed a stadialist view that conditions of human life were determined by environmental circumstances and the moral and dispositional adaptations to them. While Somerville could speak a medicalised discourse of race and physiognomy, this did not constitute his entire intellectual framework.

Conclusion

Both Park and Somerville's ethnographic writings are offered here as examples of the complicated and entangled intellectual inheritance of late Enlightenment natural history. These Scotsmen were the product of an education that equipped them to apply European forms of knowledge to non-European peoples and lands. This could be seen, as Pratt suggests, as a form of intellectual colonisation. We have argued, however, that this was not solely due to the application of a Linnaean or Buffonian framework of natural history, but to the complicated melding of a variety of methods, theories and conjectures conveyed to them through their education at the University of Edinburgh. Park and Somerville here stand as examples of a much wider body of work produced by fellow students before and after them; they serve here only as indicators of the conceptual shifts they and their former fellow students' work can be used to chart.

93. Somerville, *William Somerville's narrative*, p.136.

The shift toward a more rigorously deterministic notion of race was being articulated throughout this time in Scotland by thinkers such as the antiquarian John Pinkerton (1758-1826) who set out to debunk the image of the supposedly 'noble savage' and in particular the romanticisation of the Scottish Highlander popularised by James Macpherson's (1736-1796) mythic poems of Ossian. So far as Pinkerton was concerned, neither ancient nor contemporary 'Celtic' Highlanders were noble in character, nor were they capable by themselves of progress or improvement. He perceived the Celts as an inferior race who had failed to stand up to invading 'Goths' whom he idealised as the noble ancestors of most Europeans bar the Highlanders and 'Laplanders'. To gain an understanding of what the Celts were really like as a historic people, Pinkerton argued, one had to look at the manners of contemporary 'Laplanders' who like the Celts were indigenous to Europe and to the 'Hottentots' in Africa. Denigrating all of these indigenous peoples, European and African, as merely 'savage', Pinkerton thus lumped them together as natural victims of the 'stronger' Gothic 'invaders' whose racial destiny he argued was to conquer and rule.[94]

Pinkerton's ideas reflect the increasing presence of racial theories with explicitly racist implications in late Enlightenment Scottish thought. From the 1820s, such views were hammered into scientific orthodoxy by Edinburgh intellectuals such as Robert Knox (1793-1862) and John Crawfurd (1783-1868), and Edinburgh educated figures such as James Cowles Prichard (1786-1848), who argued that all of humanity was divided into racial groups whose relative civilisation, political status and future fate were all seen as products of their largely unalterable racial endowments. Although it is anachronistic to suggest that Linnaeus and his early eighteenth-century followers could have foreseen how racial studies were to develop, it is equally important not to deny the clear relationship that exists between classification and racism. What Linnaeus and other European naturalists such as Buffon and Blumenbach had in common, despite their different taxonomies, was the drive, as described by Yuri Slezkine, to search for order within the 'family of man' and to organise 'People into peoples'.[95] Their 'typological thinking', in which plants, minerals and animals (humans included) were classified after their supposed essence, strengthened the concept of solid

94. John Pinkerton, *An Enquiry into the history of Scotland preceding the reign of Malcolm III, or The Year 1056*, 2 vols (London, John Nicols, 1789), vol.1, p.17. William Ferguson, *The Identity of the Scottish nation: an historic quest* (Edinburgh, 1998), p.259.

95. Quoted in Colin Kidd, *The Forging of the races: race and scripture in the protestant Atlantic world, 1600-2000* (Cambridge, 2006), p.81-82. See also Andersson Burnett, 'Northern noble savages?', p.176.

and static differences between different humans.[96] The defining characteristics assigned to each human variety were as a result, by the end of the eighteenth century, perceived as more static and unlikely to change.[97] We have argued here that this development should not be understood as a clear intellectual transition from one discourse to another. It was rather a process of gradual transformation within entangled and entwined discourses, and global circuits of knowledge, reflected in the teaching of both natural history and moral philosophy at the University of Edinburgh.

96. Stepan discussed in Paul Stock, '"Almost a separate race": racial thought and the idea of Europe in British encyclopedias and histories, 1771-1830', *Modern intellectual history* 8:1 (2011), p.3-29 (8).
97. Wheeler, *Complexion of race*, p.32.

7. Negotiating people, plants and empires: the fieldwork of Johann Gerhard König in South and South East Asia (1768-1785)

NIKLAS THODE JENSEN

Introduction

This chapter explores a known yet understudied example of how Linnaean natural history was practised, structured and exchanged across colonial South and South East Asia in the second half of the eighteenth century. As in the chapter on Pehr Löfling by Nyberg and Lucena Giraldo, the main character here is one of Carl Linnaeus's travelling students. His name was Johann Gerhard König (1728-1785), but contrary to Löfling, König has never been included as one of the Linnaean 'apostles'. Indeed, he is rarely mentioned at all in research on Linnaeus's students.[1] This is curious since standard works on the history of science in India recognise him as the first person to introduce Linnaean taxonomy to India and as the first formally employed naturalist of the powerful English East India Company (EIC).[2] König's influence on Indian botany is an important element in recent research which suggests that the informal scientific networks established by less significant colonial powers like Denmark-Norway and Sweden may provide the key to a re-evaluation of current views of 'colonial science'.[3]

In accordance with the general theoretical framework outlined in the Introduction, the focus of this chapter is on how König practised Linnaean natural history during his travels in South and South East Asia 1768-1785. The first part examines how he built a career as a naturalist by constantly shifting his allegiances between several patrons and nations, from centres to peripheries and back again. This overview of his career also establishes the background and parameters for his work in the colonial field. The second part investigates how König applied

1. Bengt Jonsell, 'Apostlarnas resor och gärningar', in *Ljus över landet? Upplysning som drivkraft i 1700-talets svenska vetenskap ock vitterhet*, ed. Paul Hallberg (Gothenburg, 2005), p.79-98.
2. David Arnold, *Science, technology and medicine in colonial India* (Cambridge, 2000), p.46-47; Michael Sterll, 'Life and adventures of Johann Gerhard König (1728-1785): a phantom of the herbaria', *Rheedea* 18:2 (2008), p.111-29; Isaac Henry Burkill, *Chapters on the history of botany in India* (Kolkata, 1965), p.12-17, 37.
3. Mark Harrison, 'Science and the British Empire', *Isis* 96:1 (2005), p.56-63 (62-63).

Linnaeus's ideals of fieldwork in the colonial context of South and South East Asia through his field methods, and how he negotiated local informants, their languages and knowledge. It was through this application of Linnaean ideals to the local social context that he came to position himself as the man in the field: bio-prospector and mediator between the colonial field and distant Europe. The third and last part returns to a more abstract perspective by exploring König's networks of colonial and metropolitan correspondents, supporters and learned societies. Here the importance of networks of German expatriate experts is brought out alongside those consisting of academics from minor colonial powers like Denmark-Norway and Sweden.

This chapter is based mainly on archival material from three collections: the König manuscripts in the Banks Collection held at the Library and Archives of the Natural History Museum (London), the letters from König to Linnaeus in the online database *The Linnaean correspondence*, and archival records held at the archive of Danish-Halle Mission at Die Franckeschen Stiftungen (Halle and der Saale).[4]

Between patrons and plants: an overview of König's career and travels

Denmark and Sweden

Johann Gerhard König (also spelt Kønig or Koenig) was born in 1728 to German parents near the town of Krustpils (Kreutzburg) in what was then Polish Latvia. He trained as an apothecary in the nearby seaport of Riga, but in 1748 he went to Denmark where he held positions at a number of provincial pharmacies.[5] From 1757-1759, König studied natural history under Linnaeus in Uppsala and made a number of friends among Linnaeus's students who later became very important in his life, particularly Daniel Solander (1733-1782). However, König did not gain a degree at Uppsala. After his return to Denmark in 1759, König was appointed pharmacist at the new Royal Frederick Hospital in Copenhagen under Joachim Dietrich Cappel (1717-1784), a fellow pharmacist from Riga.[6]

4. *The Linnaean correspondence* is an electronic edition prepared by the Swedish Linnaeus Society, Uppsala, and published by the Centre international d'étude du XVIIIe siècle, Ferney-Voltaire, available at http://linnaeus.c18.net/ (last accessed 8 August 2017). The online catalogue of the archive of Die Franckeschen Stiftungen zu Halle is available at http://192.124.243.55/cgi-bin/dhmeng.pl (last accessed 8 August 2017). The abbreviations AFSt/M and ALMW/DHM refer to this database. The Natural History Museum's collection of König manuscripts in twenty-two volumes is hereafter referred to as NHM, König MS.
5. Sterll, 'Life and adventures', p.116-17.
6. Johann Clemens Tode, *Medicinischchirurgische Bibliothek*, 10 vols (Copenhagen, Johann Gottlob Rothe, 1775-1787), vol.1:1 (1775), p.153-54. On Cappel see Carl Frederik Bricka,

The financial security of the position at the hospital may have facilitated König's decision to start a family. Sometime before 1764 he married Mariana (née Mørch) and the couple had first a son in 1764, and then a daughter in 1766.[7] Around the same time, König undertook a journey with two other naturalists to the Danish island of Bornholm in the Baltic Sea.[8] The purpose was to investigate the natural history of the island, particularly its minerals, yet it was König's collections of plants that earned him the position as amanuensis to Georg Christian Oeder (1728-1791), professor of botany in Copenhagen and curator of the newly launched and highly prestigious royal project to publish a Flora Danica.[9] Like Cappel, Oeder belonged to the influential but controversial group of German scientific experts that dominated Danish-Norwegian science and medicine from the 1730s to the 1770s.[10] The connection to Oeder propelled König onto the next stage of his career. In October 1763, King Frederik V of Denmark-Norway wished to send him to the Danish-Norwegian colonies in the West Indies as royal botanist, but for some reason this came to nothing.[11] Instead, in the spring of 1764, Oeder sent König to Iceland, at that time a dependency of Denmark-Norway. The Danish-Norwegian Crown funded the journey, the purpose of which was to collect Icelandic plants for the Flora Danica.[12] König was the first naturalist proper to do research in Iceland.[13]

In the autumn of 1765, König returned to Copenhagen and his position at the Royal Frederik Hospital. Around this time, he started studying medicine at the University of Copenhagen, and in 1767 he obtained the degree of *candidatus medicinae*.[14] However, during these same years of 1765-1767 König's career became seriously threatened due to

---

Dansk biografisk lexikon, tillige omfattende Norge for Tidsrummet 1537-1814, 19 vols (Copenhagen, Gyldendal, 1887-1905), vol.3 (Copenhagen, 1889), p.354-55.

7. The son Gerhard Friderick, christened 4 May 1764, and the daughter Friderica Charlotte, christened 27 October 1766. Danish National Archives (DNA), Church records of Copenhagen St Nikolaj Parish, Enesteministerialbog (1694-1805), F1746-F1765 https://www.sa.dk/ao-soegesider/billedviser?bsid = 156081 and F1766-F1788 https://www.sa.dk/ao-soegesider/billedviser?bsid = 156082 (both last accessed 8 August 2017).
8. Tode, *Medicinischchirurgische Bibliothek*, p.153-54.
9. Sterll, 'Life and adventures', p.116-17.
10. Helge Kragh, *Natur, Nytte og Ånd, 1730-1850* (Aarhus, 2005), p.14-15, 46-47.
11. Johann Gerhard König to Carl Linnaeus, 26 October 1763, *The Linnaean correspondence*, http://linnaeus.c18.net/, letter L3312 (last accessed 8 August 2017).
12. Johan Zoëga, 'Tilhang om de Islandske Urter', in Eggert Ólafsson and Bjarni Pálsson, *Vice-Lavmand Eggert Olafsens og Land-Physici Biarne Povelsens Reise igiennem Island foranstaltet af Videnskabernes Sælskab i Kiøbenhavn*, vol.2 (Sorøe, Jonas Lindgrens Enke, 1772), p.1-2.
13. Sterll, 'Life and adventures', p.116.
14. Bricka, *Dansk biografisk lexikon*, 19 vols (Copenhagen, Gyldendal, 1887-1905), vol.9 (1895), p.633-34.

the severance of his relationship with Oeder, his *patronus* in the world of botany. Oeder discovered that König had sent duplicates of specimens from Iceland to Linnaeus in Uppsala and thus contravened his duty, as a royal botanical collector, to send all material to Oeder in Copenhagen. This was all the more serious because Oeder disagreed with Linnaeus's sexual system and because the purpose of the Flora Danica was to earn prestige for the absolutist monarchy of Denmark-Norway in the field of natural history – a field where its ancient rival Sweden was gaining immense prestige from Linnaeus.[15]

Fortunately for König, his Linnaean network in Copenhagen managed to save his career by securing an appointment as the new doctor to the Lutheran (pietist) Danish-Halle Mission in the Danish-Norwegian colony of Tranquebar in South India. The old mission doctor had died in late March 1767 but, by this time, the government Board of Missions (Missionskollegiet) in Copenhagen had already hatched the idea of having the new mission doctor investigate Indian natural history based on the Linnaean system.[16] This combination of mission doctor and natural historian was nothing new as, since 1730, it had been an integral part of the mission doctor's job description to investigate the local natural history.[17] What was new, however, was that the investigation was based on the Linnaean system. The appointment of a new mission doctor had previously been the privilege of the other centre of leadership of the Danish-Halle Mission, the Francke Foundations (Die Franckesche Stiftungen) in Halle an der Saale. Nevertheless, this time it seems that the Board of Missions seized the opportunity to appoint its own candidate, namely König. Central to this manœuvre was Christian Friis Rottbøll (1727-1797), professor of medicine at the University of Copenhagen. Rottbøll was also a student of Linnaeus, a friend of König and an old rival of Oeder.[18] König was appointed mission doctor and the first royal Danish botanical collector of Tranquebar and, in December 1767, he embarked on the voyage accompanied by his wife Mariana.

15. Sterll, 'Life and adventures', p.118.
16. AFSt/M1B56:5 (28 March 1767); ALMW/DHM9/17:45a (17 November 1766).
17. Niklas Thode Jensen, 'Making it in Tranquebar: science, medicine and the circulation of knowledge in the Danish-Halle Mission, *c.*1732-1744', in *Beyond Tranquebar: grappling across cultural borders in South India*, ed. Esther Fihl and A. R. Venkatachalapathy (Delhi, 2014), p.325-51.
18. ALMW/DHM8/14:28 (20 November 1765); ALMW/DHM9/17:45a (17 November 1766); AFSt/M1F2:118 (14 November 1767); AFSt/M2E38:1 (25 November 1767); ALMW/DHM3/3c:32 (4 March 1768); Sterll, 'Life and adventures', passim; Anders Nørgaard, 'Secretaries of the Mission Board in Copenhagen 1714-1868', in *Halle and the beginning of protestant Christianity in India*, vol.3, ed. Andreas Gross, Y. Vincent Kumaradoss, Heike Liebau (Halle, 2006), p.1566.

On 5 July 1768 the ship anchored off Tranquebar on the Coromandel Coast.[19]

The Coromandel, the Mission and the Nawab of Arcot

Tranquebar was the administrative headquarters of the Danish-Norwegian (and from 1814 just the Danish) East Indies from 1620 to 1845. When König arrived, the colony consisted of the fortified trading town itself, comprising approximately 3000 inhabitants, and about thirty-two square kilometres of surrounding agricultural lands containing a few villages.[20] The colony remained roughly this size throughout most of its existence, and thus never deviated from the mode of the small colonial trading post of the seventeenth and early eighteenth centuries.[21] The other Danish-Norwegian possessions in the East Indies were the trading town of Serampore in Bengal (near Calcutta) and the Nicobar Islands in the Bay of Bengal.[22] Until 1777, the colonies were controlled by the Danish Asiatic Company (DAC) and subsequently became Crown colonies.[23]

What set Tranquebar apart from other small trading colonies in India, however, was its status as the headquarters of the Danish-Halle Mission. From the early years of the Mission in the first decade of the eighteenth century, its university-trained missionaries and mission doctors had carried out a range of scientific investigations. This work reached its first peak during the 1730s and 1740s with the production of a series of herbaria that made the mission in Tranquebar a South Indian node for the production and circulation of natural history.[24] Accordingly, when König arrived, natural history in Tranquebar was the traditional domain of the Mission, but not the concern of the DAC, as it later became with the British EIC, or the concern of the Crown, as in the case of the French and the Spanish colonies.[25] This was why the position as mission doctor

19. Johann Gerhard König to Carl Linnaeus, 2 December 1767 and 26 February 1769, *The Linnaean correspondence*, http://linnaeus.c18.net/, letter L4008 and L4181 (last accessed 8 August 2017); AFSt/M1B58:3 (30 July 1768).
20. Kamma Struwe, 'Dansk Ostindien, 1732-1777', in *Vore gamle tropekolonier*, ed. Johannes Brøndsted, vol.6 (Copenhagen, 1966), p.10-11.
21. Niels Brimnes, *Constructing the colonial encounter: right and left hand castes in early colonial South India* (Richmond, 1999), p.240-42.
22. Struwe, 'Dansk Ostindien, 1732-1777', p.106, 162-96, 227-35.
23. Aage Rasch, 'Dansk Ostindien, 1777-1845', in *Vore gamle tropekolonier*, ed. Johannes Brøndsted, vol.7 (Copenhagen, 1966), p.7-21.
24. Jensen, 'Making it in Tranquebar', p.327-31.
25. For the EIC, see below; for the French, see James E. McClellan III and François Regourd, *The Colonial machine: French science and overseas expansion in the old regime* (Turnhout, 2011), p.11-47; for the Spanish, see chapter 8 by Nyberg and Lucena Giraldo in this volume.

in Tranquebar was interesting to König, to his Linnaean friends in Copenhagen and to the Danish-Norwegian Crown.

Relocating to South India opened up great research opportunities to the now forty-year-old König, but it also created new and grave problems. Firstly, his wife Mariana died from disease during the journey. König tried to remedy the loss by marrying his second wife, a local European woman named Catharina Elisabeth de Tittoh (-1777).[26] Secondly, some of the missionaries in Tranquebar did not like König and complained repeatedly to Copenhagen and Halle of his alleged lack of interest in his job as mission doctor.[27] Thirdly, it quickly dawned on König that his salary was not equal to the high cost of working as a naturalist in the European colonies in India. This meant that he could not afford to go on field trips which were essential to his research. The two latter problems were so frustrating to König that after half a year in Tranquebar he asked Linnaeus to apply on his behalf for a new job with the Empress of Russia, Catherine the Great (1729-1796).[28] Though nothing came of this idea, it was not entirely fanciful. König's native Latvia had now become part of Russia and both the Empress and German scientists like himself had significant influence on the Russian Academy of Sciences in St Petersburg.[29]

While he was stuck in Tranquebar König nonetheless managed to send a steady stream of seeds, dried plants and zoological specimens back to Copenhagen to Rottbøll, to the Royal Chamber of Arts and Sciences (Det Kongelige Kunstkammer), and to the Board of Missions.[30] He also sent specimens and descriptions to Linnaeus, even though he was still officially forbidden from sending material anywhere else but Copenhagen.[31] As he could not afford to move more than ten miles away from Tranquebar, he had to rely on other travellers, like captains and merchants, to bring him specimens from other locations, such as the

26. Church records of the Church of Zion in Tranquebar. DNA, archive no. EC-777, Tranquebar Ny Jerusalemskirke og Zionskirke. Kontraministerialbog for Zionskirken, 1767-1845. Micro fiche no. 3/777-5, 26 December 1768.
27. Arno Lehmann, 'Hallesche Mediziner und Medizinen am Anfang deutsch-indischer Beziehungen', *Wissenschaftliche Zeitschrift der Martin-Luther Universität Halle-Wittenberg* 5:2 (1955), p.117-32 (131).
28. Johann Gerhard König to Carl Linnaeus, 26 February 1769 and 26 July 1769, *The Linnaean correspondence*, http://linnaeus.c18.net/, letter L4181 and L4232 (last accessed 8 August 2017).
29. Alexander Vucinich, *Science in Russian culture: a history to 1860* (Stanford, CA, 1963), p.75-124.
30. Johann Gerhard König to Carl Linnaeus, 26 February 1769, *The Linnaean correspondence*, http://linnaeus.c18.net/, letter L4181 (last accessed 8 August 2017).
31. Johann Gerhard König to Carl Linnaeus, 6 February 1771, *The Linnaean correspondence*, http://linnaeus.c18.net/, letter L4477 (last accessed 8 August 2017).

Nicobar Islands and Malacca.[32] Such financial difficulties were a recurring theme in the travels of several Linnaean 'apostles', for instance Carl Peter Thunberg (1743-1828) who travelled in the service of the Dutch East India Company (VOC) to the Cape, Batavia and Japan in 1771-1779.[33]

Despite his efforts to find a new job, König remained with the Mission until 1774. By this time, his prospects of finding another position had improved markedly because in February 1773 Rottbøll had awarded him a doctoral degree in medicine *in absentia* based on his doctoral thesis.[34] Thus, it was not the aging Linnaeus, but the Linnaean network in Copenhagen that again interceded on König's behalf. With his credentials secured, in September 1774 König travelled to Madras (Chennai), the South Indian headquarters of the prosperous EIC. This was a deliberate attempt to make a new career in British service. Since the EIC's decisive victory in the North over the Nawab of Bengal and his French allies at the Battle of Plassey in 1757, and in the South over the French in the Third Carnatic War (1756-1763), British power in South Asia had been increasing rapidly.

On arrival at Madras, König stayed at the garden (country house) of the influential merchant Andrew Ross (-1793), with whom he had started a correspondence concerning botany at least half a year earlier.[35] Through Ross's network, König now gained access to influential people in the EIC administration and high society in India who could further his career. These included Lady Anne Monson (1726-1776), a well-known British botanist, correspondent of Linnaeus and wife of Colonel George Monson (1730-1776) of the Bengal army, and Dr Alexander Boswell, first physician to the local Indian ruler, the Nawab of Arcot, until 1776.[36] Moreover, König also started corresponding with two individuals already introduced in the chapter by Hodacs in this volume: his old fellow student Daniel Solander in London, at this time Keeper of the Natural

32. Johann Gerhard König to Carl Linnaeus, 26 July 1769 and 1 July 1772, *The Linnaean correspondence*, http://linnaeus.c18.net, letter L4232 and L4708 (last accessed 8 August 2017).

33. Marie-Christine Skuncke, *Carl Peter Thunberg, botanist and physician: career-building across the oceans in the eighteenth century* (Uppsala, 2014), p.30.

34. Johann Gerhard König, *Dissertationem inauguralem de Remediorum Indigenorum ad Morbos cuivis regioni endemicos expugnandos efficacia* (Copenhagen, Gerhardus Giese Salicath, 1773).

35. Concerning Ross, see Henry Davidson Love, *Vestiges of old Madras 1640-1800*, vol.3 (New Delhi, 1913), p.53, 473, n.7. NHM, König MS, Journal, vol.3, p.177-79, letter from Johann Gerard König to Daniel Solander, 30 September 1774. NHM, König MS, List of plants, vol.1, p.49.

36. NHM, König MS, Journal, vol.13, p.255-56. On Lady Monson, see J. Browne, 'Monson, Lady Anne (*c.*1727-1776)', in *Oxford dictionary of national biography* (Oxford, 2004), http://www.oxforddnb.com/view/article/57839 (last accessed 8 August 2017). On Alexander Boswell, see D. G. Crawford, *A History of the Indian medical service, 1600-1913*, vol.2 (London, 1914), p.225.

History Department of the British Museum, and a close associate of the influential Sir Joseph Banks (1743-1820), President of the Royal Society in London and scientific advisor to the EIC.[37] König repeatedly appealed to Solander and Banks to help him obtain a position with the British in India.[38] In February 1775, König was finally granted a new position as naturalist with the Nawab of Arcot, Muhammad Ali Khan Wallajah (1717-1795).[39] According to König, this was arranged by Boswell who was also an aficionado of herb lore, but the Nawab was also very interested in natural history.[40] However, it quickly became apparent that the position with the Nawab did not solve König's problems: the prince was heavily indebted to the EIC and could not pay his bills. Still, the Nawab assigned König to a post in the city of Tanjore, which gave him the opportunity in early 1776 to travel on specimen-gathering trips around Tanjore and south to the Palk Strait between India and Sri Lanka.[41] Later that year, König went first on a short trip to the Palliacatti Mountains, that is, the Nagari Hills north-east of Madras, and then later, on a trip lasting several months, south via Tanjore to the city of Jaffnapatnam in the northern part of Dutch Ceylon (Sri Lanka).[42]

The journey to Ceylon seems to have become particularly important for König's future research, because it opened up a new Dutch branch in his network and supplied him with much needed resources. König had long planned to go to Jaffnapatnam because he wanted to visit Dr Christian Rose (1719-1786), the commandant there.[43] Dr Rose was German, had a doctoral degree in medicine from the University of Halle,

37. John Gascoigne, *Science in the service of empire: Joseph Banks, the British state and the uses of science in the age of revolution* (Cambridge, 1998).
38. NHM, König MS, Journal, vol.3, p.177-79, letter from Johann König to Daniel Solander, 30 September 1774.
39. NHM, König MS, Journal, vol.3, p.181-84, letter from Johann König to Daniel Solander, 24 January 1775. NHM, König MS, Journal, vol.3, p.185-87, letter from Johann König to Daniel Solander, 7 February 1775.
40. Johann Gerhard König to Carl Linnaeus, 14 February 1775, *The Linnaean correspondence*, http://linnaeus.c18.net/, letter L5086 (last accessed 8 August 2017). Savithri Preetha Nair, 'The Museum in colonial India (1770-1936): a history of collecting, exhibiting and disciplining knowledge', doctoral dissertation, School of Oriental and African Studies, University of London, 2002, p.41.
41. NHM, König MS, Journal, vol.3, p.189-90, letter from Johann König to Daniel Solander, 17 February 1776; NHM, König MS, Journal, vol.3, p.191-94, letter from Johann König to Daniel Solander, 24 February 1776.
42. NHM, König MS, Journal, vol.3, p.195-200, letter from Johann König to Daniel Solander, 14 July 1776; NHM, König MS, Journal, vol.1, p.149-56, letter from Johann König to Daniel Solander, 6 March 1777. A description in Danish of the two journeys was published as: Otto Frederik Müller, 'Doctor Koenigs Reise fra Trankebar til Zeylon i et Brev til Etatsraad Müller af 15 Februari 1777', *Videnskabernes selskabs skrifter* 12 (1779), p.383-402.
43. NHM, König MS, Journal, vol.3, p.185-87, letter from Johann König to Daniel Solander, 7 February 1775.

was a member of the Berlin Society of Friends of Natural Science (Gesellschaft Naturforschender Freunde zu Berlin), and was well known to the Danish-Halle Mission in Tranquebar as a generous benefactor.[44] He was a member of the extensive but often ignored network of Germans who served both the VOC and the EIC as doctors, engineers, soldiers and so on.[45] König obtained support from this network on many occasions. From Dr Rose, König gained not only indispensable assistance from the VOC for his specimen gathering trips around northern Ceylon, but also a magnificent gift of botanical books from Rose's private library, such as all twelve volumes of van Rheede's *Hortus Malabaricus*, all the works of Rumphius and many more.[46] These formed priceless resources to which very few researchers in the East Indies had access, and were a great asset for König's future work.

When König returned from Jaffnapatnam in October 1776, his problems continued. The financial difficulties of the Nawab remained the same, and in early 1777 König's wife Catharina died. This left König in deep sorrow and confusion, but he did manage to go on a short research trip in August 1777 to Vandalur (Wændeluhr) and Chengalpattu (Chinglæ Pattæ) just south of Madras. In December that year, König was discharged from the Nawab's service, which seems to have been a relief to him as he could now at least travel where he pleased. Accordingly, he went to the Palk Strait again and visited a mountainous region close to the city of Tiruchirappalli. As the Nawab still owed König a great deal of money, the funding for these trips seems to have come from local and overseas friends, such as Ross and the Dowager Duchess of Portland, Margaret Cavendish Bentinck (1715-1785). As mentioned in chapter 3 by Hodacs in this volume, the duchess owned one of the largest natural history collections in Great Britain. She sent König a gift through Daniel Solander and, in return, König promised to send her specimens of natural history, in the same way that he had been shipping crates of specimens throughout to Solander, Linnaeus and a range of other correspondents (see p.205-208 below).[47]

44. Christianus Rose, *Dissertationem Sollemnem De Augusto Contraria Medicina Curato* (Halæ Magdeburgicæ, Hilligerus, 1741); Lorenz Spengler, 'Auzug aus einem Schreiben des Herrn D. Johann Gerhard König', *Beschäftigungen der Berlinischen Gesellschaft Naturforschender Freunde* 3 (1777), p.427-30 (430). AFSt/M2D48:1, 22 June 1771.

45. Iris Bruijn, *Ship's surgeons of the Dutch East India Company: commerce and the progress of medicine in the eighteenth century* (Leiden, 2009), p.155-67; Roelof van Gelder, *Das Ostindische Abenteuer: Deutsche in Diensten der Vereinigten Ostindischen Kompanie der Niederlande (VOC), 1600-1800* (Hamburg, 2004), p.42-56; Chen Tzoref-Ashkenazi, *German soldiers in colonial India* (Oxford, 2015).

46. NHM, König MS, Journal, vol.1, p.149-56, letter from Johann König to Daniel Solander, 6 March 1777.

47. NHM, König MS, Journal, vol.3, p.209-12, letter from Johann König to Daniel Solander,

In the service of the East India Company

König finally gained a position with the EIC Madras Government in July 1778. According to König, this appointment was achieved through another friend, Dr Gilbert Pasley (-1781), Senior Surgeon of the Madras Presidency.[48] Nonetheless, the correspondence of Banks shows that he too had a hand in it.[49] This appointment meant that König, in effect, became the first naturalist formally employed by the EIC.[50] At this time, the EIC had neither any scientific institutions nor, indeed, a centralised structure of colonial and metropolitan scientific institutions similar to the French so-called 'colonial machine'.[51] It was only in 1783 that the British Government began to apply this scientific approach worldwide to compete with the French.[52] König became one of the frontrunners in this development. When he was employed by the EIC, much of the scientific efforts undertaken in the name of the company took place on the margins of institutions established for other purposes, because they were still being carried out in the form of private pursuits of individuals who were employed in these institutions such as doctors, engineers, lawyers and merchants. As the EIC was a joint-stock company, its directors were not interested in science for its own sake but as a means to find 'useful knowledge', or in other words anything that might have commercial value or be used to solve the countless practical problems of governing the Company's vast and growing possessions in India.[53]

This new position with the EIC released König from the financial difficulties he had been mired in since his arrival in India ten years earlier, and sent him in a new direction – quite literally. On 8 August

5 August 1778. On the trip to Vandalur, see NHM, König MS, Journal, vol.7, p.165-75. On the death of his wife, see letter from Johann König to Linnaeus, 15 February 1777, *The Linnaean correspondence*, http://linnaeus.c18.net/, letter L5190, last page, last note (last accessed 8 August 2017). On the Duchess of Portland, see P. Rogers, 'Bentinck, Margaret Cavendish [Lady Margaret Cavendish Harley], duchess of Portland (1715-1785)', in *Oxford dictionary of national biography* (Oxford, 2004); http://www.oxforddnb.com/view/article/40752 (consulted 8 August 2017).

48. NHM, König MS, Journal, vol.3, p.209-12, letter from Johann König to Daniel Solander, 5 August 1778. On the friendship with Pasley, see NHM, König MS, Journal, vol.5, p.73. On Pasley, see Crawford, *A History of the Indian medical service*, vol.2, p.152, 234, 304, 365 *et* passim.

49. Neil Chambers (ed.), *The Indian and Pacific correspondence of Sir Joseph Banks, 1768-1820*, vol.1, *Letters 1768-1782* (London, 2008), p.248, letter from Sir George Wombwell at East India House to Joseph Banks, 31 March 1779.

50. Arnold, *Science, technology and medicine in colonial India*, p.46.

51. McClellan and Regourd, *The Colonial machine*, p.41-47.

52. Richard Drayton, *Nature's government: science, imperial Britain and the 'improvement' of the world* (New Delhi, 2005), p.106-24.

53. Arnold, *Science, technology and medicine in colonial India*, p.19-33.

1778 the EIC sent him on a voyage of more than a year that took him to Malacca, Siam (Bangkok, Chanthiburi), Selangor, Phuket Island and the Nicobar Islands. The exact purpose of this journey is not quite clear. However, as the EIC was very interested in these areas, and later expanded into some of them, recent research has suggested that König was sent as a spy or prospector of natural resources.[54] If this is correct, the EIC employed König as an agent of colonisation in a somewhat similar way to how the Spanish Crown had employed Löfling some twenty years earlier.[55] In any case, König's work reveals that he investigated all aspects of natural history – mineralogy, botany and zoology – as well as a range of new information on culture, mores and customs.[56] He returned to Madras in December 1779 and seemingly spent the following months researching, ordering, preserving and shipping parts of the huge amount of material he had collected. In April and May 1780 König visited Ceylon again, this time the area around the harbour city of Trincomalee on the east coast.[57] Half a year later, in January 1781, he returned to the island, but this time via Tranquebar. By that time, the Second Anglo-Mysore War (1780-1784), which also involved France, had broken out in South India. Consequently, it was dangerous to undertake research trips in the British-held areas, but Danish-Norwegian Tranquebar remained neutral, and Dutch Ceylon was not yet involved in the war. The journey took König through the south-western region of Dutch Ceylon, along the coast from Colombo to Tangalle and deep into the country to a mountain called Adam's Peak (Sri Pada). This was only possible because of logistical and financial support from the government of Dutch Ceylon, in particular Governor Iman Willem Falck (1736-1785).[58] In fact, the governor offered König a very lucrative position as *Landphysicus*, or Surgeon General of Dutch Ceylon, but König declined out of respect for his British patrons.[59]

In early May 1781, König left Ceylon and returned to Tranquebar. The Fourth Anglo-Dutch War (1780-1784) had broken out and this escalation of warfare not only blocked König from going anywhere else but neutral Tranquebar, it also left him virtually bankrupt. Payments from the EIC

54. Sterll, 'Life and adventures', p.125.

55. See chapter 8 by Nyberg and Lucena Giraldo in this volume.

56. J. G. König, 'Journal of a voyage from India to Siam and Malacca in 1779', *Journal of the Straits Branch of the Royal Asiatic Society* 27 (October 1894), p.57-101, 103-25, 127-33.

57. NHM, König MS, Index, p.3.

58. August Hennings, *Gegenwärtiger Zustand der Besitzungen der Europäer in Ostindien*, vol.2 (Hamburg, Carl Ernst Bohn, 1785), p.303. Hennings reproduces the contents of a letter from König to Lorenz Spengler in Copenhagen dated 1781.

59. NHM, König MS, Journal, vol.3, p.243-44, letter from Johann König to Daniel Solander, 26 October 1780; NHM, König MS, Journal, vol.3, p.245-48, letter from Johann König to Daniel Solander, 28 October 1781.

had not been getting through for about a year and his savings had been lost in a raid. König's situation had become so desperate by February 1782 that he begged Linnaeus Jr in Uppsala to help him obtain a salary from the Russian Academy of Sciences in St Petersburg.[60] However, nothing came of this request.

König remained in Tranquebar at least until June 1782.[61] Then, in 1783, we find him back in Madras where he stayed at least until August.[62] At this time, the ongoing wars involving France (part of the American Revolutionary War, 1775-1783) and Mysore (1780-1784) were drawing to an end. This made it possible to travel again and König took the long road over land north to Calcutta in Bengal, visiting his friend and fellow natural historian Patrick Russell (1726-1805) in Visakhapatnam on the way. König spent 1784 working in Calcutta and in February 1785 his official post as EIC naturalist was transferred from the Madras Presidency to the Bengal Presidency. In order to gather and ship his belongings and collections from Madras and Tranquebar to the new station, König travelled south again in March and April 1785.[63] However, on 26 June 1785 he died of dysentery in Jagannadhapuram by Kakinada while in the care of his friend and fellow botanist William Roxburgh (1751-1815).[64]

Negotiating the field

Linnaean field methods in the East Indies

König's travel accounts and field notes show beyond doubt that his methods of investigating the world while travelling emulated Linnaeus's practices and ideals of fieldwork. König indicated his reliance on Linnaean ideals in 1778 when he offered to make a compendium of the natural history of India in accordance with Linnaeus's *Instructio peregrinatoris* for the EIC.[65] The *Instructio* is an apodemic text, that is it belongs to the genre of travel manuals, and it contains a set of principles divided into sixteen articles of how to travel as a natural historian, what to investigate and how to preserve the collected material during and

60. Johann König to Carl Linnaeus the Younger, 1 February 1782, *The Linnaean correspondence*, http://linnaeus.c18.net/, letter L6023 (last accessed 8 August 2017).
61. Nair, 'The Museum in colonial India', p.66.
62. NHM, König MS, Journal, vol.18, p.238.
63. Nair, 'The Museum in colonial India', p.67-69. For his itinerary from Kolkata to Ganjam 23 March to 6 April, see NHM, König MS, Journal, vol.15, p.204-12.
64. Sterll, 'Life and adventures', p.120.
65. Nair, *The Museum in colonial India*, p.43. The *Instructio* was published in 1759 as a doctoral dissertation by Linnaeus's student Erik Andreas Nordblad, but there is a consensus that Linnaeus was the author. See Hanna Hodacs and Kenneth Nyberg, *Naturalhistoria på resande fot: om att forska, undervisa och göra karriär i 1700-talets Sverige* (Lund, 2007), p.179-81.

after the journey. In general, the focus of the *Instructio* is more on education than on research, and more on what could be 'useful' in a broader economic sense than on what was interesting from a narrow scientific perspective.[66]

A comparative analysis of the articles of the *Instructio* and König's records from his travels in India, Sri Lanka and South East Asia reveals that he did follow most of the guidelines in the instruction. If we start with the problem of how to commence one's research when arriving in a new region, König usually began by visiting local European 'learned men'. This recommendation forms part of the *Instructio*'s fourth article (and links to article 15, item 1-2) and was of great practical and scientific value. First, learned European men in the East Indies were few, but usually powerful, either because they were employed in East India companies, or because of their wealth, or both. In any case, they were capable of supplying König not only with porters and logistical support, which was indispensable and often in short supply, but also with access to local people with knowledge of nature. A few examples of such learned men are Dr Rose in Jaffnapatnam, Pieter Gerardus de Bruijn (1732-1790), the governor of Dutch Malacca who supported König during his stay there in 1779, or Governor Falck on Ceylon.[67] Furthermore, by establishing personal connections to these men, König could potentially form a powerful network of correspondents with whom he could continue to exchange specimens and information.

For the next step on a research trip, moving from a port city into the countryside, König followed the recommendations of the sixth and seventh article in the *Instructio* in describing the landscape, the state of the roads, the rest houses, villages and landmarks such as temples and statues.[68] When possible, he expanded these descriptions with data on the height of mountains, the stratification of the soil, the properties of local springs and the temperature, as recommended by article 8.[69] The following step, according to the *Instructio's* articles 9, 10 and 11, was a

66. Hodacs and Nyberg, *Naturalhistoria på resande fot*, p.199-210.

67. Frits van Dulm, *'Zonder eigen gewinne en glorie': Mr Iman Wilhelm Falck (1736-1785), gouverneur en directeur van Ceylon en Onderhorigheden* (Hilversum, 2012), p.27, n.34. On de Bruijn, see NHM, König MS, Journal, vol.3, p.231-34, letter from Johann König to Daniel Solander, 1 February 1780; Ulbe Bosma and Remco Raben, *Being 'Dutch' in the Indies: a history of creolisation and empire, 1500-1920* (Singapore, 2008), p.57.

68. See for instance König's description of his journey from Colombo to Tangalle in 1781. NHM, König MS, Journal, vol.15, p.15-133.

69. On the height of mountains, see the climb of Adam's Peak, NHM, König MS, Journal, vol.15, p.126. On stratification of the soil, see the description of the digging of a well outside Tranquebar in 1770, NHM, König MS, Journal, vol.3, p.264-65. On hot springs, see the journey to Trincomalee, NHM, König MS, Journal, vol.12, p.20-22. On measuring temperature, see König, 'Journal of a voyage from India to Siam and Malacca in 1779', p.146 (9 December 1779).

deeper examination of local nature divided into the three Linnaean kingdoms of minerals (soils, ores and fossils), plants and animals (mammals, birds, amphibians, fish, shellfish and insects). König investigated all three kingdoms during his travels and in his research. For instance, during his specimen gathering trips around Tanjore in early 1776 he did a report on white ants (termites), and later, on his journey to Trincomalee in 1780, he made detailed field notes on the local geological variation.[70] Yet, the main focus of König's investigations was always botany, as it had been on his early journeys to Bornholm and Iceland.

The twelfth article of the *Instructio* introduces the economic and utilitarian perspective on the three natural kingdoms, which was typical of the Linnaean approach to natural history. This perspective included all cultural productions as related to nature's three kingdoms, from mines, agriculture and fishponds to clothes, cooking and perceptions of productivity. In König's accounts of his travels, this economic aspect is always present and intertwined with his rigorous scientific descriptions of plants and animals. Again, the kingdom of plants takes centre stage. This is reflected in his descriptions of garden designs in Tanjore, on Chinese *Penzai* trees (similar to Bonzai trees) or 'Gummæ Guttæ' (a resin used as a yellow dye), but he also investigated the mining and smelting of tin in Malacca and the physical and moral character of the Malays and other ethnic groups.[71]

Interacting with local people and their knowledge

The *Instructio* as the framework for König's work in the field might give the impression that he found and investigated everything on his own – with his own eyes (autopsy) – and without the help of anyone else except 'learned men' of European origin. However, this was not at all the case. In fact, article 10 of the *Instructio* stipulates that if the travelling natural historian is not capable of investigating the local flora himself, he should consult local botanists in order to learn from them the location of rare plants, and that local names and uses of plants should always be added to descriptions of the same. König followed and went beyond these guidelines, which makes it interesting to explore how he collaborated

70. On the white ants, see Johann Gerhard König, 'Naturgeschichte der sogenannten weissen Ameise', *Beschäftigungen der Berlinischen Gesellschaft Naturforschender Freunde* 4 (1779), p.1-28. On geology around Trincomalee, see NHM, König MS, Journal, vol.12, p.55-58.
71. On garden designs in Tanjore, see NHM, König MS, Journal, vol.10, p.131-51. On Penzai trees, see NHM, König MS, Journal, vol.18, p.24-26. On 'Gummæ Guttæ', see NHM, König MS, Journal, vol.6, p.89. On tin smelting in Malacca, see NHM, König MS, Journal, vol.10, p.11-39. On the moral character of people in Malacca, see NHM, König MS, Journal, vol.18, p.34-43, 53-60.

with local informants and the value that he, as a Linnaean, assigned to local experts and local knowledge of nature.

König's records of his travels show that he met and collected information from a great variety of people with diverse ethnic, linguistic and religious backgrounds. Both the number and the range of informants went far beyond the assumptions in the *Instructio*. To divide this group into Europeans and non-Europeans would not capture its diversity. It is more appropriate to see this group of people as distributed along a spectrum of knowledge ranging from what König saw as purely empirical knowledge (skills), to theoretical knowledge. This range encompassed for instance the Thai boy who aided König by climbing trees in Chanthiburi (Siam), through the Armenian merchant Monsieur Sathur in Malacca who shared his special knowledge of shellac from Pegu (Burma), to formally educated European men like Dr Rose.[72]

If we start at the empirical end of this spectrum, the first people we find are those who accompanied König on his travels: palanquin-carriers, porters, sailors, guides and translators. They were indispensable for the practical aspects of his journeys and could number up to fifty individuals, but nonetheless he only rarely mentioned them.[73] When he did, he generally respected the advice of those of them who had some kind of specialised knowledge, whereas he was critical of those who did not. For instance, on his journey to the Nagari Hills in 1776, he, on the one hand, accused some of his indigenous assistants of contributing nothing but negligence and deceit, while, on the other hand, he was very interested in the practices of a local herbalist (in Danish 'Urtemand', in Tamil 'Illeri') who was acting as his guide.[74] This question of the reliability of local people's skills and knowledge also came up when König needed to send an assistant to collect plants for him. In general, he felt that indigenous people could not be trusted with this task because they did not have the proper skills, or in other words they did not preserve the specimens properly or they were not capable of distinguishing between different varieties of the same plant.[75] Conveniently enough, the alleged failure of his assistants to comply with Linnaean standards fortified König's own position as the only trustworthy local specialist of scientific natural history. However, this was not just a matter of being European or not, because at times König's entourage also contained local European assistants who were not necessarily assigned any greater credibility

72. On the Thai boy, see König, 'Journal of a voyage from India to Siam and Malacca in 1779', p.176. On Monsieur Sathur, see NHM, König MS, Journal, vol.18, p.135.
73. Müller, 'Doctor Koenigs Reise', p.393, 395.
74. Müller, 'Doctor Koenigs Reise', p.384, 386, 390.
75. Hennings, *Gegenwärtiger Zustand der Besitzungen*, p.297, 304, 310; König, 'Journal of a voyage from India to Siam and Malacca in 1779', p.64.

than that of the non-European assistants.[76] Furthermore, König held some ethnic groups, in particular the Chinese, in higher regard than ordinary colonial Europeans because the former were generally more 'industrious' (or *fleissig*, in German).[77]

Another set of informants were those local experts that König encountered during his travels. Many of them were indigenous doctors, which is hardly surprising given König's own training as a doctor. In general, he was not impressed with their knowledge of medicine. This was because, like most European doctors at the time, he saw exact anatomical knowledge, surgical skills and the performance of autopsies as the emblems of advanced medical knowledge.[78] Since he did not find such knowledge among the Siamese, Chinese and Sinhalese doctors, he had little respect for their medical skills. However, at the same time he showed great interest in their knowledge of herbs, resins and other pharmaceutical substances.[79] One instance of this took place in Tranquebar in the early 1770s, when König called in a local female specialist to cure his cook of an eye disease. While the woman prepared eye drops for the patient, König did his best to observe and note the various ingredients and their preparation. This is also an example of the fact that there were women among the local experts König encountered, though they were few in number.[80]

Studies of exchanges of knowledge between European and non-European experts in the colonial world have often noted that Europeans rarely acknowledged the names of their non-European counterparts.[81] Judging by the examples of Thunberg and Löfling, Linnaean botanists generally behaved the same way and König was no different from most of his colleagues.[82] However, there was one striking exception, which may suggest that the absence of indigenous names in König's records was not only an attempt to elevate his own expertise over that of the locals, but also a side effect of being a travelling natural history collector instead of

76. NHM, König MS, Journal, vol.14, p.147-49.
77. NHM, König MS, Journal, vol.18, p.53-60.
78. Mark Harrison, 'Medicine and orientalism: perspectives on Europe's encounter with Indian medical systems', in *Health, medicine and empire: perspectives on colonial India*, ed. Biswamoy Pati and Mark Harrison (Hyderabad, 2001), p.37-87 (41, 52-58).
79. On Siamese doctors, see König, 'Journal of a voyage from India to Siam and Malacca in 1779', p.131-32, 154. On Chinese doctors/chemists in Siam, see König, 'Journal of a voyage from India to Siam and Malacca in 1779', p.157. On Sinhalese doctors, see NHM, König MS, Journal, vol.6, p.98.
80. NHM, König MS, Journal, vol.5, p.69.
81. Mark Harrison, 'Networks of knowledge: science and medicine in early colonial India, *c.*1750-1820', in *India and the British Empire*, ed. Douglas M. Peers and Nandini Gooptu (Oxford, 2012), p.191-210 (199).
82. On Thunberg, see Skuncke, *Carl Peter Thunberg*, p.30. On Löfling, see chapter 8 by Nyberg and Lucena Giraldo in this volume.

a resident one. The exception was a Tamil doctor named Wangar-Naiki (or Vangar Naiken) in Tranquebar, who in 1780 supplied König with the Tamil names of twelve minerals.[83] The reason why König chose to name this Tamil doctor in his records was probably that he knew him well from his years of living and working in Tranquebar.[84] This would have distinguished Wangar-Naiki from almost all the other non-European doctors or local experts König encountered on his journeys. His interaction with them was always brief. Historian Mark Harrison has suggested that one of the reasons why European doctors in Madras around the same time did not mention the names of the local Indian experts of medicinal plants they did business with was due to the lack of prolonged knowledge interaction between the two groups.[85] As a travelling natural history collector, König rarely had extended interaction with individual indigenous experts. In addition, as he rarely spoke their language he consequently would not have been able to maintain a scientific exchange with them after the journey had ended. As mentioned above, this continued scientific exchange was important to König's way of practising natural history, and since it was not possible with indigenous experts, save perhaps with Wangar-Naiki, what was useful to König to record was not their names but their *positions* (doctor, priest or woodcutter, for example) as an indication of their social standing and general credibility.

On this basis, it seems that König's approach to local knowledge was ambivalent. On the one hand, he was very interested in local knowledge regardless of what ethnic group it came from. On the other hand, he rarely named his non-European informants but always his European peers and superiors. He tended to see local knowledge as empirical and thus less advanced than his own theoretical training in the Linnaean system. This view emphasised the importance of his own role as a mediator between non-European knowledge and the scientific circles in Europe. It is also resonant of Alix Cooper's conclusion that Linnaean science created a hierarchy which relegated local experts of, say, plants, to a lower position than Linnaean scholars because the latter could make knowledge move globally.[86]

83. NHM, König MS, Journal, vol.13, p.73.
84. The 1790 census of Tranquebar shows that Dr Vangar Naiken was a member of the Catholic congregation and lived in Northern Wall Street. DNA, archive no. 445:1447A-1447B, Det kgl. ostindiske Guvernement, Kolonien Trankebar, Mandtal over indbyggere i Tranquebar og landsbyerne, 1790.
85. Harrison, 'Networks of knowledge', p.199.
86. Alix Cooper, *Inventing the indigenous: local knowledge and natural history in early modern Europe* (Cambridge, 2007), p.169.

Managing languages and connecting names

Having discussed the issue of local experts and the perceived value of their knowledge, we now turn to an important dimension of local knowledge of nature: the vernacular names. As pointed out by Lisbet Koerner, and mentioned above in reference to article 10 of the *Instructio peregrinatoris*, it was important to the Linnaeans to record vernacular names of natural history objects because these names were the keys to the local use and usefulness of the object.[87] At the same time, however, local languages also proved to be major obstacles for a travelling natural history collector, like König, in the colonial world.

Even though König understood a range of European languages, it seems he never really mastered any of the languages of India or South East Asia. Consequently, he was dependent on interpreters during his journeys, which often caused frustration. On more than one occasion, he blamed the lack of useful results of his investigations on the ignorance and indolence of his interpreter.[88] He also tried to cope with the language problem by compiling phrasebooks – short dictionaries of single words, phrases and numerals – in one or more European languages and a local language. From his journey to Siam and the journeys to Ceylon, we have a number of these manuscripts translating from Latin, Portuguese, Dutch or German into Thai, Sinhalese and Malay.[89] In other words, language and names were much more important in König's work than presumed in the *Instructio*.

As regards vernacular names of natural history objects, König's records contain a great abundance of them, particularly lists of plant names. These lists reveal how König collected much of his knowledge of natural history and that the organising principle of the lists changed as his knowledge of the botanical world of the East Indies increased. When he first came to India, König collected the botanical knowledge he could find in the mission in Tranquebar and adopted the missionaries' technique of listing based on the Tamil language. This is evident from a list dated 1771, the *Plantarum Thamulica*, which is a collection of Tamil plant names from the herbaria in the library of the mission in Tranquebar collected and annotated by König.[90] It is a more or less alphabetic

87. See the Introduction to this book.
88. NHM, König MS, Journal, vol.9, p.247.
89. NHM, König MS, Journal, vol.6, p.159-65; NHM, König MS, Journal, vol.10, p.215; NHM, König MS, Journal, vol.15, p.103, 177-82; NHM, König MS, Journal, vol.18, p.107-109, 146-49.
90. NHM, König MS, Journal, vol.3, p.137-60, 'Nomina plantarum thamulica speciis herbarii vivi bibliothecæ missionis tranqvebarensis cum notis sporadicis collectoris. In catalogum redacta. Spreta in tempore gloria interdum cumulatior redit. Liv. Ab Johann Gerhardo König. 1771'.

catalogue of about 350 Tamil plant names arranged in table form. Following the first column with the Tamil name in Latin script are columns with the name in Tamil script, the binomial name in Latin and further comments in German regarding the local use of the plant. A comparison of this list with the *Plantae Malabaricae*, the only surviving herbarium vivum from the botanical research of the Tranquebar mission in the 1730s and 1740s, shows a strong correspondence in content and structure.[91] Accordingly, König expanded his first knowledge of local South Indian plants using the foundational work by the mission some thirty years earlier. His development of this tradition is visible in the column with Linnaean binomial names, which are not present in the *Plantae Malabaricae.*

In the course of the years after 1771, König changed the defining language in his lists to Latin. One instance is a manuscript from 1776 with the title 'A register of the plants which are used in medicine and are native to Ceylon'.[92] It is arranged according to its binominal name in the Linnaean system and added to this is the Sinhalese name and names of the plant in Tamil or Portuguese. The change to Latin was of course necessary to comply with the Linnaean system and to circulate the descriptions of new plants with Europe. Nevertheless, the vernacular plant name was still indispensable for further research. Without it, it was impossible to find the plant again in India and to compare with older European botanical works that generally used local names. This comparative method of collecting, cross-referencing, validating and accumulating knowledge is also clearly visible in König's records. One instance is an undated list of the plant names in Malayalam recorded in the first two volumes of the *Hortus Malabaricus* and the equivalent names in Tamil.[93]

Colonial and metropolitan networks

Just as relations with people locally in the field and between names and objects of nature were important to König's work, so were relations and exchanges with naturalists, collectors and scientific societies in the region and on the global scale. The above outline of his career shows how this network of correspondents evolved as he moved between national and imperial spheres and between patrons. This is not the place for a full study of König's network, but an analysis of his correspondence with Linnaeus, the files at the NHM and at the Franckesche

91. Jensen, 'Making it in Tranquebar', p.340-41.
92. NHM, König MS, Journal, vol.1, p.265-78, 'Verzeichniss dererjenigen pflanzen welche in der medicin gebrauchtlich und hier auf Zeylon einheimisch sind nach dem Linnæischen system'.
93. NHM, König MS, Journal, vol.7, p.213-24.

Stiftungen reveals that, during the years in South and South East Asia 1769-1785, his network consisted of two parts: a colonial network in South and South East Asia and a metropolitan network in northern and north-western Europe. It was essential to König's work to maintain both parts of this structure. The metropolitan part contained the most powerful friends, collectors and institutions with whom he could exchange his specimens and descriptions of natural history in return for scientific prestige, a position, money or valuable objects like books and instruments. The colonial part was less prestigious though equally important because it not only supplied him with specimens to exchange with the metropolitan network, but was also the immediate source that supplied him with a job, money and much needed logistical resources.

Many of König's colonial correspondents we have already met at the beginning of this chapter. For instance, the Dutch branch including Dr Rose, Governors de Bruijn and Falck, the Madras circle including Ross, Boswell, Pasley, Russell and Roxburgh, and Lady Monson representing the Calcutta connection. However, the network was, in fact, even wider and more impressive. In Madras, the circle of doctors also included Dr James Anderson (1739-1809), surgeon in Madras since 1765 and from 1786 Physician General there.[94] The class of high government officials and their spouses was represented by Lady Joanna Rumbold, the wife of Thomas Rumbold, Baronet and governor of Madras 1777-1780, by General Sir Hector Monroe (1726-1805), 8th Laird of Novar and 1778-1780 commander of the Madras Army and by Lord George Macartney (1737-1806), governor of Madras 1781-1785.[95] König's connection to high government in Calcutta was equally impressive as it included Warren Hastings (1732-1818), governor-general of British India 1773-1785.[96]

94. NHM, König MS, List of plants, vol.2, p.1, and Journal, vol.5, p.67. On Anderson, see Pratik Chakrabarti, '"Neither of meate nor drinke, but what the Doctor alloweth": medicine amidst war and commerce in eighteenth-century Madras', *Bulletin of the history of medicine* 80:1 (2006), p.1-38.

95. NHM, König MS, List of plants, vol.1, p.217, 219, 221, and vol.2, p.1, 29. On Joanna Rumbold, see M. M. Drummond, 'Rumbold, Thomas (1736-1791), of Woodhall, Herts.', in *The History of Parliament: the House of Commons 1754-1790*, ed. L. Namier and J. Brooke (London, 1964), http://www.historyofparliamentonline.org/volume/1754-1790/member/rumbold-thomas-1736-91/ (last accessed 8 August 2017). On Munro, see 'Munro, Hector (1726-1805)', in *Encyclopædia Britannica*, ed. Hugh Chisholm (New York, 1911), vol.19, https://en.wikisource.org/wiki/1911_Encyclopædia_Britannica/Munro,_Hector (last accessed 8 August 2017). On Macartney, see 'Macartney, George (1737-1806)', in *Encyclopædia Britannica*, ed. Hugh Chisholm (New York, 1911), vol.17, http://encyclopedia.jrank.org/LUP_MAL/MACARTNEY_GEORGE_MACARTNEY_EARL.html/ (consulted 8 August 2017).

96. NHM, König MS, List of plants, vol.2, p.1. On Hastings, see P. J. Marshall, 'Hastings, Warren (1732-1818)', in *Oxford dictionary of national biography* (Oxford, 2004), http://www.oxforddnb.com/view/article/12587 (last accessed 8 August 2017).

Turning to König's metropolitan network, it had its most important nodes in Uppsala with Linnaeus, in Copenhagen with Rottbøll and in London with Solander and Banks.[97] Yet, this network too was much wider. First, the high number of correspondents in Copenhagen shows that this part of the network was at least as important to König as the connection to Linnaeus and London, despite the fact that he was more dependent on the British to secure his career and livelihood in India. The network in Copenhagen included notable scholars such as J. D. Cappel, König's former employer and member of the Danish government's Board of Health (Collegium Medicum), Morten Thrane Brünnich (1737-1827), professor of natural history and economy at the University of Copenhagen, Otto Frederik Müller (1730-1784), naturalist and court official, and finally Lorenz Spengler (1720-1807), turner to the Royal Court, naturalist and head of the Royal Chamber of Arts and Sciences (Det Kongelige Kunstkammer).[98] All of these scholars were correspondents of Linnaeus, and Spengler was also the guardian of König's son, Gerhard.[99] Outside the academic world, but still in Copenhagen, König's Danish network included Niels Ryberg (1725-1804), probably the most influential and wealthy merchant and shipping magnate in Denmark-Norway at the time and member of the Board of Directors of the Danish Asiatic Company.[100]

A second tier in König's metropolitan network consisted of scholars in university towns scattered across North and North West Europe who were also correspondents of Linnaeus. Among them were Anders Jahan Retzius (1742-1821), professor of natural history and economy at the University of Lund, Johann Christian Daniel von Schreber (1739-1810), professor of botany at the University of Erlangen, David van Royen (1727-1799), professor of botany at the University of Leiden and John Hope (1725-1786), professor of botany at the University of Edinburgh.[101]

97. On Linnaeus, see NHM, König MS, List of plants, vol.1, p.43-47, 267. On Rottbøll, see NHM, König MS, List of plants, vol.1, p.20, 229, 287, 415, and vol.2, p.57. On Solander, see NHM, König MS, List of plants, vol.1, p.125, 225, 250, 297, 305, and vol.2, p.1.

98. On Brünnich, see NHM, König MS, List of plants, vol.1, p.237-38, 282-83, 331-33; Bricka, *Dansk biografisk lexikon*, vol.3, p.191-94. On Müller, see NHM, König MS, List of plants, vol.1, p.265, 335, and vol.2, p.51; Bricka, *Dansk biografisk lexikon*, vol.11, p.594-600. On Cappel, see NHM, König MS, List of plants, vol.1, p.239, 275, 281, 331-33; Bricka, *Dansk biografisk lexikon*, vol.3, p.354-55. On Spengler, see NHM, König MS, List of plants, vol.1, p.273, and vol.2, p.56, 71; Bricka, *Dansk biografisk lexikon*, vol.16, p.209-12.

99. Nair, *The Museum in colonial India*, p.46.

100. On Ryberg, see NHM, König MS, List of plants, vol.1, p.273, 331-33; Bricka, *Dansk biografisk lexikon*, vol.14, p.444-47.

101. On Retzius, see NHM, König MS, List of plants, vol.1, p.239, 247, 265, 323, and vol.2, p.55; G. Eriksson, 'Retzius, Anders Jahan', in *Svenskt biografiskt lexikon*, vol.30 (Stockholm, 2000), vol.30, http://sok.riksarkivet.se/sbl/artikel/6599/ (last accessed 8 August 2017). On Schreber, see NHM, König MS, List of plants, vol.1, p.402; E. Wunschmann, 'Schreber,

The latter has already been discussed in chapter 6 by Andersson Burnett and Buchan in this volume.

In general, König's communication with his network consisted of letters with descriptions of travels and individual species, and shipments of specimens of natural history such as plants, seeds, minerals, seashells and zoological specimens in alcohol. The fact that the network also contained influential officials and merchants in both the colonial part and the metropolitan points to the broad interest in natural history in these circles at the time and it underscores how dependent König was as a researcher in the colonial field on economic, logistical and political support.

In addition to the network of individuals, and overlapping with it in significant ways, König was also a member of a number of scholarly societies. In 1772 he became member of both the Imperial Leopoldina Carolina German Academy of Natural Scientists (Kaiserlich Leopoldinisch-Carolinische Deutsche Akademie der Naturwissenschaftler) and the Royal Norwegian Society of Sciences and Letters in Trondheim (Det Kongelige Norske Videnskabers Selskab, KNVS).[102] Later, he became member of the Physiographic Society in Lund (Physiographiska Sällskapet i Lund), and in the 1780s of the Royal Danish Academy of Sciences and Letters in Copenhagen (Det Kongelige Danske Videnskabernes Selskab, KDVS) and the Berlin Society of Friends of Natural Science (Gesellschaft Naturforschender Freunde zu Berlin, GNFB). He was also a founding member of the Asiatic Society in Calcutta (AS).[103] As a member of these societies, König gained prestige, a broader network of potential correspondents and, most importantly, a platform for publishing his research.[104]

Joh. Christian Daniel (v.)', in *Allgemeine Deutsche Biographie* (Leipzig, 1891), vol.32, http://daten.digitale-sammlungen.de/bsb00008390/images/index.html?seite=467/ (consulted 8 August 2017). On van Royen, see NHM, König MS, List of plants, vol.2, p.81; W. M. C. Regt, 'Royen, David van', in *Nieuw Nederlandsch biografisch woordenboek*, ed. P. C. Molhuysen and P. J. Blok, vol.10 (Leiden, 1937), http://www.dbnl.org/tekst/molh003nieu10_01/molh003nieu10_01_1256.php (last accessed 8 August 2017). On Hope, see NHM, König MS, List of plants, vol.1, p.240, 309, and vol.2, p.29; G. T. Bettany, 'Hope, John (1725-1786)', in *Dictionary of national biography*, ed. Sidney Lee (New York, 1891), vol.27, http://en.wikisource.org/wiki/Hope,_John_(1725-1786)_(DNB00) (last accessed 8 August 2017).

102. On the Leopoldina, see NHM, König MS, List of plants, vol.1, p.351; 'Members of the Leopoldina', http://www.leopoldina.org/de/mitglieder/mitgliederverzeichnis/member/4703/ (last accessed 8 August 2017). On the KNVS, see Olaus Schmidt, *Det kongelige norske videnskabernes selskab: Matrikkel, 1760-1960* (Trondheim, 1960), p.34.

103. On the KDVS, see Asger Lomholt, *Det kongelige danske videnskabernes selskab 1742-1942: Samlinger til selskabets historie*, vol.1 (Copenhagen, 1942), p.355-85. On the GNFB, see *Schriften der Berlinischen Gesellschaft Naturforschender Freunde* 5 (1784), p.xliv. On the AS, see *Asiatick researches*, vol.1, 5th edn (1784; London, 1806), p.489.

104. Svend-Erik Sandermann Olsen, *Bibliographia discipuli Linnaei. Bibliographies of the 331 pupils*

Conclusion

This examination of the meandering journeys of Johann Gerhard König through South and South East Asia has revealed that working and making a career as a Linnaean naturalist in the colonial field was not simply a matter of diffusing European science to the world. Making knowledge in the field was a complex undertaking that required constant negotiation. As formal structures of scientific institutions did not yet exist, or were only beginning to be established, in the Danish-Norwegian, British and Dutch colonies that he roamed, König had to rely on networking personal contacts within the larger Republic of Letters and on constant movement across national and imperial spheres to make a living despite wars and financial difficulties. In this social context, rare specimens of natural history and information from distant lands were not just resources for scientific research but also gifts to be exchanged for advances in his personal career.

Like his fellow Linnaeans Löfling and Thunberg, König was not so much a direct emissary of Linnaeus himself as of networks of knowledge brokers exercising their influence through European trading companies and monarchies. König's correspondence network consisted of a colonial and a metropolitan part, both of which he depended on for economic, logistical and political support. To König, collaboration with this network of naturalists, collectors and scientific societies was just as important as collaboration with people in the field. Nonetheless, he remained very much the man in the field: surveyor, bio-prospector, collector and organiser. In other words, he positioned himself as a go-between and nexus between the colonial field and the Linnaean networks of natural history in Europe.

König came to work in several geographical areas, from Iceland to Siam. Sometimes he worked in areas already colonised, sometimes in areas colonised by powers other than the one he worked for, and sometimes in areas not under European control. Consequently, he became an expert in negotiating the field. His methods followed Linnaeus's ideals of fieldwork. Still, a crucial social dimension of fieldwork was how to negotiate locals and their knowledge. For all his experience, König's approach to local knowledge seems to have been ambivalent. On the one hand, he was very interested in local knowledge regardless of what ethnic group it came from. On the other hand, he

of Linnaeus (Copenhagen, 1997), p.236-38; Brigitte Hoppe, 'Von der Naturgeschichte zu den Naturwissenschaften die Dänisch-Halleschen Missionare als Naturforscher in Indien vom 18. bis 19. Jahrhundert', in *Mission und Forschung: Translokale Wissensproduktion zwischen Indien und Europa im 18. und 19. Jahrhundert*, ed. Heike Liebau, Andreas Nehring and Britta Klosterberg (Halle, 2010), p.153-58, 162, 164.

rarely stated the name of his non-European informants but always that of his European peers and superiors. He tended to see local knowledge as empirical and thus as less advanced than his own theoretical knowledge based on the Linnaean system. This view conveniently emphasised the importance of his own role as a mediator between the field and scientific circles in Europe. Furthermore, while the power relations between König and non-Europeans were usually asymmetric it was also highly entangled. On the one hand, König generally assigned his non-European collaborators to inferior positions, yet on the other, he was himself a subordinate when he worked for the Nawab of Arcot.

When surveying the field and discussing objects of nature with locals, command of the vernacular languages was, of course, a central issue to König. However, vernacular names were also crucial in the parallel work of cross-referencing local names with older botanical literature in order to validate them and harness valuable information from them. It was a process of combining information from many sources, new and old, non-European and European, and in several languages, into the new Linnaean structure.

As indicated at the beginning of this chapter, recent research has suggested that informal scientific networks of less globally significant colonial powers like Denmark-Norway and Sweden may have the potential to revise current views of 'colonial science'. From the examples of Löfling, it is clear that Linnaeus's apostles did in fact participate in colonisation carried out by other states and empires. König, however, is an even stronger example that shows the full potential of the above suggestion. He was not just an emissary of networks in Denmark-Norway but also a member of the large group of expatriates from the German lands who worked as scientific experts for European states like Denmark-Norway and Russia and companies like the VOC and EIC. It seems that it was not just his status as a Linnaean naturalist but also this status as an expert from an inconspicuous country with no colonies of its own that allowed König to be employed in Denmark-Norway, by the EIC, to receive a job offer from the VOC and move freely around VOC territory. Yet, with the rise of national science at the expense of the transnational network of the Republic of Letters, this flexibility was rapidly disappearing.

8. Lives of useful curiosity: the global legacy of Pehr Löfling in the long eighteenth century

KENNETH NYBERG *and* MANUEL LUCENA GIRALDO

Introduction

Processes of (re)production of knowledge are usually examined within a particular type of context or through a particular aspect of the circulation of knowledge. In this chapter the goal is rather to explore a whole range of such contexts and aspects by studying one single individual's life geography and global legacy in the long eighteenth century. That individual is Pehr Löfling (1729-1756), one of Carl Linnaeus's travelling students or 'apostles'.[1] While our point of departure is the general theoretical framework laid out in the introduction to this volume, we shall not dwell on its basic elements here but rather focus on previous scholarship in Swedish and Spanish, concerning the apostles in general and Löfling in particular, which may be little known outside the Swedish-Spanish context.

An important first point in this regard is that the traditional narrative of Linnaeus and his rise to fame is very much part of a Swedish historiographical tradition that, in turn, takes a deeply Eurocentric perspective for granted. In other words, Linnaeus in the role of a great man of science serves the purpose of lending prestige to Sweden within a wholly European context of national rivalries, given added relevance by

Kenneth Nyberg is grateful to Riksbankens Jubileumsfond, http://rj.se (last accessed 30 March 2017) for supporting the research on which his contribution to this chapter is based. Manuel Lucena Giraldo has had the support of the research project 'History and philosophy of experience: cultural history of wellbeing' (FFI2016-78285-R), funded by the Spanish Ministry of Economy, Industry and Competitiveness through the Spanish National Research Council, CSIC, http://csic.es (last accessed 30 March 2017).

1. For 'life geography', see David N. Livingstone, *Putting science in its place: geographies of scientific knowledge* (Chicago, IL, 2003), p.182-83. For the apostles, see Sverker Sörlin, 'The apostles', in *The Linnaeus apostles: global science and adventure*, vol.1, *Introduction*, ed. Lars Hansen (Whitby, 2010), p.151-79; Kenneth Nyberg, 'Linnaeus's apostles and the globalisation of knowledge, 1729-1756', in *Global scientific practice in an age of revolutions, 1750-1850*, ed. Patrick Manning and Daniel Rood (Pittsburgh, PA, 2016), p.73-89; Sverker Sörlin and Otto Fagerstedt, *Linné och hans apostlar* (Stockholm, 2004); Hanna Hodacs and Kenneth Nyberg, *Naturalhistoria på resande fot: om att forska, undervisa och göra karriär i 1700-talets Sverige* (Lund, 2007).

Sweden's recent loss at that time (with the conclusion of the Great Northern War, 1700-1721) of great power status. Further, this narrative is to an extraordinary degree and quite literally the product of Linnaeus's own view of his accomplishments and his role in the history of science, since a great deal of the sources for it emanate directly or indirectly from him and have not always been approached with the amount of critical analysis that they deserve.[2]

Ironically, however, in those very sources we can also find visible support for the interpretation of Linnaean science offered in this volume, especially the notion of it being a collaborative effort which rests not only on the ideas of one man but on the contributions and interactions, the work and the practices, of many individuals and institutions.[3] This is made very clear in the prefaces to several of Linnaeus's major works, where he used lists and tables to demonstrate the overwhelming amount of learning and evidence upon which these works were based. First in these lists were always Linnaeus's own previous books and papers, his travels and collections, but this was followed by external contributions: the writings and the collections of his travelling students; the gardens and herbaria he had visited or consulted; the other prominent botanists he had corresponded with or whose works he had studied, and so on.[4] In other words, it is obvious that Linnaeus wanted to demonstrate beyond any shadow of doubt that his work was based on the entire body of existing (European) knowledge about the topics he was covering. But his point in doing so was to establish that he was not only the 'first among equals', but rather that he was the one who created order out of the chaos of all of these (otherwise meaningless) efforts. Further, he claimed to do so on the basis of ideas and principles that were uniquely his own achievements and for which no one else could take credit. Today, we can reject such a claim and turn it on its head, taking Linnaeus's own word for it that his 'reformation' of botany was ultimately a collaborative effort.

An important part of that effort, in interpretations old and new, was the travels by the seventeen or eighteen young men who were referred to as his apostles, although the exact nature of their contribution to

2. For a discussion of the historiography, see the introduction to this volume and Nyberg, 'Linnaeus's apostles and the globalisation of knowledge', p.73-79.
3. See especially chapter 2 by Dietz in this volume; see also Bettina Dietz, 'Contribution and co-production: the collaborative culture of Linnaean botany', *Annals of science* 69:4 (2012), p.551-69; and Marie-Christine Skuncke, *Carl Peter Thunberg, botanist and physician: career-building across the oceans in the eighteenth century* (Uppsala, 2014).
4. See, for example, the preface and various opening materials in Carl Linnaeus, *Species plantarum, exhibentes plantas rite cognitas ad genera relatas, cum differentiis specificis, nominibus trivialibus, synonymis selectis, locis natalibus, secundum systema sexuale digestas* (Stockholm, Lars Salvius, 1753).

Linnaeus's main publications and ideas is open to debate.[5] It should also be stressed that while they have often tended to be treated as a group – again, largely the result of the impact of Linnaeus's perspective – there were great individual differences between them in terms of scientific achievements as well as other respects.[6] However, most of them had this in common: they spent significant amounts of time studying under Linnaeus, and then travelled through Europe and other parts of the world for scientific and/or educational purposes.

This means that, while the apostles have often played an important role in diffusionist accounts of Linnaean science, in at least some cases they were in fact the very embodiment of the circulation of knowledge, of knowledge brokers and go-betweens.[7] In recent scholarship the latter epithets have often been reserved specifically for individuals who were not themselves trained scientists but rather facilitators and intermediaries between scientists and institutions in Europe and/or indigenous people and other local agents outside Europe.[8] As will be apparent below, however, this is very much the role played by many of Linnaeus's apostles in practice, and sometimes there were interesting tensions between that type of position and their identity or persona as a trained scientist in their own right.[9]

Whether these travellers qualify as go-betweens or not, however, is less important than the simple fact that studying them allows us to (at least in some respects) much more directly examine the global nature of the

5. See, for instance, Sörlin, 'The Apostles'; Bengt Jonsell, 'Apostlarnas resor och gärningar: Linnélärjungarnas roll i upptäckten av världen', in *Ljus över landet? Upplysningen som drivkraft i 1700-talets svenska vetenskap och vitterhet*, ed. Paul Hallberg (Gothenburg, 2005), p.79-98; Lisbet Koerner, *Linnaeus: nature and nation* (Cambridge, MA, 1999), p.49-52, 113-18 and 146-49; Mariette Manktelow and Kenneth Nyberg, 'Linnaeus's apostles and the development of the *Species plantarum*', in *'Species plantarum' 250 years: proceedings of the 'Species plantarum' symposium*, held in Uppsala 22-24 August 2003, ed. Inga Hedberg (Uppsala, 2005), p.73-80; Kenneth Nyberg, 'Linnaeus's apostles, scientific travel and the East India trade', *Zoologica scripta* 38, Supplement 1 (2009), p.7-16.

6. Jonsell, 'Apostlarnas resor och gärningar', p.80, and Bengt Jonsell, 'Thunberg, Linné och det linneanska', in *Carl Peter Thunberg: linnean, resenär, naturforskare 1743-1828*, ed. Bertil Nordenstam (Stockholm, 1993), p.41-55 (41).

7. This is one of the main themes in Skuncke, *Carl Peter Thunberg*, which builds on and further develops the concepts and arguments in works such as *The Brokered world: go-betweens and global intelligence, 1770-1820*, ed. Simon Schaffer *et al.* (Sagamore Beach, MA, 2009); Kapil Raj, *Relocating modern science: circulation and the construction of knowledge in South Asia and Europe, 1650-1900* (Houndmills, 2007); and Neil Safier, 'Itineraries of Atlantic science: new questions, new approaches, new directions', *Atlantic studies* 7:4 (2010), p.357-64.

8. See, for instance, Neil Safier, 'Spies, dyes and leaves: agro-intermediaries, Luso-Brazilian couriers and the worlds they sowed', in *The Brokered world: go-betweens and global intelligence, 1770-1820*, ed. Simon Schaffer *et al.* (Sagamore Beach, MA, 2009), p.239-65.

9. Lorraine Daston and H. Otto Sibum, 'Introduction: scientific personae and their histories', *Science in context* 16:1-2 (2003), p.1-8.

circulation of knowledge than if we study Linnaeus himself. The notion of 'centres', in the Latourian sense and others, has been much discussed, but to the extent that such centres existed Linnaeus can be identified with, or as, one of them.[10] He was, indeed, stationary for much of his career, in Uppsala, while his students travelled. Whether they did so for him, for themselves or others is an issue we will return to later. The point here is that they were on the move, and by following them we can follow the circulation of knowledge and what that actually entailed in practice. Most of all, what we can do with one single life, in this case Pehr Löfling's, is to study closely how Linnaean science was constantly negotiated and renegotiated in different contexts.

The purpose of what follows here, then, is to explore how the global circulation of Linnaean science impacted on, and was influenced by, individual lives on different levels by following a Linnaean naturalist around, through changing contexts of negotiation during his lifetime and after.[11] We hope to show that, while the idea of diffusion is not entirely irrelevant to understanding these issues, those who saw themselves as 'spreading' Linnaean knowledge were also affected and changed by the contexts in which they found themselves; they were in a constant process of negotiating new identities, relationships and allegiances over time (and space). The chapter consists of four sketches of such (re)negotiations of Linnaean science in different contexts, all connected to one scientific life and its changing geography: Uppsala, Madrid, colonial Venezuela and newly emancipated Spanish America. Since the last sketch is centred around Löfling's posthumous legacy, we have chosen to talk of not only one life, but lives, of 'useful curiosity'.

Becoming a naturalist

Pehr Löfling was born in 1729 as the son of a bookkeeper at the iron works of Tolvfors in the north of Sweden.[12] At the age of fourteen he

10. David Philip Miller, 'Joseph Banks, empire, and "centres of calculation" in late Hanoverian London', in *Visions of empire: voyages, botany and representations of nature*, ed. David Philip Miller and Peter Hanns Reill (Cambridge, 1996), p.21-37. See also chapter 4 in this volume by Van Damme.
11. See Mary Terrall, 'Following insects around: tools and techniques of eighteenth-century natural history', *British journal for the history of science* 43:4 (2010), p.573-88. The approach here is similar, but applied on a broader global scale while focusing more narrowly on one single person's movements.
12. Unless otherwise noted, the outline of Löfling's biography here is based primarily on Linnaeus's narrative in his preface to Pehr Löfling's *Iter Hispanicum* (Stockholm, Lars Salvius, 1758) and its recounting in later scholarship, especially Stig Rydén, *Pehr Löfling: en linnélärjunge i Spanien och Venezuela 1751-1756* (Stockholm, 1965); Manuel Lucena Giraldo, *Laboratorio tropical: la expedición de límites al Orinoco, 1750-1767* (Caracas, 1993); and Kenneth Nyberg, *Pehr Löfling's letter-book in the archives of the Real Jardín Botánico in Madrid: a catalogue*

matriculated at the University of Uppsala, where he intended to study theology in order to become a clergyman. At some point during the next few years he came into contact with Carl Linnaeus and the field of natural history, especially botany. Although there are hints that Löfling had previously encountered Linnaean botany in the form of one of the earliest apostles, Pehr Kalm, it was only in Uppsala that he abandoned his plans for a career in the state Church in favour of becoming a naturalist in training, as well as a member of the Linnaean circle.[13] That circle included not only the teacher himself and his students, but also his family members and other dependents.

Löfling apparently did not make an instant impression on his teacher, but in 1748 Linnaeus assigned him a dissertation topic on tree buds that he completed in a way that impressed his seniors; he then seems to have made very rapid progress as a botanist. Unlike other students Löfling probably wrote his academic 'trial dissertation' *Gemmae arborum* (1749) himself, or at least most of it,[14] and around this time or soon after he became a private tutor to Linnaeus's son, also named Carl. This meant that he lived in Linnaeus's house and spent much of his time there, and he also helped his teacher with secretarial services from time to time. Notably, this included Linnaeus dictating much of the manuscript of the *Philosophia botanica* (1751) to Löfling, who, Linnaeus claims, constantly asked him about its contents so that he became firmly acquainted with the principles and practices articulated there.[15]

Both the *Gemmae arborum* and the *Philosophia botanica* played a role in the development of binomial nomenclature, one of the main features of

(Gothenburg, 2008). The best concise summary is in Swedish; see Olle Franzén, 'Löfling, Pehr', in *Svenskt biografiskt lexikon*, vol.24 (Stockholm, 1982-1984), http://sok.riksarkivet.se/sbl/artikel/10050 (last accessed 1 March 2017). Some rhetorical flourishes aside, as far as can be determined from other sources, including the few fragments of biographical data left by Löfling himself, Linnaeus's account is more or less correct.

13. Pehr Löfling to Pehr Kalm, Madrid, 15 May 1752, Archivo del Real Jardín Botánico, Madrid (hereafter RJB), II.2.1. Unless otherwise noted, all dates are in the Gregorian calendar used in much of Europe at this time, rather than the Julian calendar used in Sweden before 1 March 1753. Compare with a slightly different story in Löfling's speech to a 'nation' of students in Uppsala, 24 May 1749, RJB, II.1.1.8, later published by Stig Rydén and Arvid Hj. Uggla as 'Pehr Löflings oration i Medelpado-Jämtländska Nationen den 24 maj 1749', *Norlandica, Norrlands Nations skriftserie* 5 (Uppsala, 1961), p.98-123.

14. Gustaf Drake af Hagelsrum, *Linnés disputationer: en översikt* (Nässjö, 1939), p.192.

15. Carl Linnaeus, *Philosophia botanica in qua explicantur Fundamenta Botanica cum definitionibus partium, exemplis terminorum, observationibus rariorum* (Stockholm, Gottfried Kiesewetter, 1751); the manuscript is now (partially) preserved in the Library of the Linnean Society of London, LM/LP/BOT/3/3/1. Around the same time, Löfling also compiled the index to Linnaeus's account of a journey in the Swedish province of Skåne in 1749; see Löfling to E. G. Lidbeck, Uppsala, 16 November 1750, Lund University Library, Brevsamlingen.

Linnaean botany which still remains in use to this day.[16] This is just one example of a broader, often understudied theme that is important in the context of the circulation of knowledge – in other words, the relationship between Linnaeus's scientific work and his role as a teacher.[17] That relationship has often been described as one where Linnaeus's teaching was based on his research; this direct connection, not very common in Swedish universities at the time, was one of the main reasons why he became such a popular teacher in the first place. As Skuncke points out, his classes and botanical excursions with students 'had the excitement of presenting research in the making'.[18]

However, the opposite is also true: his teaching and other interactions with the students, in particular the apostles, played a role in the development of his scientific thinking. Certainly this was true during the 1740s and 1750s, especially in the crucial years around 1750 leading up to the publication of both *Philosophia botanica* and *Species plantarum* (1753), the latter often being seen as the culmination of the crucial phase of his scientific life that had begun with the *Systema naturae* in 1735. His notions on the sexual system, in other words, had nothing to do with his teaching, but for the binary nomenclature as well as some of the methodological and practical rules established in the *Philosophia botanica* it seems clear that his students' struggles to preserve, organise and keep track of their specimens played an important role.[19]

Pehr Löfling was among the handful of more prominent students in that respect, both as an inquisitive secretary to Linnaeus during the writing of the *Philosophia botanica* and later as a very active correspondent of his during the final stretch of work on the *Species plantarum*. While the relationship between Linnaeus and his apostles varied greatly, all the extant sources suggest that Löfling closely identified with Linnaeus's view of the world and the Linnaean system's place within it.[20] In every sense of the word, then, he was a 'Linnaean', but that did not prevent him from developing an independent mind in systematics where he dared, on occasion, to question Linnaeus himself in a way that very few of the other

16. Manktelow and Nyberg, 'Linnaeus's apostles', p.73-80, and William T. Stearn, 'An introduction to the *Species plantarum* and cognate botanical works of Carl Linnaeus', in Carl Linnaeus, '*Species plantarum*': *a facsimile of the first edition 1753*, vol.1 (London, 1957), p.1-176 (66-71).

17. Hodacs and Nyberg, *Naturalhistoria på resande fot*, especially p.9-10; see also Hanna Hodacs, 'Linnaeans outdoors: the transformative role of studying nature "on the move" and outside', *The British journal for the history of science* 44:2 (2011), p.183-209 (187).

18. Skuncke, *Carl Peter Thunberg*, p.229.

19. Kenneth Nyberg and Mariette Manktelow, 'Linnés apostlar och tillkomsten av *Species plantarum*', *Svenska Linnésällskapets årsskrift* (2002-2003), p.9-30 (21-24); Koerner, *Linnaeus*, p.49.

20. Jonsell, 'Apostlarnas resor och gärningar', p.85-86.

students did (at least openly). In fact, one of the interesting themes to follow in Löfling's correspondence after he left Uppsala for Madrid in 1751 is the rapid evolution of his independence in general, as well as in relation to Linnaeus more specifically.[21]

That brings us to the question of Löfling's move from Sweden to Spain: what does it say about the Linnaean enterprise? It is interesting not least because it illustrates so clearly the flaws of previous interpretations of the apostles and how Linnaeus sent them out into the dangers of the world. There are somewhat contradictory versions about how Löfling's journey to Spain came about, but even if it was exclusively Linnaeus's initiative, as seems likely at first, the power to realise it – and even more so the continuation to South America a few years later – was entirely in the hands of the Spanish authorities. Their motives were not necessarily the same as those of Linnaeus, Löfling or other interested parties, a fact that helps illustrate the negotiated nature of how Linnaean science travelled.

According to preserved diplomatic correspondence, it was Linnaeus who approached the Spanish ambassador to Stockholm, the Genoese Jerónimo de Grimaldi, in the autumn of 1750, and suggested that one of his students could be sent to Spain in order to advance the study of Spain's flora according to the latest (that is, his own) scientific methods. He also asked the Spanish government to fund the work since it would not be possible to raise money in Sweden for it. Grimaldi supported the proposal and sent it to Madrid, where it was approved. After being notified in early January 1751, Linnaeus named Löfling as his choice to undertake the assignment.[22]

It is unclear whether the prospect of going to America was already raised at this early stage. Although the Spaniards may not have con-

21. See the exchange with Linnaeus below about naming some plants after Spanish botanists, where Löfling does not refrain from correcting Linnaeus (however politely). This stands in stark contrast to the tone in an early letter where he complains of his difficulties in 'making real differentials' between species; Löfling to Linnaeus, Porto, 7 August 1751, Library of the Linnean Society of London, Linnaeus correspondence, vol.9, p.339-40; draft in RJB, II.2.1. In *The Linnaean correspondence*, http://linnaeus.c18.net/, this letter's reference number is L1299. For Linnaean letters mentioned below, whenever possible the *Linnaean correspondence* number has been given in addition to the source reference.

22. Reports from Jeronimo de Grimaldi, Spanish ambassador in Sweden, to his superiors in Madrid from 23 October 1750 to 5 February 1751, Archivo General de Simancas (hereafter AGS), Estado, leg. 6678 and 6679. Linnaeus also wrote a letter of thanks to the Spanish authorities, dated Uppsala 25 January 1751 and composed in Latin, after receiving word that his request had been granted (AGS, Estado, leg. 6679); this seems to be identical to L1222 in *The Linnaean correspondence* where Spanish prime minister José de Carvajal y Lancaster is identified as the recipient. See also Lucena Giraldo, *Laboratorio tropical*, p.118-20, and Rydén, *Pehr Löfling*, p.19.

sidered it that way, Linnaeus himself probably saw the journey to Spain as a chance to put one of his students on a path that, with any luck, might lead on to the New World. This was very interesting territory from his perspective, and the only way to gain access to most of it for scientific investigation was through the Spanish authorities.[23] While his apostles' travels may not have been as systematically planned an enterprise as has often been assumed, Linnaeus was very adept at recognising and grasping opportunities to explore other parts of the world when they presented themselves.[24] This was one of the most important of such opportunities, as Spanish America was more difficult to access than other areas of almost equally keen interest to Linnaean botanists, such as the Cape Colony in southern Africa. Here, as elsewhere, it is Linnaeus's own version of events that has come to prevail, in which Löfling travelled to Spain out of loyalty to his teacher and his love of science, while Spanish officials were driven by the desire to advance the state of botany in their country to invite a close associate of the great Linnaeus. There are probably elements of truth in both of those assumptions, but it is interesting that Löfling's private letters to friends reveal that he was less than enthusiastic about going – an indication in itself, perhaps, that initially there was only talk of Spain and not of America.[25]

If we try to free ourselves from Linnaeus's perspective as regards not only Löfling but also the Spanish government, what could their motives have been for agreeing to employ him? In part, they probably did see the benefit that a student of Linnaeus might be able to give to Spanish botany. Having said that, it is difficult not to assume that Löfling's potential role in an American expedition later on might also have been a consideration. In January 1750, Spain and Portugal had reached an agreement which finally settled long-standing disputes over the borders between their territories in South America. While the expedition that Löfling joined had several objectives and did not depart until early 1754, formally it constituted one of the Spanish border commissions organised in accordance with that treaty. All indications show that these commissions were several years in the making, suggesting

23. The example of Daniel Rolander, who spent seven months in the Dutch colony of Surinam in 1755-1756, illustrates that other opportunities did exist to travel to South America; see James Dobreff, 'Daniel Rolander: the invisible naturalist', in *Systema naturae 250: the Linnaean ark*, ed. Andrew Polaszek (Boca Raton, FL, 2010), p.11-28. But the vast majority of the continent's territory was Spanish and could only be explored with Spanish approval and, realistically, through their logistical support, as seen in the geodesic mission to the Equator (1735-1748).

24. Sörlin, 'The apostles', p.164; Hodacs and Nyberg, *Naturalhistoria på resande fot*, p.178.

25. Löfling to Pehr Bjerchén, Uppsala, 29 January 1751; copy in Royal Swedish Academy of Sciences (hereafter RSAS), Bergianska brevsamlingen, vol.3, p.545.

that this is the context in which to view the original agreement between Linnaeus and the Spanish government.[26]

A Linnaean among Tournefortians

Following a few months of preparations, Löfling departed from Stockholm in May 1751 on a ship bound for Porto, Portugal, and after some time spent there and elsewhere along the Portuguese Atlantic coast he travelled to the Spanish capital, where he arrived in October. He stayed in Madrid and the surrounding area for two years, making some short excursions but never being permitted to explore other parts of the country, to the great disappointment of his and Linnaeus's initial hopes. These spatial restrictions would become a source of steadily increasing irritation, until it was finally resolved that he would travel to South America.[27] Perhaps this is another indication that the interest of the Spanish authorities never really lay in Löfling's potential contributions to the flora of Spain, but rather in how he could serve their efforts to more effectively mobilise and exploit the natural resources of their American territories.

Löfling's situation in Madrid differed quite dramatically from what it had been in Uppsala. In some ways, he was now more independent and free than ever before, the journey having transformed him from a student to a scientist in his own right. In other ways, there were now also new constraints due to the fact that he was seen, and also saw himself, as Linnaeus's representative in an environment where Linnaean botany was still treated with scepticism – most of the leading Spanish botanists being adherents of Tournefort's older system.[28] Löfling was not only a foreigner in the Spanish capital but, not insignificantly in his own view at least, also a Protestant among Catholics. He was forced, at twenty-two years of age, to quickly develop skills as both a politician and a diplomat in his dealings with leading Spanish politicians, court officials and botanists.[29] He seems to have been highly aware of the necessity of

26. See Lucena Giraldo, *Laboratorio tropical*, p.17-18, 117-21.
27. See for example Löfling to Lars Montin, Madrid, 3 January 1752 (draft, RJB, II.1.4.1). In Löfling to Pehr Kalm, Madrid, 15 May 1752 (draft, RJB, II.2.1), he wrote: 'I'd prefer sooner rather than later to get out of Spain to Am[erica as I have been promised]' ['jag ju förr ju häldre ser, att jag kommer ut af Spanien till Am[erica som mig blifvit lofvadt']; translation by the authors. It is interesting that as early as mid-1752 he considered the American journey a 'promise'.
28. Löfling to Pehr Kalm, Madrid, 15 May 1752 (draft, RJB, II.2.1), mentions some irritation among Spanish botanists over Linnaeus's remarks about the state of botany in Spain in his *Bibliotheca botanica* (Amsterdam, Salomon Schouten, 1736).
29. In a letter to Georg Svahn from Madrid, 18 September 1752 (draft, RJB, II.2.1), Löfling wrote that while he himself was a Linnaean, 'since my arrival here I have studied T[ournefort's system] so well that no Spaniard can trump me there. So I understand

not provoking anyone, especially the botanists, and from the very start he developed strategies to mollify them and engender their trust. One of his first steps in that direction was to write to Linnaeus and others back in Sweden, requesting that they not print any of his letters without his explicit consent since the opinions he expressed in them might upset his hosts.[30]

This is an interesting example of the contradictory demands between common practices in the scientific community – where letters would be printed and circulated in order to disseminate results as well as to increase personal recognition – and the considerations necessitated by political and diplomatic realities. Löfling had all the more reason, then, to employ the full range of other network-building activities in his interactions with his local counterparts in Madrid. He became an intermediary, quite literally a go-between, between Linnaeus (and, more rarely, other Swedes) and prominent Spanish naturalists, as shown in his preserved correspondence.[31] They also communicated directly with Linnaeus, but the majority of these contacts were managed through Löfling who brokered greetings, news, questions and answers, specimens and books in both directions. Spanish government officials also took part in the exchange in a few cases that involved plants with commercial potential, but this was a rare occurrence.[32]

It is quite striking how Löfling worked very actively to build bridges between the Linnaeans and the Tournefortians, efforts in which he frequently solicited and usually received the energetic support of his old teacher. Linnaeus, who was by this time in the final stages of work on the *Species plantarum*, in the autumn of 1752 named three related plants after some Spanish botanists and Löfling himself. His intention was 'to signify in future your conversation with these gentlemen at Madrid'.[33]

both [systems], and know the flaws of both.' ['sedan min hitkomst, har jag så braf studerat T. att ingen Spanior skal slå mig på fingrarna deruti. Jag förstår således bägge, och vet bägges fel.']; translation by the authors.

30. Löfling to Lars Salvius, [Porto] July 1751 (draft, RJB, II.2.1).

31. Linnaeus's letters to Löfling from 1751 to 1753 in RJB, II.1.6.6-II.1.6.23. Löfling's letters to Linnaeus in the Library of the Linnean Society of London, Linnaeus correspondence, vol.9; drafts of most of these as well as of Löfling's other correspondence in RJB, II.2.1. Some letters are also preserved, in original or copy, in Stockholm, Uppsala, Lund and elsewhere. For online summaries of much of the correspondence between Löfling and Linnaeus, see the already mentioned *The Linnaean correspondence*, http://linnaeus.c18.net/ (last accessed 2 March 2017).

32. One well-known example was Löfling being instructed to translate parts of a treatise by Linnaeus's student Pehr Kalm on some useful North American plants into Spanish, so it could be read in Madrid. The draft of the translation is preserved in RJB, II.2.3.6.

33. Linnaeus to Löfling, Uppsala, 3 October 1752: 'det beteknar framdeles Eder conversation med desse herrar vid Madrit'; RJB, II.1.6.15; L1477, translation by the authors.

Unfortunately, Linnaeus seems to have mixed up the names of two of these botanists, Quer and Velez, in creating a *Querezia* (along with the *Minuartia* and *Loeflingia*), an error Löfling was quick to point out in his next letter. Politely but quite firmly he urged Linnaeus to correct the mistake, probably fearing that the botanists in Madrid might otherwise be insulted.[34] Although Linnaeus had already printed the sheets in question for the *Species*, he heeded Löfling's advice and had them reprinted with modified names. In his letter informing Löfling about this, he enclosed samples of the revised pages to show the Spanish botanists. Löfling did so, and later noted their gratitude for Linnaeus's courtesy towards them.[35]

A similar example was Löfling's request that Linnaeus nominate the French astronomer Louis Godin as a member of the Royal Swedish Academy of Sciences in Stockholm. Löfling had met Godin, a recently returned member of the French geodesic mission to the Equator, in Lisbon and then travelled in his company to Madrid. Throughout the journey and upon arrival there Godin had been very helpful, introducing Löfling to various officials in Madrid, giving advice and being supportive in general. While Godin did not have a position of immediate influence over his work in Spain, Löfling was very anxious to have Linnaeus thank him for his assistance and get him admitted as a foreign member of the Academy.[36] Linnaeus seems to have been unwilling or unable to accomplish this, but instead Godin was elected a fellow of the Uppsala Society of Sciences (of which Linnaeus was then Secretary), and Löfling was able to reconnect with him in Cadiz before leaving for Venezuela in early 1754.[37]

Perhaps it was a measure of his success as a network builder that, during his stay in Madrid, Löfling increasingly faced the problem of how to balance the requirements and expectations of his many patrons and allies both in Spain and in Sweden. First and foremost among these was

34. Löfling to Linnaeus, Madrid, 13 November 1752, Library of the Linnean Society of London, Linnaeus correspondence, vol.9, p.356-59; draft in RJB, II.2.1; L1507.
35. Linnaeus to Löfling, Uppsala, 13 March 1753 (RJB, II.1.6.17; L1554), and Löfling to Linnaeus, Madrid, 14 May 1753, Library of the Linnean Society of London, Linnaeus correspondence, vol.9, p.367-69 (draft in RJB, II.2.1; L1586); see also Stearn, 'An introduction to the *Species plantarum*', p.136-42.
36. Löfling to Linnaeus, Madrid, 15 November 1751, Library of the Linnean Society of London, Linnaeus correspondence, vol.9, p.398-99 (draft in RJB, II.2.1). Godin is also mentioned in several other letters written from Madrid around this time, such as to Isaac Biberg, Jonas Hollsten and his own parents (all undated), as well as to the Secretary of the RSAS, Pehr Wargentin, 28 November 1751 (drafts in RJB, II.2.1).
37. Löfling to Louis Godin, informing him of his election to the Uppsala Society, 16 January 1753 (draft in RJB, II.2.1), and to Linnaeus, Puerto de Santa María, 18 December 1753, Library of the Linnean Society of London, Linnaeus correspondence, vol.9, p.385-87 (draft in RJB, II.2.1; L1683).

always Linnaeus himself, which is very clear and often stated explicitly also in his private letters to close friends. Second, once he had arrived in Spain, was the Spanish government, ultimately the king himself, but in practical terms the influential botanist José Ortega and officials like Grimaldi. In addition to these, Löfling was also careful to keep in touch with the secretary of the Swedish Academy of Sciences, Pehr Wargentin, and the leading physician Abraham Bäck who in late 1752 was appointed president of the Collegium medicum (Royal College of Physicians).

In the Swedish context Löfling had a number of other correspondents, ranging from former fellow students to aristocratic and royal benefactors that covered a broad spectrum of scientific, economic and social stature. What they all had in common was an interest in his collections: plants, insects, fish and minerals. Further, locally, in Madrid, as discussed above he had to build relationships with Spanish botanists, which was partly done through sharing plant specimens, descriptions and other scientific intelligence.[38] Taken all together, there were many demands to satisfy, and in several letters Löfling discussed how to balance them out, who took priority as recipients of his collections and so on.[39] His papers in Madrid also include what seems to be a preserved list of correspondents, with notes for each one about the kind of topics they were interested in or that Löfling should remember to emphasise in letters to them. If such an interpretation is correct, he thought very carefully and systematically about how to build up lasting social relations with those who might play a role for his future prospects.[40]

In a larger context, while Löfling undoubtedly established a connection between Linnaean natural history and Spanish botanists that came to have long-lasting effects, one might ask how and to what degree it actually bridged the scientific differences between them or advanced the 'cause' of Linnaean botany in Spain. Gómez de Enterría has argued that the introduction of Linnaean systematics in the Spanish Empire was delayed by Löfling's death. In fact, it does not seem that any of the Spanish botanists with whom he interacted in Madrid actually abandoned the Tournefortian system, at least not immediately. It was only

38. Löfling gathered plants in the garden at Ciempozuelos, in Aranjuez and elsewhere, but he also received specimens 'from the garden of [José] Quer' ['ex hortis Quer'] (translation by the authors), RJB, II.1.7.2, and rock samples from Cristóbal Vélez, RJB, II.2.2.5.

39. Löfling to Lars Montin, Madrid, 10 July 1752 and 11 September 1752, and Löfling to Pehr Bjerchén, Madrid, 15 May 1752 (drafts in RJB, II.2.1). As in Carl Peter Thunberg's case (see Skuncke, *Carl Peter Thunberg*, p.294), Löfling's practice of ranking echoed Linnaeus himself, whose systematic work to a significant degree built upon principles of ranking and listing plants, books and botanists alike.

40. Undated sheet of notes in Löfling's 'Letter-book', RJB, II.2.1; see also Nyberg, *Pehr Löfling's letter-book*, p.71-72, no.168.

several decades later that new generations of botanists there would adopt Linnaean principles of taxonomy and nomenclature.[41]

As Lafuente and Valverde have emphasised, however, the Linnaean system was, indeed, used in the many botanical expeditions sent out by Spain in the second half of the eighteenth century, which makes the Orinoco expedition of which Löfling was a member an interesting forerunner of future developments.[42] To what extent there is a direct lineage between Löfling and the later expeditions is open to debate and interpretation, but, as we will see below, Löfling did leave a legacy partly independent of Linnaeus in the later South American context in a way that was not really mirrored in Sweden. The prominent Spanish-Colombian botanist José Celestino Mutis (1732-1808), for instance, owned a copy of Löfling's *Iter Hispanicum* that survives to this day, and the work was also included in a list of officially recommended natural history books issued for the expedition of Ruiz and Pavón as it prepared to leave for Peru in 1777.[43] Even more telling, perhaps, is the fact that the expedition papers contain a copy of Löfling's bill from January 1754 for ink and other supplies for his draughtsmen, suggesting that it served as a guide in travel preparations almost a quarter of a century later.[44]

Botanist of the Catholic king

While it is not possible to prove what the original intentions of the Spanish authorities were in inviting Löfling to Madrid, in the summer of 1753 they decided that he would join the expedition led by José de Iturriaga to determine the border between Spain and Portugal in the Orinoco and Amazon basins in what is today Venezuela, Colombia and Brazil. In October, Löfling left Madrid for Cadiz, and after a few months of preparation and local excursions, the whole expedition departed in February 1754, arriving in Cumaná, eastern Venezuela, in early April. Here Löfling would spend less than two years working scientifically, interrupted with increasing frequency and duration by bouts of ill

41. Josefa Gómez de Enterría, 'The reception of Swedish scientific and technological innovations in Spain and their lexical contributions in botany, metallurgy and mining', in *Commerce and navigation between Spain and Sweden throughout history*, ed. Enrique Martínez Ruiz and Magdalena de Pazzis Pi Corrales (Madrid, 2000), p.283-99 (286-87).
42. Antonio Lafuente and Nuria Valverde, 'Linnaean botany and Spanish imperial biopolitics', in *Colonial botany: science, commerce and politics in the early modern world*, ed. Londa Schiebinger and Claudia Swan (Philadelphia, PA, 2005), p.134-47 (136).
43. Mutis's copy of *Iter Hispanicum* is kept in the Biblioteca Nacional de Colombia (Bogotá), Fondo Mutis, 2996, and the list of natural history books for Ruiz and Pavón can be found in the Museo Nacional de Ciencias Naturales, Madrid (hereafter MNCN), Flora Peruana y Chilense (Ruiz y Pabón), ref.7, appendix 1.
44. MNCN, Flora Peruana y Chilense (Ruiz y Pabón), ref.7, appendix 3.

health, until he died from a fever in the Capuchin mission of San Antonio del Caroní on 22 February 1756.

Linnaean science as an instrument of colonial exploration and exploitation is often associated with the late eighteenth and early nineteenth centuries, for example in the work and travels of Joseph Banks and his associates, including prominent students of Linnaeus such as Daniel Solander and Anders Sparrman.[45] (See also chapter 7 by Thode Jensen in this volume.) Pehr Löfling's participation in the Orinoco expedition, however, is one of the earliest examples of the direct mobilisation of Linnaean science as part of a colonial enterprise. It is true that there were other Linnaean naturalists active in the region at the same time, but not in Spanish service and not as directly involved in colonial enterprises as Löfling.[46] His work was even part of an actively colonising effort in the sense that the area where he worked was not yet fully under Spanish control, and new towns like San Fernando de Atabapo or San Carlos de Ríonegro were only established later, in 1759 and 1760. As noted above, the Orinoco expedition became a precedent for a great number of later Spanish expeditions that entailed botanical exploration based on Linnaean systematics. In some respects, this potentially gives the expedition and Löfling's contributions to it a special significance in the early modern global history of science.

Pehr Löfling's presence in Cumaná and the Orinoco was one of a European scientist in the colonial 'periphery', adding yet another layer of contexts and identities to his interactions. He was still a Swede among Spaniards, but also a white European among Americans from different ethnic origins and a scientist among (mostly) military men. His mission as a botanist of the Catholic king became increasingly important, as did his authority, not least since he now had Spanish naturalists (Benito Paltor and Antonio Condal) in training and two draughtsmen (Bruno Salvador Carmona and Juan de Dios Castel) to assist him. He could work more

45. For just a few examples see Miller, 'Joseph Banks', p.21-37; Hanna Hodacs, 'Local, universal and embodied knowledge: Anglo-Swedish contacts and Linnaean natural history', in *Global scientific practice in an age of revolutions, 1750-1850*, ed. Patrick Manning and Daniel Rood (Pittsburgh, PA, 2016), p.90-104; and, more generally, Staffan Müller-Wille, 'Walnuts at Hudson Bay, coral reefs in Gotland: the colonialism of Linnaean botany', in *Colonial botany: science, commerce and politics in the early modern world*, ed. Londa Schiebinger and Claudia Swan (Philadelphia, PA, 2005), p.34-48.

46. One example of this was the previously mentioned Daniel Rolander who visited Surinam in the mid-1750s; another, perhaps better known, case was that of Nikolaus Joseph von Jacquin, the Dutch botanist in the service of the Austrian monarchy who travelled in the Caribbean region between 1754 and 1759. Jacquin's travels culminated in the publication of a famous work, *Selectarum Stirpium Americanarum Historia* in 1763, five years after Löfling's *Iter Hispanicum*; see Santiago Madriñán, *Nikolaus Joseph Jacquin's American plants: botanical expedition to the Caribbean (1754-1759) and the publication of the 'Selectarum Stirpium Americanarum Historia'* (Boston, MA, 2013), p.21-45.

independently as a qualified scientist here than ever before, while at the same time being more clearly situated within a hierarchy of command where he was expected to perform certain duties in relation to the duties performed by others. While this was his daily reality, he still had obligations to Linnaeus and other benefactors back in Sweden as well as having to maintain his relationships with the Spanish botanists in the now distant Madrid.

At first glance one might assume that Löfling and his fellow expedition members operated from a position of strength, representing as they did the full force of the Spanish Empire in an area that had been under its control for a long period. While Löfling himself does not say much on the matter, other sources, however, give a rather different impression of realities on the ground. From 1728 the Royal Guipuzcoan Company of Caracas had a monopoly on the trade and production of cacao, the main export crop in the region. Essentially, the interior of both Caracas and Cumaná provinces were only under very patchy and intermittent Spanish control, with frequent outbursts of conflict between certain indigenous populations allied to the Dutch, mainly Caribs, and small, scattered contingents of Spanish soldiers, missionaries and traders. There were a number of strategic missions run by various orders, especially Capuchins and Jesuits, but they struggled to survive and in many cases did not last long.

Finally, many personal conflicts arose during the expedition – not least between the commander, Iturriaga, and his deputy, Alvarado – and between the expedition and the permanent colonial administration in the area. One issue was that of jurisdictions and precedence; another (and probably more critical) was that of mobilising resources for an expedition of several hundred men, which clearly placed a very severe strain on the local and regional treasuries.[47] There is much evidence of these conflicts and frictions in the extensive collections of sources left behind by the expedition in the archives of Madrid, Simancas, Seville, Bogotá and Caracas. When reading them, the fragility and weakness of the Spanish Empire on the Guayana and Amazon frontier becomes very apparent.

In fact, considering its scale and composition the Orinoco expedition was not so much a group of travelling scientists and officials protected by

47. This is a recurring theme, for instance, in many letters and reports in Archivo General de Indias (Seville), Caracas, 439 and 440. See also Manuel Lucena Giraldo, 'El dorado geométrico: la expedición de límites al Orinoco, 1754-1761', in *La Comisión naturalista de Löfling en la expedición de límites al Orinoco*, ed. María Pilar de San Pío Aladrén (Madrid, 1998), p.23-40 (27-33); and Luisa Martín-Merás, 'Fondos cartográficos y documentales de la comisión de límites de Brasil en siglo XVIII en el Museo Naval de Madrid', *Terra Brasilis* 7-8-9 (2007), p.1-81 (4-5), http://terrabrasilis.revues.org/402 (last accessed 2 March 2017).

military troops as a sort of parallel administration on the move. In accordance with the Hapsburg tradition of a *visita*, in extraordinary circumstances a royal minister could be sent to a colonial outpost to execute judicial affairs and deal with abnormal situations. This aspect of the expedition became increasingly important in the years after Löfling's death, and once the expedition eventually ended in name it morphed into something different; many of its participants simply settled and turned to the task of establishing villages, towns and lines of communication further into the interior of the continent. In 1777 all the territories finally became united in a new and powerful Captaincy General of Venezuela with Caracas as its capital. While the old provincial capitals remained, new ones were founded along the rivers, hundreds of miles from the coast. Some of them did not survive long but others became more firmly established and would begin to grow rapidly in the following decades. It was only long after Löfling's death, then, that the area he and fellow members of the expedition travelled in were brought under control. The region would serve as a fundamental base of resources for both royalists and patriots in the revolutionary wars of independence from which it would later emerge as Gran Colombia.[48]

The practical and logistical challenges of conducting scientific work in these difficult and volatile conditions was by all accounts daunting, although Löfling does not address them much in his writings from this time. In general, aside from the purely scientific manuscripts, this material is rather minor and fragmentary, mostly consisting of a handful of letters written to Linnaeus and friends in Sweden and a somewhat larger number of short messages to fellow expedition members about practical matters.[49] It is mainly in these that we get a glimpse of his daily work and travels in the region, as well as in some of the sources left behind by the expedition itself: account books, orders, circulars and letters of various kinds. One question that becomes particularly interesting in this context is that of the role played by local assistants, collectors or informants in Löfling's work. Unfortunately, it is difficult to determine to what extent he depended on indigenous knowledge for his 'Flora Cumanensis' (later published in the *Iter Hispanicum*), or other writings at this time, due to the lack of sources discussing this explicitly. If there are sources elsewhere that may shed light on local knowledge of plants at the time, this could be compared to the entries in the 'Flora Cumanensis' where – alongside the Latin names – there are sometimes

48. Manuel Lucena Giraldo, *Naciones de rebeldes: las revoluciones de independencia latinoamericanas* (Madrid, 2010), p.148-64.

49. Drafts of these, some in fragmentary form, can be found in sections 2 and 4 of the Löfling collection (div. II) in the RJB. A few of the Swedish letters, in original or copy, survive in the Library of the Linnean Society of London and the RSAS.

notes about either Spanish or indigenous American names.[50] This would require very extensive analysis that is beyond the scope of this study, but from the local names that are extant it does seem likely that Löfling did have some interaction with local people, including knowledge transfer concerning plants.

However, when it comes to how this knowledge was recorded and organised there seems to have been little interaction or mutual influence, at least at a basic level. If we study the manuscript sheets of the 'Flora Cumanensis' from 1754-1755 and compare them with the corresponding sheets of plant descriptions from excursions around Uppsala in 1749-1750, they are organised in almost exactly the same way, with identical headings and the same visual layout on the page.[51] We cannot say to what extent Löfling's ideas about taxonomic groupings or the Linnaean system as a whole were affected by his work in America; although very relevant it is uncertain whether this can be studied based on the surviving sources. It should also be noted that in the draft of the 'Flora Cumanensis', for instance, Löfling did not use any form of binary nomenclature such as the one Linnaeus had introduced in the *Species plantarum* a few years earlier and that he himself had used in the *Gemmae arborum*. Instead, for each species he only identified its likely genus (sometimes with references to Linnaean or other systematic works) and gave a traditional diagnostic name in the form of a brief description of the plant or animal. Sometimes he would also note the colloquial Spanish name of a species, but only in a very few cases did he mention indigenous American nomenclature or terminology at all.[52]

We shall return to the issue of names and languages below, but this cannot help but raise the question about what we actually mean when we refer to the 'circulation of knowledge' or say that 'knowledge travels'. It is probably true that Löfling depended on indigenous and local knowledge, whether Spanish Creole or even African – according to the accounts one 'mulatto' was paid in return for collecting plants for him – in order to gather data needed to determine the place of various plants in the Linnaean system.[53] In that respect, Linnaean science depended not only on other trained naturalists than Carl Linnaeus himself, such as his

50. Draft of 'Flora Cumanensis', RJB, II.3.2 and II.3.3. In the Löfling collection in Madrid there is also a fragment of a dictionary of naturalist terms in Spanish and 'la lengua Paniagota' (or, *pariagota*) based on a manuscript by 'P_e F. Mig_l Angel de Gerona' (a capuchin), RJB, II.2.5.9. The translated words seem to be from that indigenous language.

51. Plant descriptions from Uppsala and Hillebola dated 1749-1750 are in RJB, II.1.1.3 and II.1.2; descriptions of fish on the Cumaná coast in 1754 are in II.2.6.12; drafts of the 'Flora Cumanensis' dated 1754-1755 can be found in II.3.1, II.3.2 and II.3.3.

52. Draft of 'Flora Cumanensis', RJB, II.3.2 and II.3.3.

53. Account of travelling expenses for Pehr Löfling and others, 25 July to 19 August 1754, RJB, II.2.6.21.

student Löfling, but also on the contributions of nameless local collectors and their knowledge of nature. But one might also argue that what really constituted Linnaean knowledge was the system itself, and the methods and practices associated with it, not the individual data being fed into that system. In that regard, at least in Löfling's case, there does not seem to have been much change in either the Spanish or the American context.

Yet again, however, there is still much we do not know about other aspects of this issue. The procedures for gathering plants, for preserving and transporting them – which were also important parts of the whole Linnaean (eco)system of practices – may have been changed or adapted to circumstances that were rather different from those in Uppsala. Unfortunately, the sources are largely silent in this particular case.[54]

Löfling's long and unexpected shadow

Another issue is just what do these scattered sources reveal about Löfling's continued influence on local, Spanish American Creole science after his death. It must be said that the 'grand narrative' of scientific expeditions, the journey from the centre to the periphery and the return home (alive and safe, if possible), evolutionist and positivist, as established in Victorian times, does not explain what happened. This is a fragmentary history of isolated data, but this reflects precisely how much of global history of science is being made. As stated before, Löfling was a complex human being, curious to the extreme about plants, animals and humans, first in Sweden, then in Spain, and later in Spanish America. Nobody expected him to be a neutral scientific figure; that is an invention of the nineteenth century. The idea of a terrestrial paradise in the Americas, coming from Linnaeus and ultimately the European Renaissance, was a persistent one. Löfling's letters purveyed the degree of excitement he felt upon his arrival in Spanish America. Earlier, in November 1751, he had expressed to Linnaeus his hopes of being sent to the Americas to examine nature and fulfil his aspirations. In another letter, dated August 1753, Löfling enthusiastically informed Linnaeus of a possible visit to Cuba, South America, Mexico and New Mexico, which could come to last between four and nine years.[55]

We can imagine how specimen collection worked, more through common sense than through established protocol. During the journey

54. Many years later, however, the prominent Spanish botanist Casimiro Gómez Ortega would publish an official instruction regarding these practical aspects of botanical work in 'the most distant countries', *Instrucción sobre el modo más seguro y económico de transportar plantas vivas por mar y tierra a los países más distantes* (Madrid, Joaquín Ibarra, 1779).

55. Löfling to Linnaeus, 15 November 1751 and 27 August 1753, Library of the Linnean Society of London, Linnaeus correspondence, vol.9, p.376-78, 398-99 (drafts in RJB, II.2.1).

from the Caribbean coast to the Orinoco River, or in the plains of Maturin, Löfling and his team – he had disciples too, the most important one of which was Benito Paltor – collected herbs, plants and animals. Paltor had studied with the Jesuits and obtained a degree in medicine in Valencia. Although he was not schooled in the principles of natural history he learned fast and no doubt was extremely loyal to Löfling. After his teacher's death, Paltor deserted from the expedition. In an effort to explain such an inconceivable act, the first commissioner Iturriaga wrote 'toward the end he became very thin. It may be that the drought caused him to view things differently'.[56]

Löfling's drawings and descriptions are full of names in different languages and it is difficult to identify the origin of the words. Many unanswered questions exist regarding Löfling's drawings, such as the ball game of the 'Otomaca' nation, and others, which remain unidentified.[57] The very existence of his drawings prove that his interest was not limited to botany, but included zoology (especially fish) and, at least in some cases, indigenous peoples. Nevertheless, Linnaeus's efforts to integrate American nature into his (by then) global plan are very well known now. Daniel Scheidenburg (1720-?), the chaplain of the Swedish embassy in Madrid, was a key figure who would, from time to time, convey information from the deceased Löfling to his master. In November 1757, Scheidenburg sent Linnaeus two sections of 'Observationes in plantas cumanensis' and, importantly, Spanish authorisation for it to be printed. Although Löfling was Swedish, he died as a botanist of the Spanish king. His papers, letters and drawings became state property, not the domain of private enterprise or initiative. For a time, the publication process went smoothly and easily. These copies sent to Linnaeus came to form the source for the American section of *Iter Hispanicum*, published in Stockholm in 1758 and translated into German in 1766, then into English in 1771, and into Spanish in 1801-1802. This was a case of collaborative science. Translations opened up possibilities for future research. (See also chapter 2 by Dietz in this volume.) Further, regional long-term impacts were far more important than was understood at the time.[58]

56. Letter from José de Iturriaga to the minister of State Ricardo Wall, Cabruta, 16 August 1757, AGS, Estado, leg. 7395 (translation by the authors). Paltor appeared in Barcelona in 1772 with a good fortune and still a strong interest in botany. Furthermore, he read in the Royal Academy of Natural Sciences and Arts memoirs about the cinchona from Guayana and a balsam from Copaiba. He died in 1782. Antonio Palau, translator of Linnaeus's works into Spanish and a key figure in his success, was also a member of the same Academy at the same time; see José Iglesies Fort, *Memorias de la Real Academia de Ciencias y Artes de Barcelona* no.707, 36:1 (Barcelona, 1964), p.189.
57. RJB, II, lám. 197.
58. José Antonio Amaya, *Mutis, apóstol de Linneo: historia de la botánica en el virreinato de la Nueva Granada (1760-1783)*, 2 vols (Bogotá, 2005), vol.1, p.160-67.

Of course, Linnaeus was not satisfied with the material obtained and published so far. In 1759 Scheidenburg sent him Löfling's 'Flora madrileña' and other writings. He had obtained these items through successful use of influence and lobbying. The American collections themselves remained, however, out of reach. The editorial debate in Spain was in fact an open one. Was it a wise and enlightened policy to allow free publication, or to share scientific information and secrets even with dangerous enemies? Halfway through the Seven Years War Spain also became involved, and calls for freedom of publication were no longer welcome. For the decades to come the official policy was one of caution, and the political decision was taken to only publish his works in Spain. In a moment of despair – and some naiveté – Linnaeus sent another student, Clas Alströmer, to Madrid. He spent almost a year, in 1760, trying to retrieve Löfling's collections and after a time even began to doubt the very existence of the materials he was meant to rescue. After his arrival at Cadiz in April, he was met with official silence. He visited Granada and travelled to Madrid. Angry and frustrated, he left Spain in July 1761.[59] He was probably unaware that an Irish naturalist in the service of the king of Spain, William Bowles, was in charge of publishing Löfling's works after 1762.

In another twist of fate, in July of the following year, the ambitious courtier and botanist Casimiro Gómez Ortega took over the task. Significantly, two former members of Löfling's team, who had been trained in botanical drawing by Löfling himself and Paltor, Bruno Salvador Carmona and Juan de Dios Castel, spent the next twenty years preparing the materials for publication. Their efforts resulted in forty-four descriptions, of a total of 303 hand-made Löfling descriptions at their disposal, namely four classes of the Linnaean system, being ready. Two hundred drawings, with 115 plants, fifty-one fish, seven reptiles, nine birds, eight mammals, six arthropods, two maps and the above mentioned two drawings of indigenous peoples and customs, remained unedited. In 1801, when ideas about publication had changed, the naturalist Ignacio de Asso published some of Löfling's letters in the *Anales de ciencias naturales*. It was a great success and a vindication of the Spanish aspect of his life and works. He was never forgotten.

Developments in Spanish America took a different course. Löfling's spirit was more than alive. The famous Spanish naturalist and savant José Celestino Mutis, born in Cadiz, arrived in the viceroyalty of New Granada, what is now Colombia, in 1761, as the doctor to the Viceroy

59. Joaquín Fernández Pérez, 'Tres apóstoles de Linné en Cádiz: Pehr Osbeck, Pehr Löfling y Clas Alströmer', in *Pehr Löfling y la expedición al Orinoco, 1754-1761*, ed. Francisco Pelayo (Madrid, 1990), p.51-102.

Messía de la Cerda. A restless spirit, before embarking for Cartagena Mutis had managed to meet in Cadiz with Clas Alströmer, who informed Linnaeus of his existence.[60] In 1760 he sent Linnaeus a collection of seeds and plants that he had received. From 1761 onwards, and until Linnaeus's death in 1778, they maintained their correspondence. His proposal to write a 'Natural History of America' which he sent to the Spanish Crown was undoubtedly influenced by Linnaeus's scientific ideas and institutional approach. Mutis's success can be explained by his favourable reputation and finally afforded recognition. But what matters most is how Mutis invented himself as a botanist who was continuing the work that Löfling would have undertaken in the American tropics if he had lived. Mutis sent botanical descriptions to Sweden after 1762, a first collection of 146 plants in 1773, and a second of 119 in 1778. It is clear that the importance of drawings and visual culture in Linnaean botany influenced Mutis's approach to his work.[61] Later this devotion to botanical representation attracted much criticism, because it supposedly showed how sterile and unproductive Mutis and his team were due to an obsession with depicting nature as it is. Linnaeus was almost in shock when Mutis's material arrived. In 1778 after Linnaeus's death, 'the worst thing to have happened in my life', Mutis wrote, he exchanged seven letters with Swedish correspondents, probably sending them by way of Cadiz and then on to Sweden.[62]

As a token of Linnaeus's and his successors' gratitude, in 1784 Mutis was appointed a member of the Royal Academy of Sciences in Stockholm. When Linnaeus died, his son sent Mutis a letter professing to be his friend, as a sort of heritage from his father. Mutis answered with a fourteen page letter and an acceptance from a friend living so far away. Some of Mutis's plants were, in fact, included in the *Supplementum plantarum* published by Linnaeus the Younger in 1781. After 1783, as director of the Royal Botanical Expedition, Mutis adopted a managerial approach to botany. He was no longer an amateur now that he was in charge of a program of colonial reform, and reliant on the king's favour. Of course, just like Linnaeus during his lifetime, Mutis had plenty of enemies. He was very careful with regard to his activities, but maintained an exchange of books and plants with Swedish friends and correspon-

60. 'A long time ago I sent news of you to master Linnaeus and I was delighted to receive his answer stating that he is ready to have a friend in such a distant part of the world'; letter from Clas Alströmer to José Celestino Mutis, Madrid, 2 July 1761; Amaya, *Mutis*, vol.2, p.469 (translation by the authors).

61. 'José Celestino Mutis, director of the New Granada expedition, was especially devoted to the idea of painting as botanical exploration'; Daniela Bleichmar, *Visible empire: botanical expeditions and visual culture in the Hispanic Enlightenment* (Chicago, IL, 2012), p.13.

62. Amaya, *Mutis*, vol.1, p.302.

dents: after Linnaeus's death his own son, other naturalists – such as Peter Jonas Bergius or Gustaf von Paykull, or the diplomat Hans Jacob Gahn – before and after Linnaeus the Younger's death in 1783.[63] Even more importantly, Mutis was influenced by his knowledge of how Linnaeus nurtured a circle of disciples and generated his own similar clique. Mutis lived to be an old man and a priest, dwelling amongst gifted Creoles, until he died in 1808, still at the centre of a network of competing disciples. Mutis's life in New Granada echoed that of Linnaeus a generation earlier: the great teacher surrounded by his students, reading the book of nature while they took incredible risks. The Royal Botanical expedition to New Granada turned into a sort of Creole think-tank, a university, a laboratory and a political point of reference for agitators. It is not surprising that Francisco José de Caldas, his own nephew Sinforoso Mutis and Jorge Tadeo Lozano, like other members of the expedition, would all support the patriotic faction when the War of Independence began in 1810.[64]

Conclusion

As we have tried to show here, Pehr Löfling became deeply involved in an active and on-going colonising effort involving non-European peoples and new 'tropical' territories. The Orinoco expedition was not one of pure scientific exploration, if such a thing exists, but rather a military and political enterprise with multiple aims. It was designed to reinforce Spanish control of the Orinoco and Amazon basins, to stop Dutch incursions and Carib insurgence, to settle the border with the Portuguese and to provide the necessary knowledge for a more effective exploitation of the region's natural resources. The expedition, and Löfling's role in it, is yet another reminder that the main reason for the interest of political and military leaders in natural history was its potential for economic benefit and territorial expansion, not scientific exploration for its own sake.

At the same time, we must not forget that the motives and aspirations of different parties varied. Scientists knew that they could generate support for themselves by the 'useful knowledge' that botany might

63. Amaya, *Mutis*, vol.2, 'Catálogo de los envíos de historia natural remitidos por José Celestino Mutis a Suecia', p.149-59.

64. Jaime Bernal Villegas, 'La institucionalización del conocimiento científico en el Nuevo Reino de Granada', in *Academia Mutisiana: documentos preneogranadinos de José Celestino Mutis y la promoción de sociedades científicas en la Nueva Granada*, ed. Alberto Gómez Gutiérrez *et al.* (Bogotá, 2011), p.177-232 (210-11); Mauricio Nieto Olarte, *Orden natural y orden social: ciencia y política en el semanario del Nuevo Reino de Granada* (Madrid, 2007), about Caldas, p.109; José Luis Peset, *Ciencia y libertad: el papel del científico ante la independencia americana* (Madrid, 1987), p.271-307.

provide, even when they might personally be driven by scientific curiosity as such. They too were a heterogeneous group in this regard since some tended to value curiosity more while others prioritised utility. However, it is still important to distinguish the scientists from the patrons, ministers and others who sponsored their work since their incentives might fundamentally diverge. Further, 'pure' scientific interest does not necessarily equate with 'idealism', since one of the characteristics of this period was that science was becoming a field in its own right with its own reward systems and career paths. In other words, new incentive structures connected with the values and priorities of the emergent scientific community within the new field of natural history was an increasingly important factor at play there. These new incentive structures often tended towards what we would today call basic research, while the external supporters and patrons were interested in what we might call applied research, that is useful intelligence, for example information about new crops or other plants that could generate immediate commercial or medical benefits.

Finally, it should be pointed out that circulation of knowledge does not mean that power was equally distributed among those who have figured in this study. On the contrary, there were indeed hierarchies and inequality – political, economic, social and cultural – but the point in relation to much previous research is that those patterns of inequality were not as simple as often has been assumed. Linnaean science was not the product of one single mind, whose ideas spread out into the world where they would ultimately prevail. Rather, these ideas, and the practices associated with them, depended on the efforts of many minds and hands, and all those involved had to navigate constantly shifting contexts of negotiation where knowledge did not move in one but in many different directions. As it did so, knowledge changed, and was changed by, countless go-betweens like Pehr Löfling and their lives of useful curiosity.

Summaries

1. Notebooks, files and slips: Carl Linnaeus and his disciples at work
Isabelle Charmantier

This chapter examines the writing technologies developed by Carl Linnaeus (1707-1778) which facilitated the rapid dissemination of Linnaean science, especially botany. Following an overview of Linnaeus's writing technologies, the chapter concentrates on paper slips, which Linnaeus started using in the late 1760s. The format and flexibility of paper slips enabled the propagation of Linnaean standards of plant description and efficient collaboration between naturalists. The chapter examines their use by Daniel Solander (1733-1782), Linnaeus the Younger (1741-1783) and Carl Peter Thunberg (1743-1828), and their essential role in the late eighteenth-century institutionalisation of natural history.

2. What is a botanical author? Pehr Osbeck's travelogue and the culture of collaborative publishing in Linnaean botany
Bettina Dietz

This essay explores the culture of collaborative publishing in Linnaean botany by tracing the publication trajectory of Pehr Osbeck's report of his journey to Canton in southern China from 1750 to 1752. Osbeck, a Swedish naturalist, divided his material between a number of publications that appeared successively, building on and complementing each other. This long-term strategy was aimed at making a first version of his new material available to interested readers as quickly as possible, while leaving the process of completing and correcting it to be undertaken by multiple authors and translators at a later date. As a botanical author, Osbeck was a 'composite', and his text was, visibly, an assemblage of contributions by many.

3. The price of Linnaean natural history: materiality, commerce and change
Hanna Hodacs

Focusing on trade, materiality and consumer culture, this chapter explores the trade in collections with a Linnaean provenance. An assessment of the value in money of Linnaeus's conserved plants and animals is mapped in the first section of this paper. The second section focuses on the material dimension, foremost on aspects of preservation, presentation and fashion. The third part of the chapter discusses how, in late eighteenth-century London, evaluations of specimen collections Linnaeus had used in his taxonomic work changed, and how it reflected Linnaeus's shifting status in the history of natural history. The final section explores the market and the use of binary names when selling natural history specimens at auctions in eighteenth-century Europe, particularly in Sweden and England.

4. In the name of Linnaeus: Paris as a disputed capital of natural knowledge (1730-1789)
Stéphane Van Damme

According to historiographical tradition, if Europe as a whole was quickly convinced by the Linnaean sciences, Paris was always seen as a pole of resistance. Moreover, French provinces, especially Montpellier and Lyon, shared this resistance. Focused on several botanists, intellectual history has taken Linnaeus's poor reception in Paris for granted. The long-lasting controversy between Buffon and Linnaeus is used to encapsulate this difficult reception. Thus, Paris was not a neutral European place for Linnaeus, who visited the capital in 1738 to meet Antoine and Bernard Jussieu and to attend a meeting at the Academy of Sciences; it was rather a *passage obligé* in the recognition of his system. The reception of botanical philosophy was so sensitive in Paris because it profoundly divided the Parisian naturalist world, not only in terms of intellectual framework but also in terms of practices. The paper argues that natural knowledge featured in the strategies employed to re-scale the city as a powerful and attractive metropolis, an emergence that benefitted from new ways of commodifying, consuming and interpreting nature.

5. On the use and abuse of natural history: Linnaean science in Kant's Königsberg
Jonas Gerlings

This chapter investigates the development of eighteenth-century natural history in and around Königsberg. In doing so the chapter attempts to redraw the intellectual map of the Baltic Sea region, which has predominantly highlighted Königsberg as an exponent of a metaphysical Enlightenment personified by Immanuel Kant. By situating Kant in the midst of this tradition of natural historians, I investigate and contrast two strands. On the one hand, the long development, which transforms the local tradition of natural history into Linnaean science and, on the other hand, the varying discourses on the science of man, which draws on this tradition. The chapter contrasts this long history of institutionalisation of Linnaean science with the intellectual history of contestation in debates on the nature of man, and argues that Kant, while being deeply immersed in this tradition, also formulates a critique of a purely biological understanding of man.

6. The Edinburgh connection: Linnaean natural history, Scottish moral philosophy and the colonial implications of Enlightenment thought
Linda Andersson Burnett and Bruce Buchan

In this chapter we explore how Enlightenment notions of humanity hinged on the emergence of a colonial ethnography combining Linnaean natural history with Scottish moral philosophical theories of stadial historical progress. We do this by analysing the writings of colonial 'travellers' brought together at the University of Edinburgh in the second half of the eighteenth century, where they were trained in both 'traditions'. Our argument is that the dynamism of the encounter between Linnaean natural history and Scottish moral philosophy at Edinburgh prompted new hypotheses about the nature of humanity and of human historical progress. These were communicated to colonial settings by former students whose 'ethnographic' writings are invaluable sources for the intellectual history of European colonial thought. Their writings also highlight the dynamism and conceptual fluidity that occurred when different Enlightenment traditions of thought came into contact and were transformed by colonial encounters.

7. Negotiating people, plants and empires: the fieldwork of Johann Gerhard König in South and South East Asia (1768-1785)
Niklas Thode Jensen

Linnaean natural history relied to a large extent on the exchange of information and specimens from colonised areas around the world. This chapter investigates how Linnaean natural history was practised, structured and exchanged by one of Linnaeus's travelling students in the colonial field of South and South East Asia in the second half of the eighteenth century. The student was Johann Gerhard König (1728-1785) and the chapter investigates how he applied Linnaeus's ideals of fieldwork in the colonial context while at the same time negotiating local informants, their languages and knowledge to position himself as the mediator between the colonial field and distant Europe.

8. Lives of useful curiosity: the global legacy of Pehr Löfling in the long eighteenth century

Kenneth Nyberg and Manuel Lucena Giraldo

In this chapter we examine the case of Pehr Löfling (1729-1756), one of Carl Linnaeus's travelling students or 'apostles'. After going to Madrid in 1751 and entering Spanish service, he joined an expedition to present-day Venezuela to settle the border with Portuguese territories and to map the natural resources of the Orinoco basin. Through the global trajectory of Löfling's life we can follow how Linnaean science was constantly (re)negotiated in different contexts: the academic community in Uppsala which was centred round Linnaeus himself, the politicised world of botanist courtiers in Madrid, and the colonial contact zone of Spanish America. In the latter region Löfling and his early work as a Linnaean naturalist would come to have an enduring legacy, not least through the director of the Royal Botanical Expedition to New Granada, José Celestino Mutis (1732-1808).

Bibliography of works cited

Manuscripts[1]

London, The Linnean Society of London, Library (LSL)

Linnaeus fil. MS

'Supplementum plantarum', LM/LF/BOT/2/2/2/1.

'De Liliaceis et affinibus plantis', LM/LF/BOT/3/9/1.

'Palmarum descriptiones', LM/LF/BOT/3/9/2.

'Plant diagnoses', LM/LF/BOT/3/9/3.

'Notes from England', LM/LF/PP/6/5.

'Mammalia paper slips', LM/LF/ZOO/3/6.

Linnaeus MS

'Spolia botanica', LM/LP/BOT/1/2.

'Hortus Uplandicus I' (1730), LM/LP/BOT/2/1.

'Fundamenta botanica', vols VII and VIII, LM/LP/BOT/3/1/13 and LM/LP/BOT/3/1/14.

'Manuscripta medica I', LM/LP/BIO/3/2.

'Philosophia botanica' (press manuscript), LM/LP/BOT/3/3/1.

'Species plantarum', LM/LP/BOT/3/4/1.

Linnaeus's books

Linnaeus, Carl, *Genera plantarum* (Leiden, C. and G. J. Wishoff, 1742), BL 51.

Linnaeus's correspondence

Letters from Pehr Löfling to Carl Linnaeus (1751-1754), 'Linnaeus correspondence', vol.9.

Miscellaneous authors' manuscripts

Pehr Löfling, 'Contra Kleinii animadversions' (*c.*1743), LM/MA/LOE/1.

Carl Peter Thunberg, 'Diagnoses of *Oxalis* and *Septas*', LM/MA/THU/1.

London, Natural History Museum, Library and Archives (NHM)

König MS

Autograph Journals of König's voyages: with lists and descriptions of East Indian (including Siam and Malacca) Plants, Animals and a few Minerals. List of plants (2 vols); Journal (19 vols); Index (1 vol.).

Solander MS

'Manuscript descriptions of plants', and 'Manuscript descriptions of animals written on slips and systematically arranged in accordance with Linné's "Systema naturae"', MS BANKS COLL SOL.

London, Christie's Archive Department, King Street, St James's

Sales catalogues for sales on the 3 June 1771; 24 November 1775, 10 May 1782; 21 December 1785; 30 March 1787; 26 April 1788; 23 May 1788; 29 April 1796; 2 June 1807.

1. Manuscripts are listed under their respective archives, which have been grouped into major language areas and locations and ordered according to the relative importance of their holdings for the research presented in this book. For each archive, items have been organised and listed under the call signs of the collections or series they belong to or, lacking that, in alphabetical order based on the first words of the manuscript title or item description (excluding particles).

Edinburgh, Centre for Research Collections (CRC), University of Edinburgh (EUL)

Reverend Dr John Walker, 'Natural history of the inhabitants of the Highlands', Walker Papers, Dc. 1.18.
Reverend Dr John Walker, Walker Papers, MS Dc. 1.59.
Adam Ferguson, 'Lectures in pneumatics', 3 vols, Dc. 1.84-86.
Reverend Dr John Walker's MS Lectures, Dc. 2.17 and 2.20.
Reverend Dr John Walker, 'Queries concerning the North of Scotland', Dc. 2.36.
Dugald Stewart's 'Lectures on moral philosophy', 1793-1794, Dc. 4.97.
'Lectures on moral philosophy delivered by Professor Dugald Stewart, session 1789 & 1790', vol.1, Gen. 1987-1989.
Reverend Dr John Walker, Correspondence, Letters, 1762-1779, LAIII352.1.

Edinburgh, National Library of Scotland (NLS)

Robertson-Macdonald Papers, Materials collected by Principal Robertson, NLS, MS 3954.

Paris, Muséum d'histoire naturelle, Library

Thouin MS
MS 1989/950-990.

Paris, Archives nationales (AN)

Minutier central, CVIII, 498, 9 March 1752, dépôt du testament de Claude-Joseph Geoffroy.

Paris, Bibliothèque centrale du Muséum national d'histoire naturelle (BCMNHN)

MS 77: 'Herborisations aux environs de Paris', by Joseph Pitton de Tournefort (1656-1708).
MS 301: Philibert Commerson, 'Mémoires pour servir à l'histoire du Voyage fait Autour Du Monde par les Vaisseaux Du Roi La Boudeuse et l'Estoile Pendant les années 1766-1768'.
MS 313: 'Etats des distributions végétales du Muséum' (1790-1823).
MS 1093: 'Dénombrement des plantes qui naissent aux environs de Paris, extrait du livre de Tournefort avec des additions par Sébastien Vaillant' (1704).
MS 1169: Bernard de Jussieu, 'Catalogus generum juxta ordinem in horto Trinanonesi institutum' (1759).
MS 1297: 'Statistique végétale des environs de Paris, par Duchesne' (1771); 'Avertissement', separate sheet.

Madrid, Museo Nacional de Ciencias Naturales (MNCN)

Flora Peruana y Chilense
Ruiz y Pabón, ref.7.

Madrid, Archivo del Real Jardín Botánico (RJB)

Div. II, Fondo 'Expedición de Límites al Orinoco. Pehr Löfling' (1743-1766)
II.1.1.3: Descriptions of Swedish plants.
II.1.1.8: Speech to a group of students in Uppsala, 24 May 1749.
II.1.2: Descriptions of Swedish plants.
II.1.4.1: Letter from Pehr Löfling to Lars Montin, Madrid, 3 January 1752 (draft).
II.1.6.6-II.1.6.23: Letters from Carl Linnaeus to Pehr Löfling, 1751-1753.
II.1.7.2: Lists of plants found in Portugal and Spain.
II.2.1: 'Letter-book' (drafts of correspondence, 1751-1754).

II.2.2.5: Notes on various types of rocks in Portugal and Spain.
II.2.3.6: Spanish translation of a treatise by Pehr Kalm on useful North American plants.
II.2.5.9: Dictionary of naturalist terms in 'la lengua Paniagota'.
II.2.6.12: Descriptions of fish in Cumaná.
II.2.6.21: Account of travelling expenses from 25 July to 19 August 1754.
II.3.1-II.3.3: Drafts of 'Flora Cumanensis'.
Lám. 197.

Seville, Archivo General de Indias

Caracas
439, 440.

Simancas, Archivo General de Simancas (AGS)

Estado
Leg. 6678, 6679: Reports from Jeronimo de Grimaldi, Spanish ambassador in Sweden, 23 October 1750 to 5 February 1751.
Leg. 7395: Letter from José de Iturriaga to Ricardo Wall, Cabruta, 16 August 1757.

Bogotá, Biblioteca Nacional de Colombia

Fondo Mutis
2996: José Celestino Mutis's copy of Pehr Löfling's *Iter Hispanicum* (1758).

Stockholm, Royal Swedish Academy of Sciences (RSAS)

Bergianska brevsamlingen
Vol.3: Pehr Löfling to Pehr Bjerchén, Uppsala, 29 January 1751 (copy).
Linnaeus MS
'Museum reginae', Carl von Linné, MS 1.

Stockholm, Royal Library (Kungliga biblioteket)

Sales catalogues
(Vardagstryck, ekonomi, auktionskataloger).
Förteckning, Uppå framledne stads-auctuariens Herr PETER HOFFMANS samlingar, af åtskilliga rariteter, naturalier, mineralier, malmer, mynt och gewär, som igenom offentlig auction komma at försäljas den 9 juni 1747 (Stockholm, Peter Jör. Nyström, 1747).
Förteckning på hwad i synnerhet förtjenar utmärkas, bland de saker, som förekomma wid auction på Åkeröö den 4:de och följande dagar till och med den 16:e i febrarii månad, 1771.

Uppsala, Uppsala University Library

Linnaeus MS
'Hortus Uplandicus IV' (1731), Linnésamlingen, D67b.
'Hortus Uplandicus II' (1730), Leufsta MS.
'Adonis Uplandicus' (1731), Leufsta MS.
Other
Ihre 187:1 Travel diary written by Johan Lindwall.

Lund, Lund University Library

Brevsamlingen
Pehr Löfling to E. G. Lidbeck, Uppsala, 16 November 1750.

Copenhagen, Danish National Archives (DNA)

Det kgl. ostindiske Guvernement, Kolonien Trankebar, Mandtal over indbyggere i Tranquebar og landsbyerne, 1790, archive no. 445:1447A-1447B.
København, St Nikolaj sogn, Enesteministerialbog (1694-1805), F1746-F1765 https://www.sa.dk/ao-soegesider/

billedviser?bsid = 156081 and F1766-F1788 https://www.sa.dk/ao-soegesider/billedviser?bsid = 156082 (last accessed 8 August 2017).

Tranquebar, Ny Jerusalemskirke og Zionskirke. Kontraministerialbog for Zionskirken, 1767-1845, archive no. EC-777, micro fiche no. 3/777-5.

Halle an der Salle, Archive of Die Franckeschen Stiftungen zu Halle

Archive of Die Franckeschen Stiftungen zu Halle, http://192.124.243.55/cgi-bin/dhmeng.pl (last accessed 8 August 2017).

AFSt/M: Archiv die Franckeschen Stiftungen, das Missionsarchiv mit der Indien- und der Amerikaabteilung.

ALMW/DHM: Archiv Leipziger Missionswerk, Dänisch-Halleschen Mission.

Munich, Staatsbibliotek, Library

Linnaeus MS

'Hortus Uplandicus III' (1730), Cod. Suec. 6.

On-line sources

The Linnaean correspondence, an electronic edition prepared by the Swedish Linnaeus Society, Uppsala, and published by the Centre international d'étude du XVIIIe siècle, Ferney-Voltaire, http://linnaeus.c18.net/.

Pre-1800 works

Adanson, Michel, *Histoire naturelle du Sénégal: coquillages* (Paris, Bauche, 1757).

Beckmann, Johann, 'Linneische Synonymie zu J. T. Kleins verbesserter Historie der Vögel', *Der Naturforscher* 1 (1774), p.65-78.

Bernoulli, Johann, *Reisen durch Brandenburg, Pommern, Preußen, Curland, Rußland und Polen in den Jahren 1777 und 1778*, vol.3 (Leipzig, Fritsch, 1779).

Blumenbach, Johann Friedrich, *Handbuch der Naturgeschichte*, 5th edn (Göttingen, Dietrich, 1797).

Bock, Friedrich Samuel, *Betrachtung über das Nutzbare und Anmuthige in der Naturgeschichte* (Königsberg, Zeis, 1767).

–, *Versuch einer wirtschaftlichen Naturgeschichte von dem Königreich Ost- und Westpreussen*, vol.1 (Dessau, Buchhandlung der Gelehrten, 1783).

Boulduc [Gilles-François], 'Essai d'analise des plantes', *Mémoires de l'Académie royale des sciences* (Paris, Imprimerie royale, 1734), p.101-106.

Buffon, Georges Louis LeClerc de, *Allgemeine Historie der Natur* (Hamburg und Leipzig, Grund and Holle, 1750).

–, *Histoire naturelle générale et particulière, avec la description du Cabinet du Roi*, 15 vols (Paris, Imprimerie royale, 1749-1767), vol.13 (Paris, Imprimerie royale, 1765).

Camper, Petrus, *Dissertation sur les variétés naturelles qui caractérisent la physionomie des hommes des divers climats et des différens ages* (Paris, H. J. Jansen, 1791).

A Catalogue of the Portland Museum, lately the property of the duchess dowager of Portland (London, 1786).

Dezallier d'Argenville, Antoine Joseph, *La Conchyliologie, ou Histoire naturelle des coquilles* (Paris, Guillaume De Bure, 1780).

Duchesne, Antoine Nicolas, *Manuel de botanique, contenant les propriétés des plantes utiles pour la nourriture, d'usage en médecine, employées dans les arts d'ornement pour les jardins, et que l'on trouve à la campagne aux environs de Paris* (Paris, Didot le jeune, 1764).

[Fischer, Christian Gabriel], *Vernünftige Gedanken von der Natur. Was sie sey? Daß sie ohne Gott und seine allweise Beschränkung unmächtig sey? Und wie die einige, untheilbare göttliche Kraft, in und durch die Mittelursachen, nach dem Maß ihrer verliehen Würkbarkeit oder Tüchtigkeit, hie in der Welt alles alleine thätig würke? Durch fleißiges Nachsinnen, überlegen und schließen gefasset Und zur Verherrlichung göttlicher Majestät, auch Förderung wichtiger Wahrheiten herausgegeben von einem Christlichen Gottes-Freunde* (Königsberg, Christlichen Gottes-Freunde,1743).

–, 'Ob sich nach herrn Linnäus System ein Naturalienkabinet einrichten lasse?', *Neue Gesellschaftliche Erzählungen für die Liebhaber der Naturlehre, der Haushaltungswissenschaft der Arzneykunst und der Sitten* (Leipzig, Holle, 1758), p.163-66.

Forster, Georg, 'Noch etwas über die Menschenraßen', *Teutsche Merkur* (1786), p.57-86.

Geoffroy, Etienne-Louis, *Catalogue des livres et estampes de défunts Mrs Geoffroy* (Paris, Gabriel Martin, 1754).

–, *Catalogue raisonné des minéraux, coquilles, et autres curiosités dans le Cabinet de feu M. [Claude-Joseph] Geoffroy de l'Académie Royale des Sciences* (Paris, H. L. Guerin and L. Fr. Delatour, 1753).

– , *Histoire abrégée des insectes qui se trouvent aux environs de Paris*, vol.1 (Paris, Durand, 1762).

Girtanner, Christoph, *Über das Kantische Prinzip für die Naturgeschichte: ein Versuch diese Wissenschaft philosophisch zu behandeln* (Göttingen, Vandenhoek and Ruprecht, 1796).

Gómez Ortega, Casimiro, *Instrucción sobre el modo más seguro y económico de transportar plantas vivas por mar y tierra a los países más distantes* (Madrid, Joaquín Ibarra, 1779).

Gottsched, Johann, *Flora Prussica, sive Plantae in regno Prussiae sponte nascentes* (Königsberg, sumptibus Typographiae Georgianae, 1703).

Hagen, Karl Gottfried, 'Beschreibung der Hottentottenblumen, oder einer neuen Art wieder auflebender Pflanzen', *Berlinische Sammlungen* 9 (1777), p.113-27.

–, *Lehrbuch der Apothekerkunst* (Königsberg and Leipzig, Hartung, 1778).

–, *Programma primum de plantis in Prussia cultis* (Königsberg, Hartung, 1791).

Haller, Johann-Samuel, *Naturgeschichte der Thiere in Sistematischer Ordnung. Die Vierfüssigen Thiere, welche lebendige Jungen zur Welt bringen; nebst der Geschichte des Menschen* (Berlin, Christian Friedrich Boß, 1757).

Hasselquist, Fredrik, *Iter Palaestinum eller resa til Heliga Landet 1749-1752* (Stockholm, Lars Salvius, 1757).

Hennings, August, *Gegenwärtiger Zustand der Besitzungen der Europäer in Ostindien*, vol.2 (Hamburg, Carl Ernst Bohn, 1785).

Jacquin, Joseph Nikolaus von, *Selectarum stirpium Americanarum historia, in qua ad Linnaeanum systema determinatae descriptaeque sistuntur plantae illae, quas in insulis Martinica, Jamaica, Domingo, aliisque, et in vicinae continentis parte, observavit rariores* (Vienna, Kraus, 1763).

Kalm, Pehr, *Des Herrn Peter Kalms, Professors der Haushaltungskunst in Aobo, und Mitgliedes der königlichen schwedischen Akademie der Wissenschaften Beschreibung der Reise, die er nach dem nördlichen Amerika auf den Befehl gedachter Akademie und öffentliche Kosten unternommen hat; eine Übersetzung*, 2 vols (Göttingen, Vandenhoek, 1754).

–, *En Resa til Norra America, på Kongl. Swenska Wetenskaps Academiens befattning, och publici kostnad*, 3 vols (Stockholm, Lars Salvius, 1753-1761).

–, *Travels into North America, containing its natural history, and a circumstantial account of its plantations and agriculture in general; with the civil, ecclesiastical and commercial state of the country, the manners of the inhabitants, and several curious and important remarks on various subjects, by Peter Kalm, professor of oeconomy in the University of Aobo in Swedish Finland, and Member of the Swedish Royal Academy of Sciences, translated into English by John Reinhold Forster, F. A. S.; enriched with a map, several cuts for the illustration of natural history, and some additional notes*, 3 vols (London, William Eyres, 1770-1771).

Klein, Jacob Theodor, *Summa dubiorum circa classes quadrupedum et amphibiorum in Caroli Linnaei 'Systematen aturae'* (Leipzig, J. F. Gleditschium, 1743).

König, Johann Gerhard, 'Naturgeschichte der sogenannten weissen Ameise', *Beschäftigungen der Berlinischen Gesellschaft Naturforschender Freunde* 4 (1779), p.1-28.

–, *Dissertationem inauguralem de Remediorum Indigenorum ad Morbos cuivis regioni endemicos expugnandos efficacia* (Copenhagen, Gerhardus Giese Salicath, 1773).

Linnaeus, Carl, 'An oration concerning the necessity of travelling in one's own countrey, made by Dr. Linnaeus at Upsal, Oct. 17, anno 1741, when he was admitted to the royal and ordinary profession of physic', in B. Stillingfleet, *Miscellaneous tracts relating to natural history, husbandry and physick* (London, 1759).

–, *Bibliotheca botanica* (Amsterdam, Salomon Schouten, 1736).

–, *Caroli Linnæi Materia medica; editio secunda auctior, curante Jo. Chr. Dan. Schrebero* (Leipzig and Erlangen, Wolfgang Walther, 1772).

–, *Des Herrn Archiaters und Ritters von Linné Reisen durch einige schwedische Provinzen [...]; aus dem Schwedischen übersetzt* [by Johann Christian Daniel Schreber], 2 vols (Halle, Johann Jakob Curts, 1764-1765), vol.1, *Reisen durch Oeland und Gothland*, vol.2, *Reisen durch Westgothland.*

–, *Fauna Suecica Sistens Animalia Sveciae Regni: quadrupedia, aves, amphibian, pisces, insecta, vermes, distribute per classes & ordines, genera & species, cum differentiis specierum, synonymis auctorum, nominibus incolarum, locis natalium, descriptionibus insectorum* (Stockholm, L. Salvius, 1746).

–, *Genera plantarum, eorumque characteres naturales secundum numerum, figuram, situm, et proportionem omnium fructificationis partium* (Stockholm, L. Salvius, 1737).

–, *Genera plantarum, eorumque characteres naturales secundum numerum, figuram, situm, et*

proportionem omnium fructificationis partium (Leiden, C. and G. J. Wishoff, 1742).
–, *Genera plantarum [...], editio octava post Reichardianam secunda [...] auctior atque emendatior, curante D. Jo. Christiano Dan. Schreber* (Frankfurt am Main, Varrentrapp and Wenner, 1789-1791).
–, *Herbationes Upsalienses* (Uppsala, Laur. M. Hojer, 1753).
–, *Museum [...] Ludovicæ Ulricæ reginæ Suecorum, in quo animalia rariora, imprimis insecta & conchilia describuntur* (Stockholm, L. Salvius, 1764).
–, *Philosophia botanica in qua explicantur Fundamenta Botanica cum definitionibus partium, exemplis terminorum, observationibus rariorum* (Stockholm, Kiesewetter, 1751).
–, *Species plantarum, exhibentes plantas rite cognitas ad genera relatas, cum differentiis specificis, nominibus trivialibus, synonymis selectis, locis natalibus, secundum systema sexuale digestas* (Stockholm, Lars Salvius, 1753).
–, *Systema naturæ, sive regna tria naturæ systematice proposita per classes, ordines, genera, & species* (Lugduni Batavorum [Leiden], Theodorum Haak, 1735).
–, *Termini botanici* (Uppsala, 1762).
Linnaeus fil., Carl, *Supplementum plantarum* (Braunschweig, Impensis Orphanotrophei, 1781).
Löfling, Pehr, *Iter Hispanicum* (Stockholm, Lars Salvius, 1758).
Lösel, Johann, *Plantae in Borussia sponte nascentes* (Königsberg, typis Paschalii Mensenii, 1654).

Mémoires de la Société d'histoire naturelle de Paris 1 (1799).
Metzger, Johann Daniel, *Skizze einer pragmatischen Literärgeschichte der Medicin*, vol.1 (Königsberg, Nicolovius, 1792).
Müller, Otto Frederik, 'Doctor Koenigs Reise fra Trankebar til Zeylon i et Brev til Etatsraad Müller af 15 Februari 1777', *Videnskabernes selskabs skrifter* 12 (1779), p.383-402.

Osbeck, Pehr, *A Voyage to China and the East Indies, by Peter Osbeck, Rector of Hasloef and Woxtorp, Member of the Academy of Stockholm, and of the Society of Upsal; [...] translated from the German by John Reinhold Forster; to which are added a Faunula and Flora Sinensis*, 2 vols (London, Benjamin White, 1771).
–, *Dagbok öfwer en ostindisk resa åren 1750, 1751, 1752* (Stockholm, Lor. Ludv. Grefing, 1757).
–, *Herrn Peter Osbeck, Pastors zu Haßlöf und Woxtorp, der Königl. Schwedischen Akademie zu Stockholm und der Kön. Gesellschaft zu Upsala Mitgliedes, Reise nach Ostindien und China, nebst O. Toreens Reise nach Suratte und C. G. Ekebergs Nachricht von der Landwirthschaft der Chineser*, translated from the Swedish by J. G. Georgi (Rostock, Johann Christian Koppe, 1765).

Pinkerton, John, *An Enquiry into the history of Scotland preceding the reign of Malcolm III, or The Year 1056*, 2 vols (London, John Nicols, 1789).
Poivre, Pierre, *Voyages d'un philosophe, ou Observations sur les moeurs et les arts des peuples de l'Afrique, de l'Asie et de l'Amérique* (Lyon, J. de Ville and L. Rosset, 1769).
Pope, Alexander, *An Essay on man: in epistles to a friend. Epistle II* (London, J. Wilford, 1733).

Réaumur [René-Antoine Ferchault de], *Mémoires pour servir à l'histoire des insectes, par M. de Réaumur*, 6 vols (Paris, Imprimerie royale, 1734-1742).
Rose, Christianus, *Dissertationem Sollemnem De Augusto Contraria Medicina Curato* (Halæ Magdeburgicæ, Hilligerus, 1741).

Schriften der Berlinischen Gesellschaft Naturforschender Freunde 5 (1784).
Senebier, Jean, *Die Kunst zu beobachten* 1 (Leipzig, Weygand, 1776).
Sömmerring, Samuel Thomas von, *Über die körperliche Verschiedenheit des Negers vom Europäer* (Frankfurt and Mainz, Barrentrapp Sohn and Wenner, 1785).
–, *Über die körperliche Verschiedenheit des Mohren vom Europäer* (Mainz, 1784).
Sonnerat, Pierre, *An Account of a voyage to the Spice-Islands and New Guinea* (London, W. Green, 1781).
Sparrman, Anders, *A Voyage to the Cape of Good Hope*, 2nd edn (London, G. G. J. and J. Robinson, 1786).
Spengler, Lorenz, 'Auzug aus einem Schreiben des Herrn D. Johann Gerhard König', *Beschäftigungen der Berlinischen Gesellschaft Naturforschender Freunde* 3 (1777), p.427-30.
Stewart, Dugald, *Elements of the philosophy of the human mind* (London, Strahan and Cadell, 1792).
Stillingfleet, Benjamin, *Miscellaneous tracts relating to natural history, husbandry and physick* (London, 1759).
Stoever, Dietrich Heinrich, *The Life of Sir Charles Linnaeus [...] to which is added, a [...] list of his works, and a biographical sketch of the life of his son [...] translated from the [...] German by Joseph Trapp* (London, White, 1794).

Thunberg, Carl Peter, 'Beschryving van twee nieuwe soorten van palmboomachtige gewassen, Uti Japan en va de Kaap des Goede Hope', *Verhandelingen, uitgegeeven door de Hollandsche Maatschappye der Weetenschappen, te Haarlem* 20:2 (Haarlem, J. Bosch,1782), p.419-34.
–, *Flora Japonica* (Leipsig, I. G. Müller, 1784).
–, *Nova Acta Regiae Societatis Scientiarum Upsaliensis* 4:40 (1783).
Tode, Johann Clemens, *Medicinischchirurgische Bibliothek*, 10 vols (Copenhagen, Johann Gottlob Rothe, 1775-1787).
Tournefort, [Joseph] Pitton de, *Relation d'un voyage du Levant fait par ordre du Roy* (Paris, Imprimerie royale, 1717).

Zoëga, Johan, 'Tilhang om de Islandske Urter', in Eggert Ólafsson and Bjarni Pálsson, *Vice-Lavmand Eggert Olafsens og Land-Physici Biarne Povelsens Reise igiennem Island foranstaltet af Videnskabernes Sælskab i Kiøbenhavn*, vol.2 (Sorøe, Jonas Lindgrens Enke, 1772).

Post-1800 and digital works

Algazi, Gadi, 'Scholars in households: refiguring the learned habitus, 1480-1550', *Science in context* 16:1-2 (2003), p.9-42.
Allen, David Elliston, *The Naturalist in Britain: a social history* (London, 1976).
–, *The Naturalist in Britain: a social history* (Princeton, NJ, 1994).
Amaya, José Antonio, *Mutis, apóstol de Linneo: historia de la botánica en el virreinato de la Nueva Granada (1760-1783)*, 2 vols (Bogotá, 2005).
Andersson Burnett, Linda, 'An eighteenth-century ecology of knowledge: patronage and natural history', in *Culture unbound* 6 (2014), p.1275-97.
–, 'Northern noble savages? Edward Daniel Clarke and British primitivist narratives on Scotland and Scandinavia *c.*1760-1822',

unpublished doctoral dissertation, University of Edinburgh, 2012.

–, 'Selling the Sami: Nordic stereotypes and participatory media in Georgian Britain', in *Communicating the North: media structures and images in the making of the Nordic region*, ed. Jonas Harvard and Peter Stadius (Farnham, 2013), p.171-96.

Appadurai, Arjun, *Modernity at large: cultural dimensions of globalisation* (Delhi, 1997).

Arnold, David, *Science, technology and medicine in colonial India* (Cambridge, 2000).

Asiatick researches, vol.1, 5th edn (1784; London, 1806).

Asma, Stephen T., *Stuffed animals and pickled heads: the culture and evolution of natural history museums* (Oxford, 2001).

Aurivilius, Per Olof Christopher, 'Recensio critica lepidopterorum Musei Ludovicae Ulricae quae descripsit Carolus A Linné', *Kongl. Svenska Vetenskaps Academiens handlingar* 19 (1882), p.1-189.

Baigent, Elizabeth, 'Solander, Daniel (1733-1782)', in *Oxford dictionary of national biography* (Oxford, 2004), http://www.oxforddnb.com/view/article/25985 (last accessed 15 December 2016).

Balayé, Simone, 'La bibliothèque du Roi, première bibliothèque du monde (1664-1789)', in *Histoire des bibliothèques françaises*, vol.2, *Les Bibliothèques sous l'ancien régime, 1530-1789*, ed. Claude Jolly (Paris, 1988), p.209-34.

Bandau, Anja, Marcel Dorigny, and Rebekka von Mallinckrodt (ed.), *Les Mondes coloniaux à Paris au XVIIIe siècle: circulation et enchevêtrement des savoirs* (Paris, 2010).

Banks, Joseph, *Scientific correspondence of Sir Joseph Banks, 1765-1820*, ed. Neil Chambers, 6 vols (London, 2007), vol.4, *The Middle period 1785-1799* (London, 2007).

Beinart, William, 'Men, science, travel and nature in the eighteenth- and nineteenth-century Cape', *Journal of southern African studies* 24:4 (1998), p.775-99.

Bender, John, and Michael Marrinan (ed.), *Regimes of description: in the archive of the eighteenth century* (Stanford, CA, 2005).

Benedict, Barbara M., *Curiosity: a cultural history of early modern inquiry* (Chicago, IL, 2002).

Bénichou, Paul, *Le Sacre de l'écrivain* (Paris, 1973).

Beretta, Mario, and Alessandro Tosi (ed.), *Linnaeus in Italy: the spread of a revolution in science* (Sagamore Beach, MA, 2007).

Bernal Villegas, Jaime, 'La institucionalización del conocimiento científico en el Nuevo Reino de Granada', in *Academia Mutisiana: documentos preneogranadinos de José Celestino Mutis y la promoción de sociedades científicas en la Nueva Granada*, ed. Alberto Gómez Gutiérrez *et al.* (Bogotá, 2011), p.177-232.

Bernasconi, Robert, 'Who invented the concept of race? Kant's role in the Enlightenment construction of race', in *Race*, ed. Robert Bernasconi (Malden, 2001), p.11-36.

Bertoloni Meli, Domenico, 'Authorship and teamwork around the Cimento Academy: mathematics, anatomy and experimental philosophy', *Early science and medicine* 6:2 (2001), p.65-95.

Bettany, G. T., 'Hope, John (1725-1786)', in *Dictionary of national biography*, ed. Sidney Lee (New York, 1891), vol.27, http://en.wikisource.org/wiki/Hope,_John_(1725-1786)_(DNB00) (last accessed 8 August 2017).

Biagioli, Mario, 'Rights or rewards? Changing frameworks of scientific authorship', in *Scientific authorship: credit and intellectual property in science*, ed. Mario Biagioli and Peter Galison (New York and London, 2003), p.253-79.

Blair, Ann, 'Note-taking as an art of transmission', *Critical inquiry* 31:1 (2004), p.85-107.

–, *Too much to know: managing scholarly information before the modern age* (New Haven, CT, 2010).

Blanck, Dag, 'Att bli svensk-amerikan. Framväxten av en svensk-amerikansk identitet i USA', in *Främlingar: ett historiskt perspektiv*, ed. Anders Florén and Åsa Karlsson (Uppsala, 1998), p.59-69.

Bleichmar, Daniela, 'The geography of observation: distance and visibility in eighteenth-century botanical travel', in *Histories of scientific observation*, ed. Lorraine Daston and Elizabeth Lunbeck (Chicago, IL, 2011), p.373-95.

–, 'Training the naturalist's eye in the eighteenth century: perfect global visions and local blind spots', in *Skilled visions: between apprenticeship and standards*, ed. Cristina Grasseni (New York, 2007), p.166-90.

–, *Visible empire: botanical expeditions and visual culture in the Hispanic Enlightenment* (Chicago, IL, 2012).

–, and Peter C. Mancall (ed.), *Collecting across cultures: material exchanges in the early modern Atlantic world* (Philadelphia, PA, 2011).

Blumenbach, Johann Friedrich, *The Correspondence of Johann Friedrich Blumenbach*, vol.3 (Göttingen, 2010).

Blunt, Wilfrid, with William T. Stearn, *The Compleat naturalist: a life of Linnaeus* (London, 1971).

–, with William T. Stearn, *Linnaeus: the compleat naturalist* (Princeton, NJ, 2001).

Bödeker, Hans Erich, 'Georg Forsters Entwurf einer "Wissenschaft vom Menschen"', *NTM International journal of history and ethics of natural sciences, technology and medicine* 18 (2010), p.137-67.

–, 'On the origins of the "statistical gaze": modes of perception, forms of knowledge and ways of writing in the early social sciences', in *Little tools of knowledge: historical essays on academic and bureaucratic practices*, ed. Peter Becker and William Clark (Ann Arbor, MI, 2001), p.169-96.

Bosma, Ulbe, and Remco Raben, *Being 'Dutch' in the Indies: a history of creolisation and empire, 1500-1920* (Singapore, 2008).

Boulger, G. S., 'Smith, Sir James Edward (1759-1828)', revised by Margot Walker, *Oxford dictionary of national biography* (Oxford, 2004), http://www.oxforddnb.com/index/25/101025825/ (last accessed 5 May 2016).

Bourguet, Marie-Noëlle, 'A portable world: the notebooks of European travellers (eighteenth to nineteenth centuries)', *Intellectual history review* 20:3 (2010), p.377-400.

–, 'La collecte du monde: voyage et histoire naturelle (fin XVIIe siècle-début XIXe siècle)', in *Le Muséum au premier siècle de son histoire,* ed. Claude Blanckaert *et al.* (Paris, 1997), p.163-96.

–, 'Measurable difference: botany, climate and the gardener's thermometer in eighteenth-century France', in *Colonial botany: science, commerce and politics in the early modern world*, ed. Londa Schiebinger and Claudia Swan (Philadelphia, PA, 2005), p.270-86.

–, Christian Licoppe, and Heinz Otto Sibum (ed.), *Instruments, travel and science: itineraries of precision from the seventeenth to the twentieth century* (London, 2002).

Brandt, Reinhard, Werner Euler and Werner Stark, *Studien zur Entwicklung preussischer Universitäten* (Wiesbaden, 1999).

Bricka, Carl Frederik, *Dansk biografisk lexikon, tillige omfattende Norge for Tidsrummet 1537-1814*, 19 vols (Copenhagen, Gyldendal, 1887-1905).

Brimnes, Niels, *Constructing the colonial encounter: right and left hand castes in early colonial South India* (Richmond, 1999).

Brixius, Dorit, 'French empire on the ground: plants, peoples, and knowledge in the service of eighteenth-century Isle de France', doctoral dissertation, European University Institute, 2017.

Broberg, Gunnar, *Homo sapiens L.: Studier i Carl von Linnés naturuppfattning och människolära* (Stockholm, 1975).

–, 'The broken circle', in *The Quantifying spirit in the eighteenth century*, ed. Tore Frängsmyr, J. L. Heilbron and Robin E. Rider (Berkeley, CA, 1998).

Browne, J., 'Monson, Lady Anne (*c.*1727-1776)', in *Oxford dictionary of national biography* (Oxford, 2004), http://www.oxforddnb.com/view/article/57839 (last accessed 8 August 2017).

Bruijn, Iris, *Ship's surgeons of the Dutch East India Company: commerce and the progress of medicine in the eighteenth century* (Leiden, 2009).

Burkill, Isaac Henry, *Chapters on the history of botany in India* (Kolkata, 1965).

Bycroft, Trevor Michael, 'Physics and natural history in the eighteenth century: the case of Charles Dufay', unpublished doctoral dissertation, Cambridge University, 2013.

Cañizares-Esguerra, Jorge, *Nature, empire and nation: explorations of the history of science in the Iberian world* (Stanford, CA, 2006).

Carrillo Castillo, Jesús María, *Naturaleza e imperio: la representación del mundo natural en la Historia general y natural de las Indias de Gonzalo Fernández de Oviedo* (Madrid, 2004).

Carteret, Xavier, 'Michel Adanson au Sénégal (1749-1754): un grand voyage naturaliste et anthropologique du siècle des Lumières', *Revue d'histoire des sciences* 65:1 (2012), p.5-25.

–, *Michel Adanson (1727-1806) et la méthode naturelle de classification botanique* (Paris, 2014).

Cassirer, Ernst, *The Problem of knowledge: philosophy, science and history since Hegel* (New Haven, CT, 1978).

Chakrabarti, Pratik, '"Neither of meate nor drinke, but what the Doctor alloweth": medicine amidst war and commerce in eighteenth-century Madras', *Bulletin of the history of medicine* 80:1 (2006), p.1-38.

Chalmers-Hunt, J. M., 'Entomological sales', in *Natural history auctions 1700-1972: a register of sales in the British Isles*, compiled by J. M. Chalmers-Hunt (London, 1976), p.3-14.

Chambers, Neil (ed.), *The Indian and Pacific correspondence of Sir Joseph Banks, 1768-1820*, vol.1, *Letters 1768-1782* (London, 2008).

Chappey, Jean-Luc, *Des naturalistes en révolution: les procès-verbaux de la Société d'histoire naturelle de Paris (1790-1798)* (Paris, 2009).

Charle, Christophe (ed.), *Le Temps des capitales culturelles, XVIIIe-XXe siècles* (Seyssel, 2009).

–, and Daniel Roche (ed.), *Capitales culturelles, capitales symboliques: Paris et les expériences européennes, XVIIIe-XXe siècles* (Paris, 2002).

Charmantier, Isabelle, 'Carl Linnaeus and the visual representation of nature',

Historical studies in the natural sciences 41:4 (2011), p.365-404.
–, and Staffan Müller-Wille, 'Carl Linnaeus's botanical paper slips (1767-1773)', *Intellectual history review* 24:2 (2014), p.215-38.
–, and Staffan Müller-Wille, 'Lists as research technologies', *Isis* 103:4 (2012), p.743-52.
Chartier, Roger, 'Foucault's Chiasmus', in *Scientific authorship: credit and intellectual property in science*, ed. Mario Biagioli and Peter Galison (New York and London, 2003), p.15-31.
–, *The Author's hand and the printer's mind: transformations of the written word in early modern Europe* (Cambridge, 2014).
Clark, William, *Academic charisma and the origins of the research university* (Chicago, IL, 2006).
Cook, Alexandra, *Jean-Jacques Rousseau and botany: the salutary science* (Oxford, 2012).
Cook, Harold, *Matters of exchange: commerce, medicine and science in the Dutch Golden Age* (New Haven, CT, 2007).
Cooper, Alix, 'Homes and households', in *The Cambridge history of science*, vol.3, *Early modern science*, ed. Katharine Park and Lorraine Daston (Cambridge, 2006), p.224-37.
–, *Inventing the indigenous: local knowledge and natural history in early modern Europe* (Cambridge, 2007).
–, '"The possibilities of the land": the inventory of "natural riches" in the early modern German territories', in *Oeconomies in the age of Newton*, ed. Margaret Schabas and Neil De Marchi (Durham, NC, 2003), p.129-53.
Crais, Clifton, C., 'The vacant land: the mythology of British expansion in the Eastern Cape, South Africa', *Journal of social history* 25:2 (1991), p.255-75.
Crane, Diana, *Invisible colleges: diffusion of knowledge in scientific communities* (Chicago, IL, 1972).
Crawford, D. G., *A History of the Indian medical service, 1600-1913*, vol.2 (London, 1914).

Dance, Peter S., *A History of shell collecting*, revised edn (Leiden, 1986).
–, 'Shells', in *Natural history auctions 1700-1972: a register of sales in the British Isles*, compiled by J. M. Chalmers-Hunt (London, 1976), p.45-51.
Daston, Lorraine, 'Observation and enlightenment', in *Scholars in action: the practice of knowledge and the figure of the savant in the eighteenth century*, ed. André Holenstein, Hubert Steinke and Martin Stuber (Leiden, 2013), p.657-78.
–, 'The sciences of the archive', *Osiris* 27:1 (2012), p.156-87.
–, 'Type specimens and scientific memory', *Critical inquiry* 31:1 (Autumn 2004), p.153-82.
–, and Katharine Park, *Wonders and the order of nature, 1150-1750* (New York, 1998).
–, and H. Otto Sibum, 'Introduction: scientific personae and their histories', *Science in context* 16:1-2 (2003), p.1-8.
Daugeron, Bertrand, *Collections naturalistes entre science et empires (1763-1804)* (Paris, 2009).
Davids, Karel, 'The scholarly Atlantic circuits of knowledge between Britain, the Dutch Republic and the Americas in the eighteenth century', in *Dutch Atlantic connections, 1680-1800: linking empires, bridging borders*, ed. Gert Oostindie and J. Roitman (Brill, 2014), p.224-49.
Delbourgo, James, 'Listing people', *Isis* 103:4 (2012), p.735-42.
–, 'Sir Hans Sloane's milk chocolate and the whole history of the

cacao', *Social text* 106 29:1 (2011), p.71-101.

Descartes, René, *The Passions of the soul*, translated by Stephen H. Voss (Indianapolis, IN, 1989).

Dierig, Sven, Jens Lachmund and Andrew Mendelsohn (ed.), *Science and the city*, *Osiris* 18 (Chicago, IL, 2003).

Dietz, Bettina, 'Contribution and co-production: the collaborative culture of Linnaean botany', *Annals of science* 69:4 (2012), p.551-69.

–, *Das System der Natur: die kollaborative Wissenskultur der Botanik im 18. Jahrhundert* (Cologne and Vienna, 2017).

–, 'Linnaeus's restless system: translation as textual engineering in eighteenth-century botany', *Annals of science* 73:2 (2016), p.143-56.

–, 'Making natural history: doing the Enlightenment', *Central European history* 43:1 (2010), p.25-46.

–, and Thomas Nutz, '*Collections curieuses*: the aesthetics of curiosity and elite lifestyle in eighteenth-century Paris', *Eighteenth-century life* 29:3 (2005), p.44-75.

Diment, Judith A., and Alwyne Wheeler, 'Catalogue of the natural history manuscripts and letters by Daniel Solander (1733-1782), or attributed to him, in British collections', *Archives of natural history* 11:3 (1984), p.457-88.

Dobreff, James, 'Daniel Rolander: the invisible naturalist', in *Systema naturae 250: the Linnaean ark*, ed. Andrew Polaszek (Boca Raton, FL, 2010), p.11-28.

Douglas, Bronwen, 'Philosophers, naturalists and antipodean encounters, 1748-1803', *Intellectual history review* 23:3 (2013), p.387-409.

Drake af Hagelsrum, Gustaf, *Linnés disputationer: en översikt* (Nässjö, 1939).

Drayton, Richard, 'Knowledge and empire', in *The Oxford history of the British Empire*, vol.2, *The Eighteenth century*, ed. Peter J. Marshall (Oxford, 1998), p.231-52.

–, *Nature's government: science, imperial Britain and the 'improvement' of the world* (New Delhi, 2005).

Drouin, Jean-Marc, 'Rousseau, Bernardin de Saint-Pierre et l'histoire naturelle', *Dix-huitième siècle* 33:1 (2001), p.507-16.

–, 'Rousseau', in *Dictionnaire culturel des sciences*, ed. Nicolas Witkowski (Paris, 2001), p.368.

–, *L'Herbier des philosophes* (Paris, 2008).

–,'Les herborisations savantes du promeneur solitaire', in *L'Accident de Ménilmontant*, ed. Anouchka Vasak (Paris, 2015), p.41-55.

Drummond, M. M., 'Rumbold, Thomas (1736-1791), of Woodhall, Herts.', in *The History of Parliament: the House of Commons 1754-1790*, ed. L. Namier and J. Brooke (London, 1964), http://www.historyofparliamentonline.org/volume/1754-1790/member/rumbold-thomas-1736-91 (last accessed 8 August 2017).

Duris, Pascal, *Linné et la France (1780-1850)* (Geneva, 1993).

Duyker, Edward, *Nature's argonaut: Daniel Solander 1733-1782* (Melbourne, 1998).

Easterby-Smith, Sarah, *Cultivating commerce: cultures of botany in Britain and in France, 1760-1815* (Cambridge, 2017).

Eddy, Matthew D., *The Language of mineralogy: John Walker, chemistry and the Edinburgh Medical School, 1750-1800* (Aldershot, 2008).

Edwards, Bryan, *The History, civil and commercial, of the British colonies in the West Indies*, 3rd edn (1793; London, 1801), vol.2.

Edvinsson, Rodney, 'Foreign

exchange rates in Sweden 1658-1803' in *Exchange rates, prices and wages, 1277-2008*, ed. Rodney Edvinsson, Tor Jacobson and Daniel Waldenström (Stockholm, 2010), p.238-90.
Eliasson, Pär, *Platsens blick: vetenskapsakademien och den naturalhistoriska resan 1790-1840* (Umeå, 1999).
Embrey, Peter G., 'Minerals', in *Natural history auctions 1700-1972: a register of sales in the British Isles*, compiled by J. M. Chalmers-Hunt (London, 1976), p.39-44.
Emerson, Roger L., 'The founding of the Edinburgh Medical School', *Journal of the history of medicine and allied sciences* 59:2 (2004), p.183-218.
Endersby, Jim, *Imperial nature: Joseph Hooker and the practices of Victorian science* (Chicago, IL, 2008).
Eriksson, Gunnar, *Botanikens historia i Sverige intill år 1800* (Uppsala, 1969).
–, 'Retzius, Anders Jahan', in *Svenskt biografiskt lexikon*, vol.30 (Stockholm, 2000), http://sok.riksarkivet.se/sbl/artikel/6599/ (last accessed 8 August 2017).
Ewan, Joseph, and Nesta Ewan, *Benjamin Smith Barton: naturalist and physician in Jeffersonian America* (St Louis, MO, 2007).
Evans, Robert John Weston, and Alexander Marr (ed.), *Curiosity and wonder from the Renaissance to the Enlightenment* (Aldershot, 2006).
Eze, Emanuel C., *Achieving our humanity: the idea of the post-racial future* (New York, 2013).

Fara, Patricia, *Sex, botany and empire* (Cambridge, 2004).
Ferguson, Adam, *An Essay on the history of civil society* (1767; Edinburgh, 1967).
–, *Institutes of moral philosophy* (1769; Basil, James Decker, 1800).
Ferguson, William, *The Identity of the Scottish nation: an historic quest* (Edinburgh, 1998).
Fernández Pérez, Joaquín, 'Tres apóstoles de Linné en Cádiz: Pehr Osbeck, Pehr Löfling y Clas Alströmer', in *Pehr Löfling y la expedición al Orinoco, 1754-1761*, ed. Francisco Pelayo (Madrid, 1990), p.51-102.
Feuerstein-Herz, Petra, *'Die große Kette der Wesen': Naturgeschichte und ordnung in der Frühen Neuzeit* (Wiesbaden, 2007).
Findlen, Paula, 'Natural history', in *The Cambridge history of science*, vol. 3, *Early modern science*, ed. Katharine Park and Lorraine Daston (Cambridge, 2006), p.435-68.
Flikschuh, Katrin, and Lea Ypi (ed.), *Kant and colonialism* (Oxford, 2014).
Forsstrand, Carl, 'Linnés ekonomi och hans kvarlåtenskaps öden', *Svenska Linnésällskapets årsskrift* (1922), p.70-89.
Forster, Georg, *Georg Forster's 'Sämmtliche Schriften'*, vol.7 (Leipzig, 1843).
Foucault, Michel, *The Order of things: an archaeology of the human sciences* (New York, 2005).
–, 'Qu'est-ce qu'un auteur', *Bulletin de la société française de philosophie* 63:3 (1969), p.73-104.
Frängsmyr, Tore (ed.), *Linnaeus: the man and his work*, new edn (Canton, MA, 1994).
Franzén, Olle, 'Carl Linné, von', *Svenskt biografiskt lexikon*, vol.23 (Stockholm, 1980-1981), http://sok.riksarkivet.se/sbl/artikel/10736 (last accessed 4 May 2016).
–, 'Hur Linnébilden formades 1778-1850', *Svenska Linnésällskapets årsskrift* (1963), p.5-41.
–, 'Löfling, Pehr', in *Svenskt biografiskt lexikon*, vol.24 (Stockholm, 1982-1984), http://sok.riksarkivet.se/sbl/artikel/10050 (last accessed 1 March 2017).

Freedberg, David, *The Eye of the lynx: Galileo, his friends and the beginning of modern natural history* (Chicago, IL, 2002).

Fries, Theodor Magnus, *Linné: lefnadsteckning*, part 2 (Stockholm, 1903).

Fur, Gunlög, *Colonialism in the margins: cultural encounters in New Sweden and Lapland* (Leiden, 2006).

Galison, Peter, 'The collective author', in *Scientific authorship: credit and intellectual property in science*, ed. Mario Biagioli and Peter Galison (New York and London, 2003), p.325-55.

–, and Mario Biagioli (ed.), *Scientific authorship: credit and intellectual property in science* (New York, 2003).

Gardiner, Brian, and Mary Morris, *The Linnaean collections, The Linnean special issue* 7 (Oxford, 2007).

Garrioch, David, *The Making of Revolutionary Paris* (Berkeley, CA, 2002).

Gascoigne, John, 'Crossing the Pillars of Hercules: Francis Bacon, the Scientific Revolution and the New World', in *Science in the age of Baroque*, ed. Ofer Gal and Raz Chen-Morris (Dordrecht, 2013), p.217-37.

–, *Joseph Banks and the English Enlightenment: useful knowledge and polite culture* (Cambridge, 2003).

–, *Science in the service of empire: Joseph Banks, the British state and the uses of science in the age of revolution* (Cambridge, 1998).

Gause, Fritz, *Die Geschichte der Stadt Königsberg in Preussen*, vol.2 (Graz, 1968).

Gómez de Enterría, Josefa, 'The reception of Swedish scientific and technological innovations in Spain and their lexical contributions in botany, metallurgy and mining', in *Commerce and navigation between Spain and Sweden throughout history*, ed. Enrique Martínez Ruiz and Magdalena de Pazzis Pi Corrales (Madrid, 2000), p.283-299.

Goodman, Dena, and Kathryn Norberg (ed.), *Furnishing the eighteenth century: what furniture can tell us about the European and American past* (New York, 2006).

Gottsched, Johann Christoph, *Briefwechsel, Historisch-kritische Ausgabe*, ed. Detlef Döring and Manfred Rudersdorf, vol.1 (Berlin, 2007).

Grove, Richard H., *Green imperialism: colonial expansion, tropical island Edens and the origins of environmentalism, 1600-1860* (New York, 1995).

Guichard, Charlotte, *Les Amateurs d'art à Paris au XVIIIe siècle* (Seyssel, 2008).

–, 'Taste communities: the rise of the amateur in eighteenth-century Paris', *Eighteenth-century studies* 45:4 (2012), p.519-47.

Hagen, Karl Gottfried, *Chloris Borussica* (Königsberg, 1819).

–, *Preußens Pflanzen* (Königsberg, 1818).

Hannaford, Ivan, *Race: the history of an idea in the West* (Baltimore, MD, 1996).

Hansen, Carlo, and Anne Fox Maule (ed.), 'Linnés korrespondance med Pehr Osbeck 1750-1753: med inledning og anmærkninger', *Svenska Linnésällskapets årsskrift* (1972-1974), p.75-145.

–, and Anne Fox Maule, 'Pehr Osbeck's collections and Linnaeus's *Species plantarum* (1753)', *Botanical journal of the Linnaean Society* 67:3 (1973), p.189-212.

Hansen, Lars (ed.), *The Linnaeus apostles: global science and adventure*, vol.6, *Europe, southern Africa, East, southern and South East Asia: Carl Peter Thunberg* (London, 2007).

Hargraves, Neil, 'Beyond the savage character: Mexicans, Peruvians and the "imperfectly civilised" in William Robertson's *History of America*', in *The Anthropology of the Enlightenment*, ed. Larry Wolff and Marco Cipollini (Stanford, CA, 2007), p.103-18.

Harrison, Mark, 'Medicine and orientalism: perspectives on Europe's encounter with Indian medical systems', in *Health, medicine and empire: perspectives on colonial India*, ed. Biswamoy Pati and Mark Harrison (Hyderabad, 2001), p.37-87.

–, 'Networks of knowledge: science and medicine in early colonial India, *c.*1750-1820', in *India and the British Empire*, ed. Douglas M. Peers and Nandini Gooptu (Oxford, 2012), p.191-210.

–, 'Science and the British Empire', *Isis* 96:1 (2005), p.56-63.

Haug, Christine, '"Diese Arbeit unterhält mich, ohne mich zu ermüden": Georg Forsters Übersetzungsmanufaktur in Mainz in den 1790er Jahren', *Georg-Forster-Studien* 13 (2008), p.99-128.

Haynes, Clare, 'A "natural" exhibitioner: Sir Ashton Lever and his Holosphusikon', *Journal for eighteenth-century studies* 24:1 (2001), p.1-13.

Heckscher, Eli F., 'Linnés resor: den ekonomiska bakgrunden', *Svenska Linnésällskapets årsskrift* 25 (1942), p.1-11.

Heesen, Anke te, and Emma C. Spary (ed.), *Sammeln als Wissen: das Sammeln und seine wissenschaftsgeschichtliche Bedeutung* (Göttingen, 2001).

Herder, Johann Gottfried, *Ideen zur Philosophie der Geschichte der Menschheit* (Leipzig, 1841).

Hippel, Theodor Gottlieb von, *Sämtliche Werke*, vol.13 (Berlin, 1838).

Hodacs, Hanna, 'Circulating knowledge on nature: travellers and informants, and the changing geography of Linnaean natural history', in *Travel, agency and the circulation of knowledge*, ed. Gesa Mackenthun, Andrea Nicholas and Stephanie Wodianka (Münster, 2017), p.75-97.

–, 'In the field: exploring nature with Carolus Linnaeus', *Endeavour* 34:2 (2010), p.45-49.

–, 'Linnaeans outdoors: the transformative role of studying nature "on the move" and outside', *The British journal for the history of science* 44:2 (2011), p.183-209.

–, 'Local, universal and embodied knowledge: Anglo-Swedish contacts and Linnaean natural history', in *Global scientific practice in an age of revolutions, 1750-1850*, ed. Patrick Manning and Daniel Rood (Pittsburgh, PA, 2016), p.90-104.

–, and Kenneth Nyberg, *Naturalhistoria på resande fot: om att forska, undervisa och göra karriär i 1700-talets Sverige* (Lund, 2007).

Hoppe, Brigitte, 'Von der Naturgeschichte zu den Naturwissenschaften die Dänisch-Halleschen Missionare als Naturforscher in Indien vom 18. bis 19. Jahrhundert', in *Mission und forschung: Translokale Wissensproduktion zwischen Indien und Europa im 18. und 19. Jahrhundert*, ed. Heike Liebau, Andreas Nehring and Britta Klosterberg (Halle, 2010).

Hoquet, Thierry, *Buffon: histoire naturelle et philosophie* (Paris, 2005).

–, *Buffon-Linné: éternels rivaux de la biologie?* (Paris, 2007).

– (ed.), *Les Fondements de la botanique: Linné et la classification des plantes* (Paris, 2005).

Hulth, Johan Markus, *Linnés första utkast till Species plantarum: några anteckningar* (Stockholm, 1912).

Hunter, Ian, 'Kant's regional cosmopolitanism', *Journal of the history of international law/Revue d'histoire du droit international* 12:2 (2001), p.165-88.

–, *Rival Enlightenments: civil and metaphysical philosophy in early modern Germany* (Cambridge, 2001).

Iglesies Fort, José, *Memorias de la Real Academia de Ciencias y Artes de Barcelona* no.707, 36:1 (Barcelona, 1964).

Jacobsson, Roger (ed.), *Så varför reser Linné?: perspektiv på Iter Lapponicum 1732* (Stockholm, 2005).

Januszajtis, Andrzej, 'Societas physicae experimentalis, the first physics society in Poland: the origins of the society', *Task quarterly* 6:2 (2002), p.323-30.

Jardine, Nicholas, James A. Secord and Emma C. Spary (ed.), *Cultures of natural history* (Cambridge, 1996).

Jensen, Niklas Thode, 'Making it in Tranquebar: science, medicine and the circulation of knowledge in the Danish-Halle Mission, *c.*1732-1744', in *Beyond Tranquebar: grappling across cultural borders in South India*, ed. Esther Fihl and A. R. Venkatachalapathy (Delhi, 2014), p.325-51.

Johannisson, Karin, 'Naturvetenskap på reträtt: en diskusssion om naturvetenskapens status under svenskt 1700-tal', *Lychnos* 29 (1979-1980), p.109-54.

Johnson, David, 'Representing the Cape "Hottentots": from the French Enlightenment to post-apartheid South Africa', *Eighteenth-century studies* 40:4 (2007), p.525-52.

Jonsell, Bengt, 'Apostlarnas resor och gärningar: Linnélärjungarnas roll i upptäckten av världen', in *Ljus över landet? Upplysningen som drivkraft i 1700-talets svenska vetenskap och vitterhet*, ed. Paul Hallberg (Gothenburg, 2005), p.79-98.

–, 'Thunberg, Linné och det linneanska', in *Carl Peter Thunberg: Linnean, resenär, naturforskare 1743-1828*, ed. Bertil Nordenstam (Stockholm, 1993), p.41-55.

Jonsson, Fredrik Albritton, *Enlightenment's frontier: the Scottish Highlands and the origins of environmentalism* (New Haven, CT, 2013).

–, 'Rival ecologies of global commerce: Adam Smith and the natural historians', *The American historical review* 115:5 (2010), p.1342-63.

Kant, Immanuel, *Critique of pure reason* (Cambridge, 1998).

–, *Critique of the power of judgment* (Cambridge, 2000).

–, *Der Streit der Facultäten*, in *Kant's gesammelte Schriften*, vol.7 (Berlin, 1917), p.1-116.

–, 'Entwurf und Ankündigung eines Collegii der physischen Geographie nebst dem Anhange einer kurzen Betrachtung über die Frage Ob die Westwinde in unsern Gegenden darum feucht seien, weil sie über ein großes Meer streichen', in *Kant's gesammelte Schriften*, vol.2 (Berlin, 1912), p.1-12.

–, *Kant's Briefwechsel*, in *Kant's gesammelte Schriften*, vol.10 (Berlin, 1922).

–, *Kant's handschriftlicher Nachlaß: Anthropologie*, in *Kant's gesammelte Schriften*, vol.15 (Berlin, 1923).

–, *Kritik der reinen Vernunft*, in *Kant's gesammelte Schriften*, vol.4 (Berlin 1911), p.1-252.

–, *Kritik der Urteilskraft*, in *Kant's gesammelte Schriften*, vol.5 (Berlin, 1913), p.165-486.

–, 'M. Immanuel Kant's announcement of the programme of his lectures for the winter semester 1765-1766', in *Theoretical philosophy, 1755-1770* (Cambridge, 1992) p.287-300.

–, 'Nachricht von der Einrichtung seiner Vorlesungen in dem Winterhalbjahre von 1765-1766', in *Kant's gesammelte Schriften*, vol.2 (Berlin, 1912), p.303-14.

–, 'Of the different races of human beings', *Anthropology, history and education* (Cambridge, 2007), p.82-97.

–, 'Physical geography', *Natural science* (Cambridge, 2012), p.434-679.

–, *Physische Geographie*, in *Kant's gesammelte Schriften*, vol.9 (Berlin, 1923), p.151-438.

–, 'Plan and announcement of a series of lectures on physical geography with an appendix containing a brief consideration of the question: Whether the west winds in our regions are moist because they travel over a great sea', *Natural science* (Cambridge, 2012), p.386-95.

–, 'The conflict of the faculties', *Religion and rational theology* (Cambridge, 2001), p.233-328.

–, 'Von den verschiedenen Racen der Menschen', in *Kant's gesammelte Schriften*, vol.2 (Berlin, 1912) p.427-44.

Kapor, Vladimir, 'Translating the great maritime explorations: on Johann Reinhold Forster's translation of Bougainville's *Voyage autour du monde*', in *Travel narratives in translation, 1750-1830: nationalism, ideology, gender*, ed. Alison Martin and Susan Pickford (Abingdon, 2012), p.93-109.

Kellman, Jordan, 'Nature, networks and expert testimony in the colonial Atlantic: the case of cochineal', in *Itineraries of Atlantic science: new questions, new approaches, new directions*, ed. Neil Safier, *Atlantic studies* 7:4 (2010), p.373-95.

Kennett, Tom, *The Lord Treasurer of botany: Sir James Edward Smith and the Linnaean collections* (London, 2016).

Kenny, Neil, *The Uses of curiosity in early modern France and Germany* (Oxford, 2004).

Kidd, Colin, *The Forging of the races: race and scripture in the protestant Atlantic world, 1600-2000* (Cambridge, 2006).

Kleingeld, Pauline, *Kant and cosmopolitanism: the philosophical ideal of world citizenship* (Cambridge, 2012).

Knorr-Cetina, Karin, *Epistemic cultures: how the sciences make knowledge* (Cambridge, MA, 1999).

Koerner, Lisbet, 'Daedalus hyperboreus: Baltic natural history and mineralogy in the enlightenment', in *The Sciences in Enlightened Europe*, ed. William Clark, Jan Golinski and Simon Schaffer (Chicago, IL, 1999), p.389-422.

–, *Linnaeus: nature and nation* (Cambridge, MA, 2001).

Kontler, László, 'Mankind and its histories: William Robertson, Georg Forster and a late eighteenth-century German debate', *Intellectual history review* 23:3 (2013), p.411-29.

–, Antonella Romano, Silvia Sebastiani and Borbála Zsuzsanna Török (ed.), *Negotiating knowledge in early modern empires: a decentred view* (Basingstoke, 2015).

Kragh, Helge, *Natur, Nytte og Ånd, 1730-1850* (Aarhus, 2005).

Kuklick, Henrika, and Robert E. Kohler (ed.), *Science in the field, Osiris* 11 (Chicago, IL, 1996).

Kumar, Deepak, 'Botanical explorations and the East India Company: revisiting "plant colonialism"', in *The East India Company and the natural world*, ed. Anna Winterbottom, Vinita Damadoran and Alan Lester (Basingstoke, 2014), p.16-34.

Kusukawa, Sachiko, *Picturing the book of nature: image, text and argument in sixteenth-century human anatomy and medical botany* (Chicago, IL, 2011).
König, Johann Gerhard, 'Journal of a voyage from India to Siam and Malacca in 1779', *Journal of the Straits Branch of the Royal Asiatic Society* 27 (October 1894), p.57-101, 103-25, 127-33.

Lacour, Pierre-Yves, 'L'administration des choses naturelles: le Muséum et ses collections autour de 1800', in *Les Lieux de savoir*, vol.2, *Les Mains de l'intellect*, ed. Christian Jacob (Paris, 2011), p.246-62.
–, *La République naturaliste: collections d'histoire naturelle et Révolution française, 1789-1804* (Paris, 2014).
Lafuente, Antonio, and Nuria Valverde, 'Linnaean botany and Spanish imperial biopolitics', in *Colonial botany: science, commerce and politics in the early modern world*, ed. Londa Schiebinger and Claudia Swan (Philadelphia, PA, 2005), p.134-47.
Laissus, Yves, and Jean Torlais, *Le Jardin du Roi et le Collège royal dans l'enseignement des sciences au XVIIIe siècle* (Paris, 1986).
Lamy, Denis, 'Le dessin botanique dans la transmission des connaissances', in *Passions botaniques: naturalistes voyageurs au temps des grandes découvertes* (La Roche Jagu, 2008), p.139-55.
Larson, James L., *Interpreting nature: the science of living form from Linnaeus to Kant* (Baltimore, MD, 1994).
Latour, Bruno, *Reassembling the social: an introduction to actor-network-theory* (Oxford, 2005).
–, *Science in action: how to follow scientists and engineers through society* (Cambridge, MA, 1987).
–, and Steve Woolgar, *Laboratory life: the construction of scientific facts* (Beverly Hills, CA, 1979).
Lehmann, Arno, 'Hallesche Mediziner und Medizinen am Anfang deutsch-indischer Beziehungen', *Wissenschaftliche Zeitschrift der Martin-Luther Universität Halle-Wittenberg* 5:2 (1955), p.117-32.
Lester, Alan, 'Colonial settlers and metropole: racial discourse in the early nineteenth-century Cape Colony, Australia and New Zealand', *Landscape research* 27:1 (2002), p.39-49.
Lienhard, Luc, '"La machine botanique": zur Entstehung von Hallers Flora der Schweiz', in *Hallers Netz: ein europäischer Gelehrtenbriefwechsel zur Zeit der Aufklärung*, ed. Martin Stuber, Stefan Hächler and Luc Lienhard (Basel, 2005), p.371-410.
Lilti, Antoine, *The World of the salons: sociability and worldliness in eighteenth-century Paris*, translated by Lydia G. Cochrane (New York, 2015).
–, '"Une autre paire de manches": Buffon et la sociabilité des Lumières', *Lumières* 8 (2006), p.235-39.
Lindroth, Sten, *A History of Uppsala University 1477-1977* (Uppsala, 1977).
–, *Kungl: Svenska vetenskapsakademiens historia 1739-1818* 1:1-2, *Tiden intill Wargentins död (1783)*, (Stockholm, 1967).
–, 'Linné legend och verklighet', *Lychnos* (1965-1966), p.56-122.
–, *Svensk lärdomshistoria* 3, *Frihetstiden* (Stockholm, 1978).
–, 'The two faces of Linnaeus', in *Linnaeus: the man and his work*, ed. Tore Frängsmyr (Berkeley, CA, 1983), p.1-62.
Linnaeus, Carl, *A General system of nature through the three grand kingdoms of animals, vegetables and minerals*, translated by W. Turton (London, 1802), vol.1.
–, 'Instruktion för resande naturforskare', in Hanna Hodacs

and Kenneth Nyberg, *Naturalhistoria på resande fot: om att forska, undervisa och göra karriär i 1700-talets Sverige* (Lund, 2007), p.199-210.

–, *Linnaeus's 'Philosophia botanica'*, translated by Stephen Freer (Oxford, 2003).

–, *Linnés föreläsningar öfver djurriket: utgivna och försedda med förklarande anm. af Einar Lönnberg* (Uppsala, 1913).

–, *Linné's eigenhändige Anzeichnungen über sich selbst, mit Anmerkungen und Zusätzen von Afzelius*, translated from the Swedish by Karl Lappe (Berlin, 1826).

–, *A Selection of the correspondence of Linnaeus, and other naturalists*, ed. James Edward Smith, vol.2 (London, 1821).

–, *Vita Caroli Linnaei: Carl von Linnés självbiografier*, ed. E. Malmeström and A. Hj. Uggla (Stockholm, 1957).

Linné, Carl von, *Linnés avhandling Potus theae 1765* (Uppsala, 2002).

Lipkowitz, Elise S., 'Seized natural-history collections and the redefinition of scientific cosmopolitanism in the era of the French Revolution', *British journal for the history of science* 47:1 (March 2014), p.15-41.

Livingstone, David N., *Putting science in its place: geographies of scientific knowledge* (Chicago, IL, 2003).

Lomholt, Asger, *Det kongelige danske videnskabernes selskab 1742-1942: Samlinger til selskabets historie*, vol.1 (Copenhagen, 1942).

Love, Henry Davidson, *Vestiges of old Madras 1640-1800*, vol.3 (New Delhi, 1913).

Loveland, Jeff, *Rhetoric and natural history: Buffon in polemical and literary context*, *SVEC* 2001:03 (Oxford, 2001).

Löwegren, Yngve, *Naturaliekabinett i Sverige under 1700-talet: ett bidrag till zoologiens historia* (Uppsala, 1952).

Lucena Giraldo, Manuel, 'El dorado geométrico: la expedición de límites al Orinoco, 1754-1761', in *La Comisión naturalista de Löfling en la expedición de límites al Orinoco*, ed. María Pilar de San Pío Aladrén (Madrid, 1998), p.23-40.

–, *Laboratorio tropical: la expedición de límites al Orinoco, 1750-1767* (Caracas, 1993).

–, *Naciones de rebeldes: las revoluciones de independencia latinoamericanas* (Madrid, 2010).

Mabberley, D. J., 'Hope, John (1725-1786)', in *Oxford dictionary of national biography* (Oxford, 2004), http://www.oxforddnb.com/view/article/13730 (last accessed 8 October 2016).

'Macartney, George (1737-1806)', in *Encyclopædia Britannica*, ed. Hugh Chisholm (New York, 1911), vol. 17, http://encyclopedia.jrank.org/LUP_MAL/MACARTNEY_GEORGE_MACARTNEY_EARL.html (last accessed 8 August 2017).

Madriñán, Santiago, *Nikolaus Joseph Jacquin's American plants: botanical expedition to the Caribbean (1754-1759) and the publication of the 'Selectarum Stirpium Americanarum Historia'* (Boston, MA, 2013).

Manktelow, Mariette, and Kenneth Nyberg, 'Linnaeus's apostles and the development of the *Species plantarum*', in *'Species plantarum' 250 years: proceedings of the 'Species plantarum' symposium*, held in Uppsala 22-24 August 2003, ed. Inga Hedberg (Uppsala, 2005), p.73-80.

Marion, Michel, *Collections et collectionneurs de livres au XVIIIe siècle* (Paris, 1999).

Marouby, Christian, 'Adam Smith and the anthropology of Enlightenment: the "ethnographic" sources of economic progress', in *The*

Anthropology of the Enlightenment, ed. Larry Wolff and Marco Cipollini (Stanford, CA, 2007), p.87-91.

Marshall, John Braybrooke, 'Daniel Carl Solander, friend, librarian and assistant to Sir Joseph Banks', *Archives of natural history* 11:3 (1984), p.451-56.

–, *The Handwriting of Joseph Banks, his scientific staff and amanuenses* (London, 1978).

Marshall, P. J., 'Hastings, Warren (1732-1818)', in *Oxford dictionary of national biography* (Oxford, 2004), http://www.oxforddnb.com/view/article/12587 (last accessed 8 August 2017).

Marteau early 18th-century currency converter: a platform of research in economc history, http://www.pierre-marteau.com/currency/converter/hol-eng.html (last accessed 10 October 2016).

Martin, Alison, 'Übersetzung und die Entdeckung der Welt: Georg Forster und die Reiseliteratur', in *Übersetzung – Translation – Traduction: ein internationales Handbuch zur Übersetzungsforschung / an international encyclopedia of translation studies / une encyclopédie internationale de la recherche sur la traduction*, ed. Harald Kittel, Armin Paul Frank *et al.*, 3 vols (Berlin, 2008-2011), vol.2 (2008), p.1634-41.

Martín-Merás, Luisa, 'Fondos cartográficos y documentales de la comisión de límites de Brasil en siglo XVIII en el Museo Naval de Madrid', *Terra Brasilis* 7-8-9 (2007), p.1-81, http://terrabrasilis.revues.org/402/ (last accessed 2 March 2017).

McClellan III, James E., and François Regourd, *The Colonial machine: French science and overseas expansion in the old regime* (Turnhout, 2011).

McOuat, Gordon, 'Cataloguing power: delineating "competent naturalists" and the meaning of species in the British Museum', *The British journal for the history of science* 34:1 (2001), p.1-28.

'Members of the Leopoldina', http://www.leopoldina.org/de/mitglieder/mitgliederverzeichnis/member/4703/ (last accessed 8 August 2017).

Merrill, Elmer Drew, 'Osbeck's dagbok öfwer en ostindisk resa', *American journal of botany*, 3:10 (1916), p.571-88.

Metzger, Johann Daniel, *Über die Universität zu Königsberg: ein Nachtrag zu Arnold und Goldbeck* (Königsberg, 1804).

Miller, David Philip, 'Joseph Banks, empire, and "centres of calculation" in late Hanoverian London', in *Visions of empire: voyages, botany and representations of nature*, ed. David Philip Miller and Peter Hanns Reill (Cambridge, 1996), p.21-37.

Mornet, Daniel, *Les Sciences de la nature en France, au XVIIIe siècle: un chapitre de l'histoire des idées* (1911; Geneva, 2001).

Müller-Wille, Staffan, *Botanik und Weltweiter Handel: zur Begründung eines Natürlichen Systems der Pflanzen durch Carl von Linné (1707-1778)* (Berlin, 1999).

–, 'Collection and collation: theory and practice of Linnaean botany', *Studies in history and philosophy of science*, part C, *Studies in history and philosophy of biological and biomedical sciences* 38:3 (2007), p.541-62.

–, 'Linnaeus's herbarium cabinet: a piece of furniture and its function', *Endeavour* 30:2 (2006), p.60-64.

–, 'Names and numbers: "data" in classical natural history, 1758-1859', *Osiris* 32 (2017), p.109-28.

–, 'Nature as a marketplace: the political economy of Linnaean botany', in *Oeconomies in the age of Newton*, ed. Margaret Schabas and

Neil De Marchi (Durham, NC, 2003), p.154-72.
–, 'Vom Sexualsystem zur Karteikarte: Carl von Linnés Papiertechnologien', in *Nicht Fisch, nicht fleisch: Ordnungssysteme und ihre Störfälle*, ed. T. Bäumler, B. Bühler and S. Rieger (Zurich, 2011), p.33-50.
–, 'Walnuts at Hudson Bay, coral reefs in Gotland: the colonialism of Linnaean botany', in *Colonial botany: science, commerce and politics in the early modern world*, ed. Londa Schiebinger and Claudia Swan (Philadelphia, PA, 2005), p.34-48.
–, and Isabelle Charmantier, 'Natural history and information overload: the case of Linnaeus', *Studies in history and philosophy of science*, part C, *Studies in history and philosophy of biological and biomedical sciences* 43:1 (2012), p.4-15.
–, and Karen Reeds, 'A translation of Carl Linnaeus's introduction to *Genera plantarum* (1737)', *Studies in history and philosophy of science*, part C, *Studies in history and philosophy of biological and biomedical sciences* 38:3 (2007), p.563-72.
– and Sara T. Scharf, *Indexing nature: Carl Linnaeus (1707-1778) and his fact-gathering strategies, Working papers on the nature of evidence: how well do 'facts' travel?* 36/08, http://www.lse.ac.uk/economichistory/pdf/factspdf/3909muellerwillescharf.pdf (last accessed 10 October 2015).
'Munro, Hector (1726-1805)', in *Encyclopædia Britannica*, ed. Hugh Chisholm (New York, 1911), vol.19, https://en.wikisource.org/wiki/1911_Encyclopædia_Britannica/Munro,_Hector (last accessed 8 August 2017).
Muthu, Sankar, *Enlightenment against empire* (Princeton, NJ, 2003).
Myrin, C. G., 'Om Linnés naturhistoriska samlingar och deras bortförande till England: ett bidrag till Sveriges Litteraturhistoria', *Scandia* (1833), p.242-88.

Nair, Savithri Preetha, 'The Museum in colonial India (1770-1936): a history of collecting, exhibiting and disciplining knowledge', doctoral dissertation, School of Oriental and African Studies, University of London, 2002.
Nappi, Carla, 'Surface tension: objectifying ginseng in Chinese early modernity', in *Early modern things: objects and their histories 1500-1800*, ed. Paula Findlen (New York, 2013), p.31-52.
–, *The Monkey and the inkpot: natural history and its transformations in early modern China* (Cambridge, MA, 2009).
Natural History Museum London, 'About Paul Hermann', http://www.nhm.ac.uk/research-curation/scientific-resources/collections/botanical-collections/hermann-herbarium/about-paul-hermann/index.html (last accessed 11 June 2016).
Naum, Magdalena, and Jonas M. Nordin, *Scandinavian colonialism and the rise of modernity* (New York, 2013).
Navarro Brotóns, Víctor, and William Eamon (ed.), *Más allá de la Leyenda Negra: España y la revolución científica* (Valencia, 2007).
Neues Journal der Pharmacie für Ärzte, Apotheker und Chemiker 18 (1829).
Nieto Olarte, Mauricio, *Orden natural y orden social: ciencia y política en el semanario del Nuevo Reino de Granada* (Madrid, 2007).
Noltie, Henry. J., *John Hope (1725-1786): Alan G. Morton's memoir of a Scottish botanist* (Edinburgh, 2011).
Nørgaard, Anders, 'Secretaries of the Mission Board in Copenhagen 1714-1868', in *Halle and the beginning of protestant Christianity in*

India, vol. 3, eds. Andreas Gross, Y. Vincent Kumaradoss, Heike Liebau (Halle, 2006).

Nyberg, Kenneth, 'Från Tolvfors till Orinoco: Linnélärjungen Pehr Löflings resa och den globala historien', *Historielärarnas förenings årsskrift* (2015), p.23-39.

–, 'Linnaeus's apostles, scientific travel and the East India trade', *Zoologica scripta* 38, Supplement 1 (2009), p.7-16.

–, 'Linnaeus's apostles and the globalisation of knowledge, 1729-1756', in *Global scientific practice in an age of revolutions, 1750-1850,* ed. Patrick Manning and Daniel Rood (Pittsburgh, PA, 2016), p.73-89.

–, *Pehr Löfling's letter-book in the archives of the Real Jardín Botánico in Madrid: a catalogue* (Gothenburg, 2008).

–, and Mariette Manktelow, 'Linnés apostlar och tillkomsten av *Species plantarum*', *Svenska Linnésällskapets årsskrift* (2002-2003), p.9-30.

Oberhausen, Michael, and Riccardo Pozzo, 'The place of science in Kant's university', *History of science* 40:3 (2002), p.353-68.

–, *Vorlesungsverzeichnisse der Universität Königsberg (1720-1804)* (Stuttgart, 1999).

Ogborn, Miles, 'Discriminating evidence: closeness and distance in natural and civil histories of the caribbean', *Modern intellectual history* 11:3 (2014), p.631-53.

Ogilvie, Brian W., *The Science of describing: natural history in Renaissance Europe* (Chicago, IL, 2006).

Oitinen, Vesa, 'Linné zwischen Wolff und Kant: zu einigen Kantischen Motiven in Linnés biologischer Klassifikation', in *Kant's 'Philosophie der Natur': ihre Entwicklung im Opus Postunum un Ihre Wirkung,* ed. Ernst-Otto Jan Onnasch (Berlin, 2009), p.51-78.

Ormrod, David, *The Rise of commercial empires: England and the Netherlands in the age of mercantilism, 1650-1770* (Cambridge, 2003).

Paris, Harry S., 'La biographie de Antoine Nicolas Duchesne', in Harry S. Paris, *The Drawings of Antoine Nicolas Duchesne for his 'Natural history of the gourds'* (Paris, 2007).

Park, Katharine, and Lorraine Daston (ed.), *Early modern science* in *The Cambridge history of science,* vol.3 (Cambridge, 2003).

Park, Mungo, *Travels in the interior districts of Africa performed under the direction and patronage of the African Association in the years 1795, 1796 and 1797,* introduction and notes by Bernard Waites (1799; Ware, 2002).

Parrish, Susan Scott, *American curiosity: cultures of natural history in the colonial British Atlantic* (Chapel Hill, NC, 2006).

Peck, Robert McCracken, 'Preserving nature for study and display', and 'Alcohol and arsenic, pepper and pitch', in *Stuffing birds, pressing plants, shaping knowledge: natural history in North America 1730-1860,* ed. Sue Ann Prince (Philadelphia, PA, 2003), p.11-53.

Pelletier, Aline, 'Cabinet', in *1740, un abrégé du monde: savoirs et collections autour de Dezallier d'Argenville,* ed. Anne Lafont (Lyon, 2012), p.60-74.

Peset, José Luis, *Ciencia y libertad: el papel del científico ante la independencia americana* (Madrid, 1987).

Petitjean, Patrick, Catherine Jami and Anne Marie Moulin (ed.), *Science and empires: historical studies about scientific development and European expansion* (Dordrecht, 1992).

Pickstone, John V., *Ways of knowing:*

a new history of science, technology and medicine (Manchester, 2000).

Pietsch, Theodore W., and Hans Aili, 'Jacob Theodor Klein's critique of Peter Artedi's *Ichthyologia* (1738)', *Svenska Linnésällskapets årsskrift* (2014), p.39-84.

Pomata, Gianna, and Nancy G. Siraisi (ed.), *Historia: empiricism and erudition in early modern Europe* (Cambridge, MA, 2005).

Pratt, Mary Louise, *Imperial eyes: travel writing and transculturation* (New York, 1992).

–, *Imperial eyes: travel writing and transculturation*, 2nd edn (Abingdon, 2008).

Predeek, Albert, 'Ein verschollener Reorganisationsplan für die Universität Königsberg aus dem Jahre 1725', *Altpreußische Forschungen* 4:2 (1927), p.65-107.

Prince, Sue Ann (ed.), *Of elephants and roses: French natural history, 1790-1830* (Philadelphia, PA, 2013).

Proß, Wolfgang, '"Ein Reich unsichtbarer Kräfte": was kritisiert Kant an Herder?', *Scientia poetica* 1 (1997), p.62-119.

Quenet, Grégory, *Versailles: une histoire naturelle* (Paris, 2015).

Raj, Kapil, *Relocating modern science: circulation and the construction of knowledge in South Asia and Europe, 1650-1900* (Houndmills, 2007).

–, 'Surgeons, fakirs, merchants and craftspeople: making L'Empereur's *Jardin* in early modern South Asia', in *Colonial botany: science, commerce and politics in the early modern world*, ed. Londa Schiebinger and Claudia Swan (Philadelphia, PA, 2005), p.252-69.

Raphael, Renée J., 'Printing Galileo's *Discorsi*: a collaborative affair', *Annals of science* 69:4 (2012), p.483-513.

Rasch, Aage, 'Dansk Ostindien, 1777-1845', in *Vore gamle tropekolonier*, ed. Johannes Brøndsted, vol.7 (Copenhagen, 1966).

Rausing [Koerner], Lisbet, 'Underwriting the oeconomy: Linnaeus on nature and mind', in *Oeconomies in the age of Newton*, ed. Margaret Schabas and Neil De Marchi (Durham, NC, 2003), p.173-203.

Regt, W. M. C., 'Royen, David van', in *Nieuw Nederlandsch biografisch woordenboek*, ed. P. C. Molhuysen and P. J. Blok, vol.10 (Leiden, 1937), http://www.dbnl.org/tekst/molh003nieu10_01/molh003nieu10_01_1256.php (last accessed 8 August 2017).

Reill, Peter H., *Vitalising nature in the Enlightenment* (Berkeley, CA, 2005).

Rennie, Drummond, Veronica Yank and Linda Emmanuel, 'When authorship fails: a proposal to make contributors accountable', *Journal of the American Medical Association* 278:7 (1997), p.579-85.

Rheinberger Hans-Jörg, 'Aspekte des Bedeutungswandels im Begriff organismischer Ähnlichkeit vom 18. zum 19. Jahrhundert', *History and philosophy of the life sciences* 8:2 (1986), p.237-50.

Riedel, Manfred, 'Historizismus und Kritizismus: Kants Streit mit G. Forster und J. G. Herder', in *Deutschlands kulturelle Entfaltung: die Neubestimmung des Menschen*, ed. Bernhard Fabian, Wilhelm Schmidt-Biggemann and Rudolf Vierhaus (Munich, 1980), p.31-48.

Roberts, Lissa, 'Accumulation and management in global perspective: an introduction', *History of science* 52:3 (2014), p.227-46.

–, '"Le centre de toutes les choses": constructing and managing centralisation on the Isle de

France', *History of science* 52:3 (2014), p.319-42.
– (ed.), *Centres of accumulation and management*, special issue, *History of science* 52:3 (2014).
– (ed.), *Centres and cycles of accumulation in and around the Netherlands during the early modern period* (Berlin, 2011).
Rogers, P., 'Bentinck, Margaret Cavendish' [Lady Margaret Cavendish Harley], duchess of Portland (1715-1785)', in *Oxford dictionary of national biography* (Oxford, 2004), http://www.oxforddnb.com/view/article/40752 (last accessed 8 August 2017).
Rolfe, W. D. Ian, 'Fossils', in *Natural history auctions 1700-1972: a register of sales in the British Isles*, compiled by J. M. Chalmers-Hunt (London, 1976), p.32-38.
Rookmaaker, L. C., *The Zoological exploration of southern Africa 1650-1790* (Rotterdam, 1989).
Rosner, Lisa, 'Thistle on the Delaware: Edinburgh medical education and Philadelphia practice, 1800-1825', *Social history of medicine* 5:1 (1992), p.19-42.
Rousseau, Jean-Jacques, *Les Rêveries d'un promeneur solitaire* (Paris, Marcel Didier, 1948).
Rydén, Stig, *Pehr Löfling: en linnélärjunge i Spanien och Venezuela 1751-1756* (Stockholm, 1965).
–, and Arvid Hj. Uggla, 'Pehr Löflings oration i Medelpado-Jämtländska Nationen den 24 maj 1749', *Norlandica, Norrlands Nations skriftserie* 5 (Uppsala, 1961), p.98-123.

Safier, Neil, 'Itineraries of Atlantic science: new questions, new approaches, new directions', *Atlantic studies* 7:4 (2010), p.357-64.
–, *Measuring the new world: Enlightenment science and South America* (Chicago, IL, 2008).
–, 'Spies, dyes and leaves: agro-intermediaries, Luso-Brazilian couriers and the worlds they sowed', in *The Brokered world: go-betweens and global intelligence, 1770-1820*, ed. Simon Schaffer *et al.* (Sagamore Beach, MA, 2009), p.239-65.
–, 'The tenacious travels of the Torrid Zone and the global dimensions of geographical knowledge in the eighteenth century', *Journal of early modern history* 18:1-2 (2014), p.141-72.
Sandermann, Olsen, Svend-Erik *Bibliographia discipuli Linnaei. Bibliographies of the 331 pupils of Linnaeus* (Copenhagen, 1997).
Saraiva, Luís (ed.), *History of mathematical sciences, Portugal and East Asia II: scientific practices and the Portuguese expansion in Asia (1498-1759)* (Hackensack, NJ, 2004).
–, and Catherine Jami (ed.), *History of mathematical sciences, Portugal and East Asia III: the Jesuits, the Padroado and East Asian science (1552-1773)* (Hackensack, NJ, 2008).
Savage, Spencer, 'Description and phylogeny: on a manuscript by the Younger Linnaeus dealing with certain genera now included in the *Amaryllidaceae*', *Herbertia* 4 (1937), p.85-86, 91-94, 97-100.
Schabas, Margaret, and Neil De Marchi (ed.), *Oeconomies in the age of Newton* (Durham, NC, 2003).
Schaffer, Simon, *et al.* (ed.), *The Brokered world: go-betweens and global intelligence, 1770-1820* (Sagamore Beach, MA, 2009).
Scharf, Sara T., 'Identification keys, the "natural method", and the development of plant identification manuals', *Journal of the history of biology* 42:1 (2009) p.73-117.
Schiebinger, Londa (ed.), 'Focus: colonial science', special forum section in *Isis* 96:1 (2005), p.52-87.
–, 'Nature's unruly body: the limits

of scientific description', in *Regimes of description: in the archive of the eighteenth century*, ed. John Bender and Michael Marrinan (Stanford, CA, 2005), p.25-43.
–, *Plants and empire: colonial bioprospecting in the Atlantic world* (Cambridge, MA, 2004).
–, 'The anatomy of difference', *Eighteenth-century studies* 23:4 (1990), p.387-405.
–, 'Why mammals are called mammals: gender politics in eighteenth-century natural history', *The American historical review* 98:2 (1993), p.382-411.
–, and Claudia Swan (ed.), *Colonial botany: science, commerce, and politics in the early modern world* (Philadelphia, PA, 2007).
Schmidt, Olaus, *Det kongelige norske videnskabernes selskab: Matrikkel, 1760-1960* (Trondheim, 1960).
Sebastiani, Silvia, *The Scottish Enlightenment: race, gender and the limits of progress* (New York, 2013).
Selling, Olof H., 'Daniel Solanders naturalkabinett och dess öden', *Svenska Linnésällskapets årsskrift* (1962), p.128-37.
Shapin, Steven, 'The invisible technician', *American scientist* 77:6 (1989), p.554-63.
Simson, Clive, 'Ornithology', in *Natural history auctions 1700-1972: a register of sales in the British Isles*, compiled by J. M. Chalmers-Hunt (London, 1976), p.23-31.
Skuncke, Marie-Christine, *Carl Peter Thunberg, botanist and physician: career-building across the oceans in the eighteenth century* (Uppsala, 2014).
Sloan, Phillip R., 'The Buffon-Linnaeus controversy', *Isis* 67:3 (1976), p.356-75.
–, 'The gaze of natural history', in *Inventing human science: eighteenth-century domains*, ed. Christopher Fox, Roy Porter and Robert Wokler (Berkeley, CA, 1995), p.112-51.
Smith, James Edward, *Memoir and correspondence of the late Sir James Edward Smith*, 2 vols (London, 1832), vol.1.
Somerville, Thomas, *My own life and times 1741-1814* (Edinburgh, 1861).
Somerville, William, *William Somerville's narrative of his journeys to the Eastern Cape frontier and to Lattakoe, 1799-1802* (Cape Town, 1979).
Solander, Daniel, *Collected correspondence 1753-1782*, ed. Edward Duyker and Per Tingbrand (Melbourne, 1995).
Sörlin, Sverker, 'Apostlarnas gärning: vetenskap och offervilja i Linné-tidevarvet', *Svenska Linnésällskapets årsskrift* (1990-1991), p.75-89.
–, 'Globalising Linnaeus: economic botany and travelling disciples', *TijdSchrift voor Skandinavistiek* 29:1-2 (2008), p.117-43.
–, 'Ordering the world for Europe: science as intelligence and information as seen from the northern periphery', in *Nature and empire: science and the colonial enterprise, Osiris* 15 (2000), p.51-69.
–, 'Rituals and resources of natural history: the North and the Arctic in Swedish scientific nationalism', in *Narrating the Arctic: a cultural history of Nordic scientific practices*, ed. Michael Bravo and Sverker Sörlin (Canton, MA, 2002), p.73-122.
–, 'Science, empire and Enlightenment: geographies of northern field science', *European review of history/Revue euopéenne d'histoire* 13 (2006), p.455-72.
–, 'Scientific travel: the Linnean tradition', in *Science in Sweden: the Royal Swedish Academy of Sciences 1739-1989*, ed. Tore Frängsmyr (Canton, MA, 1989).
–, 'The apostles', in *The Linnaeus apostles: global science and adventure,*

vol.1, *Introduction*, ed. Lars Hansen (Whitby, 2010), p.151-79.
–, and Otto Fagerstedt, *Linné och hans apostlar* (Stockholm, 2004).
Spary, Emma, C. 'Codes of passion: natural history specimens as a polite language in late eighteenth-century France', in *Wissenschaft als kulturelle Praxis, 1750-1900*, ed. Hans Erich Bödeker, Peter Hanns Reill and Jürgen Schlumbohm (Göttingen, 1999), p.105-35.
–, 'The nature of Enlightenment', in *The Sciences in Enlightenment Europe*, ed. William Clark, Jan Golinski and Simon Schaffer (Chicago IL, 1999), p.272-306.
–, 'Scientific symmetries', *History of science* 42:1 (2004), p.1-46.
–, *Utopia's garden: French natural history from old regime to Revolution* (Chicago, IL, 2000).
Stafleu, Frans A., *Linnaeus and the Linnaeans: the spreading of their ideas in systematic botany, 1735-1789* (Utrecht, 1971).
Stanalker, Joanna, *The Unfinished Enlightenment: description in the age of the 'Encyclopaedia'* (Ithaca, NY, 2010).
Stark, Werner, 'Einleitung', in *Vorlesungen über Physische Geographie: Kant's gesammelte Schriften*, ed. Werner Stark, vol.26 (Berlin, 2009), p.v-lxxx.
–, 'Naturforschung in Königsberg: ein kritischer Rückblick aus den Präliminarien einer Untersuchung über die Entstehungsbedingungen von Kant's *Vorlesung über Physische Geographie*', *Estudos Kantianos* 2:2 (2014), p.29-60.
Stearn, William T., 'The background of Linnaeus's contributions to the nomenclature and methods of systematic biology', *Systematic zoology* 8:1 (1959), p.4-22.
–, 'Botanical auction sales', in *Natural history auctions 1700-1972: a register of sales in the British Isles*, compiled by J. M. Chalmers-Hunt (London, 1976), p.20-22.
–, 'An introduction to the *Species plantarum* and cognate botanical works of Carl Linnaeus', in Carl Linnaeus, '*Species plantarum': a facsimile of the first edition 1753*, vol.1 (London, 1957), p.1-176.
–, 'Introductory notes on Linnaeus's *Mantissa plantarum*', in Carl Linnaeus, '*Mantissa plantarum' 1767 & 1771: facsimile with an introduction by William T. Stearn* (Lehre, 1961).
Sterll, Michael, 'Life and adventures of Johann Gerhard König (1728-1785): a phantom of the herbaria', *Rheedea* 18:2 (2008), p.111-29.
Stock, Paul, '"Almost a separate race": racial thought and the idea of Europe in British encyclopedias and histories, 1771-1830', *Modern intellectual history* 8:1 (2011), p.3-29.
Stroup, Alice, *A Company of scientists: botany, patronage and community at the seventeenth-century Parisian Royal Academy of Sciences* (Berkeley, CA, 1990).
Struwe, Kamma, 'Dansk Ostindien, 1732-1777', in *Vore gamle tropekolonier*, ed. Johannes Brøndsted, vol.6 (Copenhagen, 1966).
Stuchtey, Benedikt (ed.), *Science across the European empires 1800-1950* (Oxford, 2005).
Sydow, Carl-Otto von, 'Den unge Linnés författarskap. En nyfunnen Linnéskrift', *Svenska Linnésällskapets årsskrift* 45 (1962), p.3-19.
–, 'Fredric Hasselquist', Svenskt biografiskt lexikon, vol.18 (Stockholm, 1961-1971), https://sok.riksarkivet.se/sbl/artikel/12645 (last accessed 1 June 2016).
–, 'Linnaeus and the Lapps', *Taxon* 25:1 (1976), p.9-14.

Terrall, Mary, *Catching nature in the act: Réaumur and the practice of natural history in the eighteenth century* (Chicago, IL, 2014).

–, 'Following insects around: tools and techniques of eighteenth-century natural history', *British journal for the history of science* 43:4 (2010), p.573-88.

–, *The Man who flattened the earth: Maupertuis and the sciences in the Enlightenment* (Chicago, IL, 2002).

Tobin, Beth Fowkes, *The Duchess's shells: natural history collecting in the age of Cook's voyages* (New Haven, CT, 2014).

Traweek, Sharon, *Beamtimes and lifetimes: the world of high energy physicists* (Cambridge, MA, 1988).

Tully, James, 'The Kantian idea of Europe: cosmopolitan and critical perspectives', in *Public philosophy in a new key*, vol.2, ed. James Tully (Cambridge, 2008), p.15-42.

Turner, G. L. E., 'The auction sales of the earl of Bute's instruments, 1793', *Annals of science* 23:3 (1967), p.213-42.

Tzoref-Ashkenazi, Chen, *German soldiers in colonial India* (Oxford, 2015).

Uggla, Arvid Hj., 'Jonas Carlsson Dryander', *Svenskt biografiskt lexikon*, vol.11 (Stockholm, 1945), https://sok.riksarkivet.se/sbl/artikel/17665 (last accessed 12 June 2016).

–, 'Linné den yngres brev till Abraham Bäck, 1778', *Svenska Linnésällskapets årsskrift* (1957), p.138-65.

–, 'Linné den yngres brev till Abraham Bäck, 2. 1779-1783', *Svenska Linnésällskapets årsskrift* (1958), p.61-100.

–, 'Ur Linnésamlingarna i London: försäljningen av Linnés mineralkabinett', *Svenska Linnésällskapets årsskrift* (1953), p.125-34.

Van Damme, Stéphane, *A toutes voiles vers la vérité: une autre histoire de la philosophie au temps des Lumières* (Paris, 2014).

–, *Paris, capitale philosophique de la Fronde à la Révolution* (Paris, 2005).

van Dulm, Frits, *'Zonder eigen gewinne en glorie': Mr Iman Wilhelm Falck (1736-1785), gouverneur en directeur van Ceylon en Onderhorigheden* (Hilversum, 2012).

van Gelder, Roelof, *Das Ostindische Abenteuer: Deutsche in Diensten der Vereinigten Ostindischen Kompanie der Niederlande (VOC), 1600-1800* (Hamburg, 2004).

van Hoorn, Tanja, *Dem Leibe abgelesen: Georg Forster im Kontext der physischen Anthropologie des 18. Jahrhunderts* (Tübingen, 2004).

Vermeulen, Han F., *Before boas: the genesis of ethnography and ethnology in the German Enlightenment* (Lincoln, 2015).

Vucinich, Alexander, *Science in Russian culture: a history to 1860* (Stanford, CA, 1963).

Wall, Cynthia, 'The English auction: narratives of dismantlings', *Eighteenth-century studies* 31:1 (1997), p.1-25.

Wallin, Sigurd, 'Urkunder rörande familjen Linnés ägodelar', *Svenska Linnésällskapets årsskrift* (1951), p.67-94.

–, 'Urkunder rörande familjen Linnés ägodelar. IV', *Svenska Linnésällskapets årsskrift* (1962), p.54-102.

Warolin, Christian, 'La dynastie des Boulduc apothicaires à Paris aux XVIIe et XVIIIe siècles (suite), Gilles-François Boulduc, 20 février 1675-17 janvier 1742', *Revue d'histoire de la pharmacie* 335 (2002), p.439-50.

Wheeler, Alwyne, 'The Gronovius fish collection: a catalogue and historical account', in *Bulletin of the British Museum (natural history), historical series* 1:5 (1953-1959) p.187-249.

–, 'Zoological specimens other than

insects, molluscs and bird', in *Natural history auctions 1700-1972: a register of sales in the British Isles*, compiled by J. M. Chalmers-Hunt (London, 1976), p.15-19.

Wheeler, Roxann, *The Complexion of race: categories of difference in eighteenth-century Britain* (Philadelphia, PA, 2000).

Winterbottom, Anna, 'Medicine and botany in the making of Madras, 1680-1720', in *The East India Company and the natural world*, ed. Anna Winterbottom, Vinita Damadoran, and Alan Lester (Basingstoke, 2014), p.35-57.

–, Vinita Damadoran and Alan Lester (ed.), *The East India Company and the natural world* (Basingstoke, 2014).

Withers, Charles W. J., *Placing the Enlightenment: thinking geographically about the age of reason* (Chicago, IL, 2007).

Wood, Paul B., 'The science of man', in *Cultures of natural history*, ed. Nicholas Jardine, James A. Secord and Emma C. Spary (Cambridge, 1996), p.197-210.

Woodmansee, Martha, 'On the author effect: recovering collectivity', in *The Construction of authorship: textual appropriation in law and literature*, ed. Martha Woodmansee and Peter Jaszi (Durham and London, 1994), p.15-28.

–, and Peter Jaszi, 'Introduction', in *The Construction of authorship: textual appropriation in law and literature*, ed. Martha Woodmansee and Peter Jaszi (Durham, 1994), p.1-13.

–, and Peter Jaszi (ed.), *The Construction of authorship: textual appropriation in law and literature* (Durham, 1994).

Wunschmann, E., 'Schreber, Joh. Christian Daniel (v.)', in *Allgemeine Deutsche biographie* (Leipzig, 1891), vol.32, http://daten.digitale-sammlungen.de/bsb00008390/images/index.html?seite = 467 (last accessed 8 August 2017).

Zammito, John H., *The Genesis of Kant's critique of judgment* (Chicago, IL, 1992).

–, *Kant, Herder and the birth of anthropology* (Chicago, IL, 2002).

Index